Twenty-sixth Edition

The Air Force Officer's Guide

By
Major General A. J. Kinney, USAF (Retired)
and
Lieutenant Colonel John H. Napier III,
USAF (Retired)

STACKPOLE BOOKS
Harrisburg, Pennsylvania

THE AIR FORCE OFFICER'S GUIDE

Published by
STACKPOLE BOOKS
Cameron and Kelker Streets
Harrisburg, Pennsylvania 17105

Continuously published and copyrighted since 1948 by The Military Service Publishing Company, Harrisburg, Pa. (1948-1957) and The Stackpole Company, Harrisburg, Pa. (1959—).

Twenty-sixth Edition.

Photographs courtesy United States Air Force.

Printed in the U.S.A.

The Library of Congress Cataloged the twenty-fifth issue of this work as follows:

Kinney, Andrew J.
The Air Force officer's guide.

Includes index.
1. United States. Air Force—Officers' handbooks.
I. Title
UG793.K56 1981 358.4'00973 80-39657
ISBN 0-8117-2143-4

CONTENTS

THE United States Air Force is an instrument available to the people of this Republic for their common defense, and for the advancement of the interests of the United States throughout the World. Insofar as this Air Force—this Instrument—is strong, sharp, and ready, its existence tends to deter the potential aggressor, giving warning of swift punishment for the lawless. A great and historic interest of the American people lies in their desire for a world of peace and justice. The United States Air Force, through prompt and efficient completion of its assigned missions, can insure the arrival of that long-hoped-for day when all human beings may walk in freedom, and in fear of none save their God . . . or, by slipshod and indifferent performance of its part, the Air Force can cause the American people to meanly lose what Abraham Lincoln called the last, best hope of earth. The issue is in the balance. You, as a member of the United States Air Force, will tip the scales for better or for worse.

The United States Air Force is symbolized by the American bald eagle with wings elevated and displayed in front of a puff cloud.

The striking power of the Air Force is from above the clouds and is represented by the winged thunderbolt taken from Mars' shield which is placed above the cloud-like base represented by the nebular partition line and white background.

There are thirteen stars surrounding the coat of arms and the Roman numerals "MCMXLVII" translated "1947."

The eagle is looking to the right, the field of honor.

Part One • You, The USAF Officer

1

Appointment and Initial Assignment

The Air Force needs you, and the nation needs you, as much as ever before.
—General David C. Jones, then Chief of Staff, USAF

SOURCES OF OFFICERS

The Air Force has a variety of sources for meeting its active-duty officer requirements, as indicated below.

The Service Academies. The Air Force Academy produces approximately 900 graduates per year. In addition, a limited number of Naval Academy and Military Academy graduates may be appointed in the Air Force.

Air Force ROTC. The Air Force conducts an ROTC program at more than 141 colleges and universities throughout the nation. This source supplies an average of 3,500 second lieutenants to the Air Force each year. The AFROTC four-year program requires completion of four academic years of ROTC studies, as well as a four-week field training. The AFROTC two-year program requires completion of two academic years of aerospace studies plus a six-week field training.

Officer Training School. The Air Force conducts an intensive three months' course of training at Lackland Air Force Base, Texas for college graduates who are not ROTC graduates. Candidates are selected on the basis of the need of the Air Force for personnel with training in the individual's academic major. Successful applicants train in the pay grade of Staff Sergeant and enter active duty in the grade of Second Lieutenant immediately upon completion of the course. This program is a valuable supplement to the Air Force ROTC program and provides an avenue of commissioning for individuals who, for various reasons, are not ROTC graduates.

Airman Education and Commissioning Program. This program is designed to select well-qualified airmen who have at least 45 semester hours of accredited college work toward degrees needed by the Air Force, and are below age 35. Those selected are sent to civilian institutions for up to 36 months to acquire their bachelor's degrees. During this period, they serve in the grade of Staff Sergeant. Upon completion of the university program, they attend Officer Training School and are commissioned Second Lieutenants upon graduation.

AFROTC Scholarships. Scholarships for enlisted members are available through the AFROTC program. Enlisted applicants who qualify for flying or technical degrees may attend college, as a member of AFROTC, after being discharged from active duty. Such scholarships may be 4, 3, or 2 years in length, depending on requirements for degrees.

APPOINTMENT OF OFFICERS

Regular Air Force Appointment. All line Reserve Officers entering active duty receive many opportunities to achieve regular status. Active duty reserve line officers are considered upon promotion to the temporary grades of captain, major, lieutenant colonel, and colonel, and also upon completion of 5 and 7 years' active commissioned service. Eligible officers are considered automatically; applications are not required nor accepted. Under present law, individuals selected are appointed in their current active duty grade.

Selection of non-line officers, i.e., chaplains, judge advocates, nurses, etc., is made by a Central Appointment Board which convenes annually. The year groups to be considered vary each year based on existing vacancies in the Regular structure. Under this program, all eligible officers in the announced year groups are considered automatically. Physicians and dentists may also submit application for appointment.

Appointment in the Air National Guard. The Air National Guard has authority to appoint officers in required specialties under State appointment criteria. Normally, such individuals are Federally recognized and are concurrently awarded appointments as Reserve of the Air Force in the same grade. Requests for information or applications for appointment should be addressed to the Adjutant General of the State National Guard.

Direct Appointments in the Air Force Reserve. At present, direct appointments in the Air Force Reserve are limited to chaplains, legal officers, and medical specialists.

INITIAL ACTIVE DUTY ASSIGNMENT

An officer newly appointed in the Air Force Reserve or other Air Force component will receive from Headquarters USAF a letter of notification with an oath of office and allied papers attached. The new appointee should execute the oath of office and such other documents as are associated with his commission, returning these to Headquarters USAF as directed. He will then await receipt of orders to active duty. When these orders to an initial tour of active duty are received, the new appointee will usually find that he has been allowed something more than thirty days in which to prepare for his initial assignment.

To all who shall see these presents, greeting:

Know Ye, that reposing special trust and confidence in the patriotism, valor, fidelity and abilities of JOHN HENRY DOE *, I do appoint* HIM SECOND LIEUTENANT *in the*

United States Air Force

to DATE *as such from the* FIFTH *day of* JULY *, nineteen hundred and* SEVENTY TWO *. This Officer will therefore carefully and diligently discharge the duties of the office to which appointed by doing and performing all manner of things thereunto belonging.*

And I do strictly charge and require those Officers and other personnel of lesser rank to render such obedience as is due an officer of this grade and position. And this Officer is to observe and follow such orders and directions, from time to time, as may be given by me, or the future President of the United States of America, or other Superior Officers acting in accordance with the laws of the United States of America.

This commission is to continue in force during the pleasure of the President of the United States of America, for the time being, under the provisions of those Public Laws relating to Officers of the Armed Forces of the United States of America *and the component thereof in which this appointment is made.*

Done at the City of Washington, this FIFTH *day of* JULY *, in the year of our Lord one thousand nine hundred and* SEVENTY TWO *and of the Independence of the United States of America the one hundred and* NINETY SEVENTH.

By the President:

Robert J. Dixon

Lieutenant General, USAF
Deputy Chief of Staff, Personnel

DEPARTMENT OF THE AIR FORCE · UNITED STATES OF AMERICA · MCMXLVII

Robert C. Seamans, Jr.

Secretary of the Air Force

OFFICER APPOINTMENT, AIR FORCE RESERVE.

Military Sponsor Program (AFR 35-35). The Air Force operates a "sponsor" program to assist new officers in getting settled into their assignments. The organization which is to receive the new officer appoints a sponsor—usually an officer of the same rank—to advise and assist the newcomer. Normal procedure is for the sponsor to get in touch with the incoming officer by telephone or letter, giving the newcomer information regarding the organization to which he or she is being assigned, the situation as to housing, the climatic conditions, and other pertinent facts. If the incoming officer has not received a contact from a sponsor shortly after receiving orders, it would be wise to telephone the organization and ask the name of the assigned sponsor, then get in touch with him or her directly. Conditions differ considerably between the various Air Force stations and there's nothing better than getting the word from someone who is already there.

Records and Orders. The officer should be certain to carry in a briefcase or other handy place all copies of orders received, all personnel records, and all other pertinent documents which bear upon his or her military service or appointment as a commissioned officer. It would also be well to carry any other papers, such as insurance records, automobile title, and income tax records, which may become necessary in conducting personal affairs in the immediate future.

Uniforms and Equipment. An explanation of Air Force uniforms and equipment used by officers is contained in Chapter 6 of this volume. Prior to departing for an initial tour of active duty, the officer should make certain to have at least a minimum of the required equipment. He or she should also make certain that this minimum will be with him or her at all times. Such a minimum might be considered to be the following: One complete service uniform, all-season, with all necessary associated components—such as shirts, socks, ties, insignia; one complete service uniform, summer; a raincoat; and necessary underwear. Additions to the military wardrobe can be obtained at most bases, but it is the best policy to come prepared. If not already in possession of the foregoing minimums, you should take care that in procuring them you secure strictly regulation articles. These can be obtained at Air Force Base Exchange Stores.

Reporting for Duty. The orders which assign an officer to a base for duty include a date for reporting. You comply with the orders if you reach the station and report to the proper official prior to midnight of the prescribed date. You should, however, make every effort to report between 0900 and 1200 of the date required. This will permit you to complete many important arrangements on the date you report.

In those cases where a large number of officers report within a short period, as at a training center, port of embarkation, or other large concentration, the reception of arriving officers may be handled by a receiving committee. In such a case the formality of reporting consists only of presenting yourself at the proper office, which is usually indicated by signs, presenting your orders, signing the register, and receiving instructions. A member of this committee usually handles quartering and messing arrangements and provides, often in a mimeographed order, the information which the newcomer requires. At a later time a meeting may be held at which the commander or his representative addresses the group for purposes of organization or orientation. When this procedure is followed be certain that you have received all of the instructions which should be in your possession and that you understand them thoroughly.

Normally, the organization to which you are assigned will designate a "spon-

HEADQUARTERS, RANDOLPH AIR FORCE BASE, TEXAS.

sor" to help you get settled in your new environment. The sponsor is an officer of your own age and grade who is familiar with the base. It is his or her duty to assist you in every way possible during your initial period at your new assignment. He or she will arrange a time to call upon the commander and arrange for you to meet your immediate commander. He or she will tell you your quarters assignment or whom to see to obtain this information, or will advise you where information may be obtained regarding available civilian housing. It is probable that he or she will arrange for the delivery of your baggage to your quarters. Request a copy of the local base regulations and, if one is available, a map of the base.

Officers Identification Card. Upon coming to active duty as a commissioned officer, each officer is given his identification card (DD Form 2 AF) printed in bleeding ink on watermarked safety paper compressed between laminated plastic sheets. It bears your photograph, fingerprints, signature, and other essential data. Carry it at all times. It establishes your identity and will serve to assist in obtaining the rights and privileges to which you are entitled. In the event of its loss (which should never occur), report the fact promptly.

Upon promotion a new card is issued.

Retired officers and officers on inactive service and dependents are issued identification cards with their status clearly shown.

Getting Established in Quarters. After reporting at the headquarters and to the subordinate commander under whom you are to serve, locate your quarters, have your baggage moved in, and get established at once. On a permanent base a set of quarters or an apartment in the bachelors' building may be assigned. Or it may be necessary to make temporary arrangements off the base, pending the assignment of quarters. It is a custom that newly joined officers are allowed sufficient time to establish themselves before undertaking their duties.

Messing Arrangements. Unless you are furnished quarters where you can provide your own meals, you may eat at the officers' open mess, commonly termed the Officer's Club. The mess officer will give you complete information concerning dues rates, prices and operating hours. It provides the social setting to widen your acquaintance with your fellow officers. The atmosphere of the officers' mess is warm and friendly, and it is a privilege to enjoy its facilities.

Local Regulations. Make a careful study of the local base regulations. They usually contain much useful information on local conditions, facilities, and conveniences, as well as requirements which will assist you in making adjustments to the new environment. Comply with them fully. These regulations answer many questions and may save embarrassment. This process will answer most of your questions; however, do not hesitate to ask others.

The Quarters Situation. The situation as to quarters for officers at most bases is generally subject to wide variation. Facilities for families of officers are certain to be limited. Private housing facilities in civilian communities adjacent to bases are sometimes scarce and rents are high. Officers are well advised to ascertain the exact conditions which confront them before moving members of their families from a location where the conditions are known to be comfortable. As to the bachelor officer, the usual quarters assignment is a room in a Bachelor Officers' Quarters (BOQ).

Learn Your Way Around. Study a map of the base and locate the important buildings, roads, training areas, and recreational facilities.

The Exchange. A visit should be paid to the Exchange soon after arrival. It is a community store operated under the supervision of the commander. At most stations the exchange consists of a general store, a tailor shop, shoe repair shop, and barber shop. It supplies other services for the benefit of officers and airmen of the base. It is noteworthy that profits resulting from the operation of the exchange revert to the organization for expenditure as provided by regulations primarily for the benefit of the airmen. Most of the immediate needs of the newly joined officer can be satisfied at the exchange, and, as it is closely supervised, the officer is assured that his purchases will meet local conditions and requirements.

Getting Set with Accounting and Finance Officer. As soon as possible after arrival at the base, you should visit the office of the Base Accounting and Finance Officer. Here you should arrange for collection of travel allowances and for receipt of monthly pay and allowances. It would be well at this time to initiate any allotments which are desired. On all of these matters, personnel at the Finance Office will provide adequate advice.

Starting to Work. The new officer arriving on active duty for a first tour will find that no one expects him or her to be expert in the performance of work, nor in knowledge of matters pertaining to the Air Force. You will also find that seniors and contemporaries will expect you to proceed at once toward bettering your technical capabilities and expanding your knowledge. It will be found that practically everyone on the base from the airman to the commander and including civilian personnel will be ready to assist the new officer in any way they can. The new officer should neither bluff nor exercise a false pride preventing him or her from seeking information and assistance. Simply proceed in a reasonable manner to learn your job and your way around.

CHANGE OF STATION

When an officer, already assigned on active duty to a base or other station, receives orders reassigning him or her to a different station, he or she must be most careful to execute certain essential pre-departure duties. The most important of these duties relates to securing release from responsibility for unit property and funds. This matter is dealt with in Chapter 11. Here we consider *personal* responsibilities when departing a station.

Be Prepared. Reassignments to new stations, including stations overseas, are routine and frequent events. Quite often such shifts from one base to another, or even from one continent to another, must be made with little warning, if any. The needs of the Air Force rather than the convenience of the individual officer must be the determinant in such events. It is therefore your duty to keep in readiness for departure from your current assignment at any time. Officers have often been recalled from leave in order to depart immediately for an oversea station. Accordingly, it is a very real duty to maintain your personal affairs and your personal equipment in such condition as will permit sudden departure from the base to which you are currently assigned without delay.

Reassignment to a New Station. (For information concerning transfer to an oversea station see Chapter 23. Attention is also invited to Chapter 21.) Upon receipt of orders for a change of station, you should proceed to the base headquarters where you can obtain a clearance form. This form will require you to secure the initials of a large number of custodial and administrative officials, such as the Club Officer, the Finance Officer, the Flight Surgeon, and many others. You must pay all outstanding bills owed to the Officers' Mess or to the Officers' Club, turn in all equipment drawn from the Base Supply Officer, and discharge many other such requirements. Generally, you can clear the base from one central location.

An officer departing a station must remember to prepare and submit any effectiveness reports on other officers which are required.

If you have outstanding debts or bill in the local community nearby the base which you are departing, you are expected to either pay these obligations or to make satisfactory arrangements for their payment with your creditors.

Before leaving, make certain that the following are on hand: copies of all official orders; record of immunization; eyeglass prescriptions, orders terminating assignment of quarters; military pay records; personnel records file. If you plan to travel by personal transportation, be certain that your orders authorize this action and also authorize the time necessary to complete such journey. Secure transportation requests if desired.

The newly-commissioned officer will find the following checklist useful to follow before leaving for his first duty station. Take these records:

(*Note:* The Armed Forces Hostess Association, The Pentagon, will provide on request considerable information on your new base, whether CONUS or overseas.)

Copies of orders
Automobile insurance policy
Automobile registration or title
Bank books
Copy of birth certificate
Driver's license
Immunization record
Insurance policies, or numbers thereof
Social security card
University transcript

If married, also take copies of marriage license, spouse's and children's birth certificates, their immunization records, and spouse's social security card.

A power of attorney and will may be drawn up at the first duty station with the aid of the Staff Judge Advocate's office.

MAVERICK MISSILE

2

Assumption of a New Assignment

Every member of the Air Force must do his job as though the safety of the entire United States depends on it: in fact it does.—General David C. Jones, former chairman, Joint Chiefs of Staff

While many of the rapidly occurring changes in the lives and careers of Air Force officers involve physical movement from one air base to another, quite often the change confronting an Air Force officer is in the nature of the assumption of a new duty. It is to this latter problem that the present chapter is addressed.

ASSUMPTION OF ASSIGNMENT OTHER THAN COMMAND OF UNITS

Command is a very infrequent duty assignment. By far the majority of the officers of the Air Force are engaged in duties which do not involve exercising command over units of the Air Force. Such work is of the most extreme importance in many cases. Every officer must be determined to perform it in an outstanding manner, despite the absence of the advantages of a command position.

Assuming a Staff Position. Upon reporting for duty on a staff assignment, the officer should report first to the executive of the staff agency to which assigned and thereafter to the chiefs of subbranches or sections as he or she may be directed by the executive. Usually a well administered headquarters will have published a manual for staff officers which sets forth the particular staff procedures desired by the commander of the headquarters. On joining a staff, an officer should secure one of these manuals and study it with care.

The next step is to become acquainted with the other members of the staff, especially with those members of staff sections closely allied to your own.

Having mastered the organization, procedures, and having become acquainted with the members of the staff to which assigned, the officer should seek acquaintance with officers in staffs senior and junior to his or her own. For example, an officer assigned to the Operations Section of a group headquarters should become acquainted with the operations personnel in the wing headquarters and also in each of the squadron headquarters within his or her own group. These personal relationships often go far in mitigating misunderstanding between staffs of various levels.

A staff officer must constantly bear in mind that he or she is not a commander and has no power of command. Frequently, however, he or she acts for the commander in matters of considerable importance. In doing so, you must strictly adhere to the policies of the commander. On the other hand, you certainly should not evade your own responsibilities by seeking to discover the views of the commander in every specific instance. You must act within the scope of the commander's authority and policies, and must act as if you yourself bore the sole responsibility for your actions. It is well to take the mental attitude of questioning yourself on the following points: (1) Would I be willing to undertake the execution of the decision I recommend were I a subordinate element commander receiving orders of this staff? (2) Would I be willing to accept the full responsibility for the effect of this decision were I the chief of this staff?

Assuming an Assignment with the Air Reserve Forces. The size and importance of the Air National Guard, Air Force Reserve, and Air Force ROTC render an assignment with one of these components both an opportunity and a challenge. Duty with these elements perhaps imposes upon the officer the maximum requirement for tact and diplomacy. As was stated to be the case in assuming a staff position, it is of great importance to become thoroughly familiar with the organization and procedures of the element to which assigned. It is of equally great importance to become acquainted with the individuals of the Air Reserve Force unit with which the officer will be associated.

Duty with the Air Force ROTC involves the special situation of dealing with students who are not members of the military forces; likewise, one must bear in mind that the universities at which these students are receiving Air Force ROTC training are operated by civilian officials whose principal interest lies in conducting civilian education. The Department of Aerospace Studies is an integral part of the college or university and must conform to the academic standards of the institution. Most host institutions give constructive credit toward degree requirements for AFROTC courses while others grant nominal credit counting only toward the grade point average.

ASSUMPTION OF A COMMAND POSITION

Each Air Force officer must anticipate being assigned without warning to assume command of an Air Force unit. This opportunity may arise as a result of combat. It may arise less dramatically as the result of a sudden need to shift personnel. The Air Force officer may find that the command position given is a larger one than normally associated with his or her grade. Accordingly, each officer should seek familiarity with the problems of command of the next higher unit.

Assumption of Command in Combat. During operations against an enemy, one's next higher superior may become wounded. It then becomes, and immediately becomes, the duty of the junior to take over the responsibility of the senior. This may be the case with a flight leader shot down in aerial combat. It may be the case of the commander of a security police squadron struck by enemy gunfire on the ground. It may be the case of the commander of a maintenance squadron becoming ill. In any case, the junior must act at once to insure that the mission of the unit continues to be performed. Such assumption of command under conditions of combat may prove to be only a temporary expedient. There must never be, however, a break in the continuity of command during combat. Every officer must be prepared to step in and take over when necessary.

Assumption of Command Under Conditions Other Than Combat. Whenever the commander of an Air Force unit is detached from command of the organization for any reason, the senior officer present for duty automatically assumes command, pending orders from higher authority. In such a situation, the officer assuming command shoulders all the responsibility and the authority previously held by the officer relieved. It is a wise policy, however, to act slowly and cautiously in such a situation. For example, it might be thought sensible to defer promotion or demotion of airmen until the new commander has been officially announced.

Transfer of Responsibilities. Other than under conditions of combat, an officer assuming command takes over all standing orders, the unit fund, all public property pertaining to the unit, and the organization records. The transfer of these responsibilities, especially for property and funds, should proceed in a careful and thorough manner with jointly acceptable conclusions being reached between the outgoing and the incoming commanders. The new commander should decline to accept responsibility until he has satisfied himself that the property, the funds, and the records of the unit are in understandable and proper order.

Do Not Hasten to Change Things. Officers assuming command of an Air Force unit sometimes are over-quick to disrupt standing operating procedures, to modify radically policies which the unit personnel have become familiar with, and generally to create confusion. The new commander should allow a reasonable period of time to study the organization, the personnel, and the practices which preceded his or her period of command before launching into major changes.

3

Military Courtesy

In dealing with others be courteous, sir; the courtesy affirms both your own good breeding and your attitude of respect toward your fellow man.

—General R. E. Lee

It is a very fine line that distinguishes between military courtesies and military customs. Both categories owe their existence to a common source: the respect for one another which is the proper attitude among military personnel. Military courtesies, however, are *required.* Omission of them can and most likely will bring immediate disciplinary action. They are a part of an officer's duty; they are a strong strand in the mesh of discipline that holds a military organization together. If military customs are ignored, whoever does so will be privately censured. If military courtesies are ignored, whoever does so will be officially clobbered.

Even though there is no choice open to an officer as to whether or not to render required military courtesies, a little thought would make this requirement no burden at all. Almost all the specifics of military courtesy parallel or are identical to those acts of civility, good breeding and thoughtfulness observed among polite people everywhere. Such courtesies improve relations between individuals, facilitate the conduct of business affairs, and add smoothness to otherwise awkward and undesirable situations. An elderly gentleman in civilian life is commonly addressed as "Sir," or with his full title, as "Mister Smith." Few, if any, people of younger age say to such a man of distinguished years, "Hey, you," or "Say, Mac. . . ." Why is this? There is no discipline in civilian life that requires it. The answer lies in a fact people have learned through experience: to render certain courtesies in certain situations is a better idea than not to do so. In the Air Force, such lessons learned through experience have been codified and made a requirement, so that one need not blunder along until experience teaches the proper reactions to all the many situations one may meet.

It has been noted that when military courtesy is lacking in a military organization, discipline becomes a casualty. For some years the Soviet Army dispensed with all forms of normal military courtesy between officers and men. The discipline of the Soviet Army promptly sank at an alarming rate. Today the Soviet Army and the Soviet Air Force are more meticulous in following military courtesies than ever before. The alternative had been tried and found disastrous. This is the kind of "experience" that only a military organization which knows the date of the next war can afford: The U.S. Air Force cannot risk such periods of disorder.

THE CORRECT USE OF TITLES

Lieutenants are addressed officially as "Lieutenant." The adjectives "First" and "Second" are not used except in written communications.

Other officers are addressed or referred to by their titles. In conversation and in non-official correspondence, Brigadier Generals, Major Generals, and Lieutenant Generals are referred to and addressed as "General." Lieutenant Colonels, under the same conditions, are addressed as "Colonel."

Senior officers frequently address juniors as "Smith" or "Jones," but this does not give the junior the privilege of addressing the senior in any way other than his proper title. If an airman is present, the senior officer should address the junior officer by his title. Officers of the same grade, when among themselves, may address each other by their given names.

Warrant officers are addressed as "Mister" or "Miss."

Chaplains are addressed as "Chaplain" regardless of their grade. A Roman Catholic chaplain may be addressed as "Father," as may be an Episcopal chaplain who prefers it.

Members of the Nurse Corps are addressed by their military titles.

Titles of Cadets. Cadets of the United States Air Force Academy and the United States Military Academy are addressed as "Cadet" officially and in written communications.

Members of the brigade at the Naval Academy have the title of "Midshipman." They are addressed by the title of "Mister" or "Miss."

Titles of Noncommissioned Officers. Noncommissioned officers are addressed by their titles. Chief master sergeants, senior master sergeants, master sergeants, technical sergeants, staff sergeants, and sergeants are all addressed simply as "Sergeant." In official communications the full title of the airman is used. Officers address airmen as "Airman Jones" or "Airman Smith."

Identification and Titles of Officers of Other Services. Courtesies are required to be observed in contacts with officers of the Army, Navy, Marine Corps, and Coast Guard in the same manner as for officers of the Air Force.

It is especially important to comply meticulously with this requirement.

The corresponding grades of commissioned officers in the Air Force, Army and Navy are shown in the illustrations in chapter 6. The grades of commissioned officers in the Marine Corps are the same as those of the Army and Air Force. In the Coast Guard the grades correspond to those of the Navy. Relative rank within grades is determined by the respective dates of commission.

THE SEVERAL MILITARY SALUTES

History of the Military Salute. Men of arms have used some form of the military salute as an exchange of greeting since the earliest times. It has been preserved and its use continued in all modern armed forces which inherit their military traditions from the Age of Chivalry. The method of rendering the salute has varied through the ages, as it still varies in form between the armed forces of today. Whatever the form it has taken, it always pertains to military personnel.

The genesis of the military salute is shrouded in the mysteries of the ages. It is known that in the Age of Chivalry the knights were all mounted and wore steel armor which covered the body completely, including the head and face. When two friendly knights met it was the custom for each to raise the visor and expose his face to the view of the other. This was always done with the right hand, the left being used to hold the reins. It was a significant gesture of friendship and confidence, since it exposed the features and also removed the right hand—the sword hand—from the vicinity of the weapon. Also, in ancient times, the freemen (soldiers) of Europe were allowed to carry arms; when two freemen met, each would raise his right hand to show that he held no weapon in it and that the meeting was a friendly one. Slaves were not allowed to carry arms and they passed freemen without the exchange of a greeting. The military salute is given to a comrade in the honorable profession of arms. The knightly gesture of raising the hand to the visor came to be recognized as the proper greeting between soldiers, and was continued even after modern firearms had made steel armor a thing of the past. The military salute is today, as it seems always to have been, a unique form of exchange of greeting between military people.

The Different Forms of the Salute. There are several forms in which the prescribed salutes are rendered. The officer uses the hand salute and the salute by

The hand salute.

Saluting the flag or National Anthem when in civilian clothes without headdress.

Saluting the flag or National Anthem when in civilian clothes with headdress.

SALUTING.

removing the civilian headdress; airmen use each of the methods and utilize also the several methods of saluting with arms, such as the salute by a sentinel armed with a carbine.

Except in formation, when a salute is prescribed the individual either faces toward the person or colors saluted or turns the head so as to observe the person or color saluted.

Covered or uncovered, salutes are exchanged in the same manner.

If a person is running, he comes to a walk before saluting.

The smartness with which the officer or airman gives the salute is held to indicate the degree of pride he or she has in the profession. A careless or half-hearted salute is discourteous.

In this chapter, unless stated otherwise, the hand salute is intended.

Methods of Saluting Used by Officers. Officers salute by using the hand salute and the salute by placing the civilian headdress over the left breast.

The hand salute is the usual method. While in most instances it is rendered while standing or marching at attention, it may be rendered while seated; *e.g.,* an officer seated at his desk who acknowledges the salute of an officer or airman who is making a report.

The salute by placing the headdress, held in the right hand, over the left breast is used under three conditions. At a military funeral, all military personnel dressed in civilian clothes use this form of the salute in rendering courtesies to the deceased. A member of the military service dressed in civilian clothes and *covered* (wearing headdress) uses the method to salute the National Anthem or *To the Color (Standard).* While in the same dress, this salute is used in paying homage to the National flag or color. Individuals in civilian clothing *uncovered* and without headdress stand at attention, holding the hand over the heart, as a courtesy to the National Anthem or the National flag or color.

Execution of the Hand Salute. Before the instant arrives to render the salute, stand or walk erectly with your head up, chin in, and pull in on the stomach muscles. Look squarely and frankly at the person to be saluted. If you are returning the salute of an airman, or of an officer junior to yourself, execute the two movements of the salute in the cadence of marching, ONE, TWO. If you are saluting a superior officer, execute the first movement and HOLD the position until the salute is acknowledged, and then complete your salute by dropping the hand smartly to your side. Do these things correctly and you will derive many rewards. Your airmen will be quick to notice it, and vie with you in efforts to outdo their officer—a particularly healthy reaction. Thus you may set the example which may then be extended to other matters. At the time of exchanging salutes, it certainly does no harm to say "Good morning" or "Good afternoon."

To execute the hand salute correctly, raise the right hand smartly until the tip of the forefinger touches the lower part of the headdress or forehead above and slightly to the right of the right eye, thumb and fingers extended and joined, palm to the left, upper arm horizontal, forearm inclined at 45°, hand and wrist straight; at the same time turn the head toward the person saluted. To complete the salute, drop the arm to its normal position by the side in one motion, at the same time turning the head and eyes to the front.

As stated above, the airman or junior officer executes the first movement, holds the position until it is returned, and then executes the second movement.

The salute is rendered within saluting distance, which is defined as the distance within which recognition is easy. It usually does not exceed thirty paces. The salute is begun when about six paces from the person or colors saluted or, in case the approach is outside that distance, six paces from the point of nearest approach.

Some of the more frequently observed errors in saluting are these: Failing to hold the position of the salute until it is returned; failure to look at the person or color saluted; failure to assume the position of attention while saluting; failure to have the thumb and finger extended and joined, a protruding thumb being especially objectionable; a bent wrist (the hand and wrist should be in the same plane); failure to have the upper arm horizontal. Gross errors include saluting with a cigarette in the right hand or in the mouth or saluting with the left hand in a pocket or returning a salute in a casual or perfunctory manner.

Uncovering. Officers and airmen under arms as a general rule do not uncover except when:

Seated as a member of or in attendance at a court or board. (Sentinels over prisoners do not uncover.)

Entering places of divine worship.

Indoors when not on duty and it is desired to remain informal.

In attendance at an official reception.

Civilians may be saluted by persons in uniform when appropriate, but the uniform hat or cap will *not* be raised as a form of salutation.

Interpretation of "Indoors" and "Outdoors." Structures such as hangars, gymnasiums, and other roofed structures when used for drill or exercise of troops are considered as "out of doors" in the application of military courtesies.

When the word "indoors" is used, it is construed to mean offices, hallways, dining halls, kitchens, orderly rooms, amusement rooms, libraries, dwellings, or other places of abode.

Meaning of the Term "Under Arms." The expression "under arms" will be understood to mean: With arms in hand or having attached to the person a hand arm or the equipment pertaining directly to the arm, such as cartridge belt, pistol holster, or automatic rifle belt.

STATUS OF WARRANT OFFICERS

Status of an Officer. Warrant officers are entitled to the salute and are extended the courtesies and respect due commissioned officers. They are addressed by the title of "Mister" or "Miss." At most bases it is customary to accord warrant officers the social status of junior commissioned officers. Warrant officers are eligible for membership in messes maintained for commissioned officers and are accepted for membership in officers' clubs.

Warrant officers rank immediately below officers (Second Lieutenant) and above all other grades.

In this chapter, the term *officers* includes warrant officers.

Identification of Warrant Officers. Warrant officers of the Air Force are authorized to wear insignia identical to that prescribed for USAF commissioned officers. Rank insignia is a bar with a colored enamel top, as shown in the illustration in chapter 6. However, there are no longer any warrant officers on active duty in the United States Air Force.

COURTESIES RENDERED BY AIRMEN TO OFFICERS

Occasions. The hand salute is required outdoors both on and off military installations whether on or off duty, except when the salute would be manifestly inappropriate or impractical. It will be required in all official greetings in the line of duty, for ceremonial occasions, and when the national anthem is played or the colors pass by.

Those entitled to the salute are commissioned officers of the Air Force, Army, Navy, Marine Corps, and Coast Guard, warrant officers, and the Nurse Corps. It is also customary to salute officers of friendly foreign countries, such as officers of the armed forces of NATO, when they are in uniform.

Members of the Air Force are urged to be meticulous in rendering salutes and in returning salutes from personnel of the sister services. Such courtesy enhances the feeling of respect which all should feel toward comrades in arms.

Covered or uncovered, salutes are exchanged in the same manner.

The salute is rendered but once if the senior remains in the immediate vicinity and no conversation takes place.

A group of airmen within the confines of military bases, and not in formation, is called to attention by the first person noticing the approach of an officer; if in formation, by the one in charge. If out of doors and not in formation, they all salute; in formation, the salute is rendered by the airman in charge. If in doors, not under arms, they uncover and come to attention.

Drivers of vehicles salute only when the vehicle is halted. Any other person in the vehicle renders the hand salute whether the vehicle is halted or in motion. A noncommissioned officer (or officer) in charge of a detail riding in a vehicle renders the hand salute for the entire detail.

An airman in charge of a detachment salutes officers by bringing the organization or detachment to attention before saluting.

Courtesies Exchanged When an Officer Addresses an Airman. In general, when a conversation takes place between an officer and an airman the following procedure is correct: Salutes are exchanged; the conversation is completed; salutes are again exchanged. *Exceptions:* An airman in ranks comes to attention and does not salute. Indoors, salutes are not exchanged except when reporting to an officer.

When NOT to Salute. Salutes are not rendered by individuals in the following cases: An airman in ranks and not at attention comes to attention when addressed by an officer.

Details (and individuals) at work do not salute. The officer or noncommissioned officer in charge, if not actively engaged at the time, salutes or acknowledges salutes for the entire detail.

When actively engaged at games, such as baseball, tennis, or golf, one does not salute.

In churches, theaters, or places of public assemblage, or in a public conveyance, salutes are not exchanged.

When carrying articles with both hands, or when otherwise so occupied as to make saluting impracticable. However, a nod and a greeting are only courteous.

No salute is rendered to persons by a member of the guard who is engaged in the performance of a specific duty, the proper execution of which would prevent saluting.

A sentinel armed with a pistol does not salute after challenging. He stands at *Raise pistol* until the challenged party has passed.

The driver of a vehicle in motion is not required to salute.

Reporting to an Officer in His Office. When reporting to an officer in his or her office, an airman (unless under arms) removes headdress, knocks, and enters when told to do so. Upon entering, he or she marches up to within about two paces of the officer's desk, halts, salutes, and reports in this manner, for example: "Sir, Airman Jones reports to Captain Smith" or "Sir, Airman Jones reports to the squadron commander." After the report, conversation is carried on in the first or second person. When the business is completed, the airman salutes, executes about face, and withdraws. An airman uncovers (unless under arms) on entering a room where an officer is present.

Procedure When an Officer Enters a Dining Hall or Mess Tent. When an officer enters the dining hall or mess tent, airmen seated at meals remain seated at ease and continue eating unless the officer directs otherwise. *Exception:* An individual addressed ceases eating and sits at attention until completion of the conversation.

Procedure When an Officer Enters a Barracks or Tent. In a barracks or tent, airmen rise, uncover (if unarmed) and stand at attention when an officer enters. If more than one person is present, the first to perceive the officer calls, "Attention." On suitable occasions the officer commands "Rest" or "At ease" when he or she expects to remain in the room and does not desire them to remain at Attention. It is not strictly correct to call officers to attention when the commander enters a room. Instead, the person nearest the door should warn of his or her approach by calling "Gentlemen, the commander!" Officers will then come to attention.

Personal Courtesies. When accompanying a senior, a junior walks or rides on the left.

Entering Automobiles and Small Boats. Military persons enter automobiles and small boats in inverse order of rank; that is, the senior enters an automobile or small boat last and leaves first. Juniors, although entering the automobile first, take their appropriate seat in the car. The senior is always on the right. In the case of aircraft, the senior usually boards first and departs first. Officers sit in the back seat of passenger cars driven by airmen unless the car is full.

COURTESIES RENDERED BY AN OFFICER TO A SENIOR OFFICER

General. The courtesies which are exchanged between officers are those which are prescribed in AFM 50-14 and, in addition, those which are observed through the force of custom as discussed in chapter 4, *Customs of the Service.* It is noteworthy that officers are enjoined to return correctly the salutes of airmen and their junior officers.

Many of the courtesies required to be rendered by airmen to officers apply with equal force and in identical manner to the officer in relations with seniors. The junior salutes first. However, in making reports at formations, the person making the report salutes first regardless of rank. An example of this is the case of a squadron commander rendering a report to the adjutant at a ceremony.

There are several courtesies required to be extended by airmen to officers which are not required between officers.

The prescribed formalities to be observed by airmen in a messroom or mess tent while at meals are not in effect.

In an officer's quarters such courtesies are not observed as are prescribed when an officer enters a barracks or tent occupied by one or more airmen.

Specific Courtesies to be Observed Between Officers. An individual officer out of doors is required to salute when meeting a person entitled to the salute, or when addressed by a senior entitled to the salute.

When reporting in an office to an officer senior in rank the junior follows the same procedure as described for airmen.

An officer (or noncommissioned officer) in charge of a detail riding in a vehicle renders the hand salute for the entire detail.

Organization or detachment commanders salute officers of higher grades by bringing the organization or detachment to attention before saluting.

The officer (or noncommissioned officer) in charge of a detail at work, if not actively engaged at the time, salutes for the entire detail.

The officer uncovers (unless under arms) on entering a room where a senior is present.

HONORS AND CEREMONIES ACCORDED VERY IMPORTANT PERSONAGES

The honors and ceremonies prescribed are designed for greeting all distinguished visitors on arrival, and are accorded to some on departure. None of the requirements will be construed to bar additional or subsequent ceremonies, such as reviews or parades. Personages entitled to honors are listed by precedence in the table of honors. The intent of honors is to extend a mark of courtesy to a distinguished visitor. Honors are rendered to individuals rather than groups. Committees and delegations are honored in the person of the senior or ranking member. Unusual questions of precedence overseas may be referred to the nearest embassy. Generally, most situations will be readily solved if it is kept in mind that the intent is to honor a distinguished guest of the Air Force (AFR 50-14).

Occasions. *a.* Unless directed otherwise, full honors will be accorded the President of the United States on arrival at and departure from any Air Force installation regardless of the day or hour.

b. Honors will be accorded officials of the same or higher level as the Chief of Staff, USAF, during hours between reveille and retreat on work days. The Chief of Staff, USAF, or commander of a major continental or oversea command may direct or request exceptions to the above.

c. Honors will be accorded officials from the grade or equivalent of full General down to and including Brigadier General or equivalent on arrival at any Air Force installation for their first visit in a calendar year during hours between reveille and retreat on work days. The Chief of Staff, USAF, or commander of a major continental or oversea air command may direct exceptions to the above.

d. Personages so entitled to honors may themselves prescribe the omission of any ceremonies otherwise due.

e. Honors will not ordinarily be accorded other commissioned officers of the United States Armed Forces unless specifically directed by the Chief of Staff, USAF. Foreign officers regardless of rank should be accorded honors on appropriate occasions.

f. When travel is by air, the airplane commander will include the pertinent part of a prescribed code in the teletype message filed with Flight Service at the departure point.

Location. The ceremonies as described may be held at any assembly area, but arrangement of distances and timing must reflect dignity and a sense of protocol. The same ceremonies may be used similarly situated when the visitor's transportation is by means other than airplane. Where facilities are limited, such honors as are possible will be rendered. The site will be controlled to avoid disturbance during the ceremony by the intrusions of spectators.

Formation. *a.* The receiving party, composed of the installation commander and appointed aide, and if necessary, the minimum number of staff personnel to properly attend high ranking members of the distinguished visitor's party, participate in the ceremony, but are not part of the honor formation. The aide and staff personnel may act as guides or escort and will render any necessary assistance throughout the visit.

b. The honor formation consists of:

(1) Honor flight commander.

(2) National flight commander.

(3) Color guard.

(4) Honor flight as prescribed.

(5) Band, if available. Recorded music of the Air Force Band may be used.

(6) Additional flights in line with band (See Diagram of Ceremony for the Secretary of the Air Force. The installation commander may use these additional flights on other special occasions).

c. Members of the honor formation will be selected for their alertness, dress, and military bearing. It is important that airmen in the honor flight be arranged according to size. Ribbons denoting decorations and awards will be worn on coats. Uniform will be service with white-covered service cap and (except for the members of the band) white gloves. Special-issue white aiguillettes and white scarves when available will be worn. Color bearers' slings will be of white material. Color guards will be the only members of the honor formation that will be armed and will carry pistols in holsters. Pistol belts and lanyards will be white.

d. Where appropriate, local civilian officials may be invited to join the installation commander in welcoming the personage.

Arrival of a Distinguished Visitor. The installation commander with his receiving party will take a post convenient to the designated point of arrival of the dignitary. Relative distances are dependent upon the location of the ceremony and the visitor's means of transportation. The honor flight will be formed as two lines of airmen in the prescribed number, at normal interval facing one another so as to form an aisle five paces wide. This aisle will begin at and extend behind and away from the line on which the installation commander and his receiving party take their posts. The honor flight commander will center himself five paces beyond the end of the aisle of airmen. The colors and color guard will form at normal interval centered four paces behind him. The band, if available, will be centered four paces to the rear of the colors and color guard. When additional flights of airmen are to be used for a Presidential ceremony or on such special occasions as considered advisable by the installation commander, they will be formed on line with and on each flank of the band. Each flight will consist of not more than 30 airmen. All members of the formation except the aisle of airmen will face the designated point of arrival of the visitor's transportation.

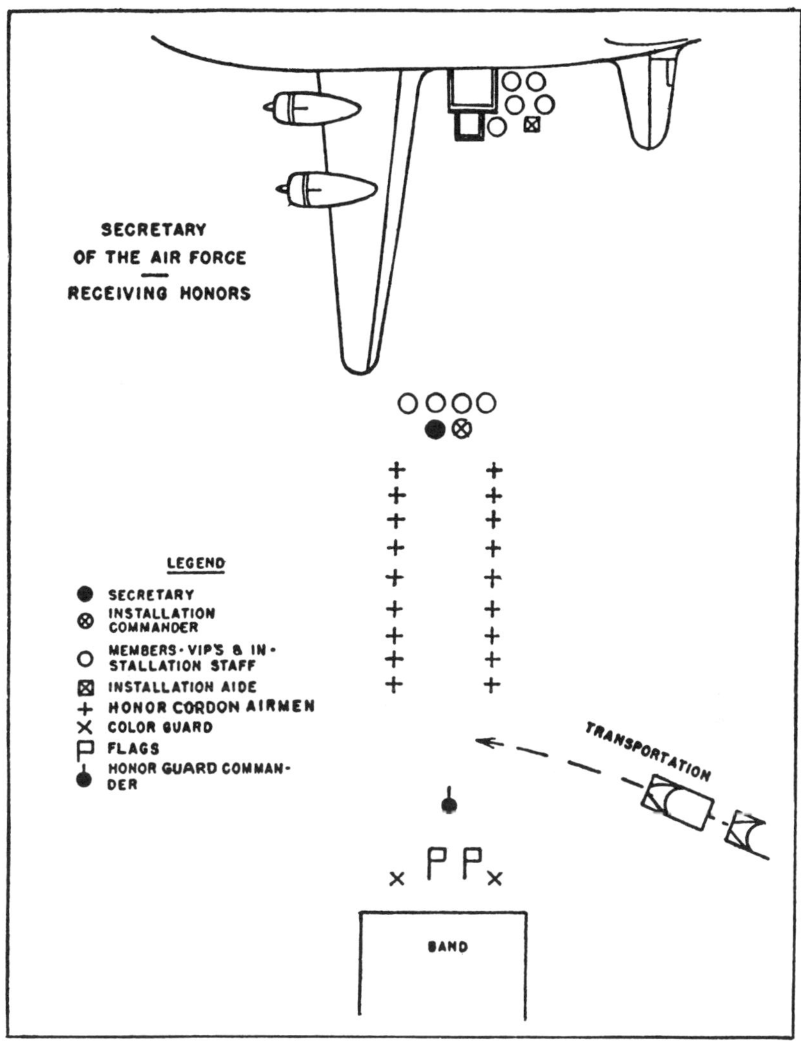

DIAGRAM OF CEREMONY FOR THE SECRETARY OF THE AIR FORCE.

The dignitary's transportation will be directed to a selected position so that his exit from it will face toward the formation. The honor flight commander will call the formation to attention as the installation commander and receiving party come forward and salute the visitor. The installation commander will extend a welcome to the base. Unless the visitor desires that certain members of his party share honors, none will participate.

The installation commander will then escort the personage to receive honors to a central position facing the colors two paces from the line at which the aisle of airmen begins and will align himself on the visitor's left. Members of the party sharing honors, if numerous, will form a line or lines behind the visitor and the installation commander. The honor flight commander, upon signal from the installation commander, will order "Present Arms." The band or music will play ruffles and flourishes and the honors march during which all present, except the national color, render salute. (A flourish is a brief trumpet fanfare; a ruffle is a roll of the drums given as the flourish is sounded.) The personage receiving honors acknowledges them throughout the music by personal salute. Personnel in uniform witnessing the ceremony render the hand salute facing towards the personage being honored. Male personnel in civilian clothing likewise face and accord the salute by placing the hat held in the right hand over the heart. Female personnel in civilian clothing and persons not wearing headgear render the salute by standing at attention. When the music stops, the honor flight commander will command, "Order ARMS," which will be executed by all present.

As the installation commander escorts the visitor forward to enter the staff car, which has been driven from the left to a point in front of the honor flight commander and between him and the aisle of airmen, the flight commander will command, "Flight," "Present ARMS." (Only the honor formation will render the

THE HAND SALUTE.

salute.) On this command, the band or music will play inspection music or a march. Members of the visiting dignitary's party receiving honors with him or her will be escorted through the aisle of airmen and will enter their staff cars. After the last of the staff cars departs from the vicinity, the formation will return to the order upon command by the flight commander when the music has stopped. The flight commander will then issue necessary orders to dismiss the formation.

Departure of a Distinguished Visitor. The ceremony on departure of a distinguished visitor entitled to such honors is substantially the same as that on arrival, but in reverse. The formation is identical. The ceremony may be modified if a visitor departs from the installation in a type of transportation different from that in which he arrived.

The ceremony described hereafter is one for departure by plane. The visitor's car is driven to a point as before in front of the honor flight commander who brings the formation to attention and salutes. The band plays inspection music or a march. The visitor, either accompanied or met there by the installation commander, proceeds to the same place at which he received honors on arrival, and both take the same positions facing the colors. The aide will meanwhile escort members of the visitor's party from their cars to the plane by a route outside the aisle of honor airmen. The honor flight commander then comes to the order and the music stops. He or she commands, "Present ARMS." All salute as before while the music sounds ruffles and flourishes and the honors march. When the music stops, the flight commander will command, "Order ARMS." The installation commander escorts the visitor to his plane and bids him goodbye. After all have embarked, the receiving party will withdraw to their posts at the beginning of the aisle of honor airmen as before. When space does not permit the honor formation to remain in position as outlined above, the flight commander should issue necessary orders to reform the honor flight at a safe distance from point of departure of the aircraft. In most instances, the closing of ranks by the honor airmen will be sufficient. The installation commander, alone, salutes as the visitor's plane moves away. If a distinguished personage departs by automobile, the ceremony previously described may be adapted according to the location by substituting his car for the plane. Drivers of staff cars for the visitor and party will not render salutes during the ceremony. They will remain in their vehicles alert to move forward. The positions of the vehicles are dependent upon the location of the ceremony.

TABLE OF HONORS

In place of gun salutes, the number of aircraft in a memorial fly-over or the number of persons in the honor cordon indicates the type of honors being accorded. The numbers in the "honor cordon" column include the honor cordon commander and airmen, but do not include additional flights used for the Presidential ceremonies, or on special occasions the installation commander considers appropriate.

Distinguished Persons	Honor Cordon	Ruffles & Flourishes	Music	Flags [10]
President	21	4	National Anthem [2] Hail to the Chief [2,3]	United States and Presidential
Ex-President	21	4	National Anthem	United States
Heads of State of foreign countries and reigning royalty	21	4	Foreign Anthem	United States and Foreign

Distinguished Persons	Honor Cordon	Ruffles & Flourishes	Music	Flags[10]
Vice President	19	4	Hail to Columbia	United States and command
Governor of a State of the United States in jurisdiction	19	4	Honors March[5]	United States and command
The Chief Justice of the United States	19	4	Honors March[5]	United States and command
Foreign prime minister or other cabinet officer, foreign ambassador, high commissioner or special diplomatic representative whose credentials give him authority equal to or greater than that of an ambassador	19	4	National or Foreign Anthem[4]	United States and command
Speaker of the House of Representatives	19	4	Honors March[5]	United States and command
Secretary of State	19	4	Honors March[5]	United States and command
The United States Representative to the United Nations	19	4	Honors March[5]	United States and command
Associate Justices of the Supreme Court of the United States	19	4	Honors March[5]	United States and command
Secretary of Defense	19[6]	4	Honors March[5]	United States and command
Cabinet Members	19	4	Honors March[5]	United States and command
Governor of a State of the United States out of jurisdiction	19	4	Honors March[5]	United States and command
United States Senators	19	4	Honors March[5]	United States and command
Members of the House of Representatives of the United States	19	4	Honors March[5]	United States and command
Deputy Secretary of Defense; Secretaries of the Army, the Navy, and the Air Force	19[6]	4	Honors March[5]	United States and command
Director of Defense Research and Engineering	19[6]	4	Honors March[5]	United States and command
Chairman of the Joint Chiefs of Staff; Chief of Staff, United States Army; Chief of Naval Operations; Chief of Staff, USAF; Commandant of the Marine Corps	19[6]	4	Honors March[5]	United States and command
General of the Army; Fleet Admiral; General of the Air Force (five-star rank)	19[6]	4	Honors March[5]	United States and command
Under Secretaries of the Cabinet, Solicitor General	17	4	Honors March[5]	United States and command
Assistant Secretaries of the Cabinet; Assistant to the Attorney General	17	4	Honors March[5]	United States and command
Assistant Secretaries of Defense and the General Counsel of the DOD and Under-secretaries of the Army, Navy, and Air Force	17[6]	4	Honors March[5]	United States and command
Generals and Admirals (four-star rank)	17[6]	4	Honors March[5]	United States and command
Assistant Secretaries of the Army, Navy, and Air Force	17[6]	4	Honors March[5]	United States and command
Lieutenant Generals, Vice Admirals	15[6]	3	Honors March[5]	United States and commands
Foreign ambassadors out of jurisdiction	15	2	Honors March[5]	United States and command

Major Generals, Rear Admirals (Upper half)	13[6]	2	Honors March[5]	United States and command
Brigadier Generals, Rear Admirals (Lower half)	11[6]	1	Honors March[5]	United States and command
Other commissioned officers	9[6]	1	Honors March[5]	United States and command

[1] Same number in honor cordon is provided on departure as on arrival for all distinguished persons.

[2] The United States Navy Band arrangements of the National Anthem and the United States Marine Corps Band arrangements of "Hail to the Chief" are designated as the official Department of Defense arrangements to be henceforth played by all service bands on appropriate occasions.

[3] The traditional musical selection "Hail to the Chief" is designated as a musical tribute to the President of the United States; as such, it will not be performed by military musical organizations as a tribute to other dignitaries. Performances of this selection will be subject to:

a. During "Hail to the Chief" by military musical organizations, military personnel in uniform, other than band personnel, will accord the same honor as they would for the National Anthem or "To the Colors."

b. If, in the course of any ceremony, honors must be performed more than once, "Hail to the Chief" may be used interchangeably with the National Anthem as honors to the President of the United States.

c. When specified by the President, the Secretary of State, the Chief of the Secret Service, or their authorized representatives, "Hail to the Chief" may be used as an opportunity for the President and his immediate party to move to or from their places while all others stand fast.

[4] When one or more foreign national anthems and the U.S. National Anthem are to be performed, the U.S. National Anthem will be performed last, except in conjunction with morning colors.

[5] Army or Air Force generals receive the Generals March; admirals, commodores, or Marine generals receive the Admirals March; all others not specified receive the last 32-bar strain of "The Stars and Stripes Forever."

[6] Foreign civilian and military officials, occupying positions comparable to these U.S. Officials, will receive equivalent honors. Foreign recipients of honors must be representatives of countries recognized by the United States.

[7] Appropriate background music is any music such as national air or a folk song favorably associated with the distinguished person or his country. If no such music is known or available, any well-sounding music of universal appeal and propriety may be used. All music performed as background should be appropriately subdued to the principal action.

[8] Appropriate inspection music may be in any meter and played so that it would not require the inspection party to conform to its cadence.

[9] Commanders of Air Force installations may obtain recorded music on discs for use in connection with honors and ceremonies by writing direct to the Audio Section, United States Air Force Band, Bolling AFB, D.C. 20332.

[10] In events honoring foreign dignitaries, the flag of the foreign country of the guest or guests being honored should be included in the color guard when available.

COURTESIES TO THE NATIONAL FLAG AND NATIONAL ANTHEM

The Flag of the United States. The term "flag" is applicable regardless of size, relative proportions, or manner of display. (See AFM 50-14.) The national flag, in its various sizes and use, is designated the Base Flag, All Purpose Flag, Ceremonial Flag, and Interment Flag.

(1) Those flown from a staff at Air Force bases are:

U.S. Base Flag: 10′ hoist by 19′ fly; used in pleasant weather.

U.S. All Purpose Flag: 5′ hoist by 9′6″ fly, for use in stormy and wind weather and as prescribed by the commander.

(2) Flags carried by a military unit are:

Ceremonial Flag: 4′4″ hoist by 5′6″ fly; carried when the Air Force ceremonial flag is appropriate.

Organizational Flag: 3′ hoist by 4′ fly. (It is not a national but a "unit" flag and is carried when the organizational flag is appropriate.)

(3) As used to cover casket:

Interment Flag: Same dimensions as all purpose flag.

Special Purpose Flags. A guidon is a swallow-tailed organizational flag carried by smaller units, such as squadrons, batteries, companies, or troops. A pennant is a triangular flag used ashore primarily for parade markers, etc. Other special purpose flags include Air Force Recruiting Service, Chapel, Chaplain, and Geneva Convention Flags.

Flag Appurtenances. The pike or staff of Air Force flags is surmounted by a silver spearhead. Other flag appurtenances may include unit streamers and silver bands, cords and tassels, and flag slings (see AFM 50-14).

Respect. Members of the military service are meticulous in observing the courtesies which are required to be rendered on prescribed occasions or circumstances with respect to the national flag or the National Anthem. Colors or standards of organizations receive the same courtesy as the national flag flown from a flagstaff. The trumpet call To the Color (Standard) receives the same courtesy as the National Anthem.

When the National Anthem is played indoors, officers and airmen will stand at attention and face the music or the flag if one is present. They do not salute unless under arms.

The same marks of respect are shown the national anthem of any other country when played upon official occasions.

Courtesies at Retreat and Escort of the Color. Retreat is a daily ceremony at bases or stations during which all personnel are required to render homage or courtesy to the flag. The ceremony may include a Retreat Parade; if organization colors or standards are carried by units on parade the courtesy is usually paid to those colors or standards by participants in the ceremony or spectators at the ceremony. The ceremony of Retreat includes the sounding of the trumpet call *Retreat* by a trumpeter and then, if the band is present, the playing of the National Anthem; in the absence of the band, field music sounds the trumpet call To the Color (Standard). The flag is lowered slowly, as if reluctantly, whereas in the morning at colors, it is raised smartly, as if eagerly. After the flag is lowered, it is folded into a triangular shape. This is supposedly commemorative of the Revolutionary War soldier's cocked hat.

At the first note of the National Anthem, or its counterpart in field music, all officers and airmen present but not in formation face the color or flag and render the prescribed salute. The position of the salute is held until the last note of the music is sounded. The prescribed salutes are: The hand salute for all officers and airmen not in formation; the prescribed salute for military personnel dressed in civilian clothes and wearing headdress, is to stand at attention, remove the headdress, and hold it over the left breast.

Vehicles in motion will be brought to a halt. Occupants remain seated in the vehicle.

Occupants of military vehicles other than passenger cars and motorcycles remain seated in the vehicle at attention, the person in charge of the vehicle dismounting and rendering the hand salute.

Salute to Passing Colors. When passing or being passed by an uncased national color, honors are rendered in the same manner as described above. For purposes of protection, the colors may be furled and then covered with a canvas case of special manufacture. When so carried honors are not required.

Courtesies to the National Anthem. Whenever or wherever the National Anthem is played or To the Color (Standard) is sounded, at the first note thereof all officers and airmen present but not in formation will face the music, stand at attention and render the prescribed salute. Exception: At the ceremony of Retreat or Escort of the Color they face the color or flag. The position of salute is held until the last note of the music is sounded. The prescribed salute is the same, in all cases, as described for the ceremonies of Retreat and Escort of the Color.

Embarrassment may be avoided by remembering this obvious rule: If the homage is to be paid to the flag, face the flag and render the correct salute; if the homage is to be paid to the National Anthem or To the Color (Standard) played by field music, face the source of the music and salute.

The method and personnel required for raising and lowering the flag on a flagstaff are prescribed in AFM 50-14.

Dipping the Flag or Colors. The national flag will not be dipped by way of salute or compliment. The organization color is dipped as a salute when the reviewing officer has the rank of a general officer. This is done by lowering the pike (as the staff of a color is called) to the front so that it makes an angle of about 45 degrees with the ground.

The organization flag salutes in all military ceremonies while the National Anthem or To the Color is being played and when rendering honors to its commander or a person of higher rank, but in no other case.

When not in use, unit flags are usually displayed in the commander's office.

Funerals. The national flag is used to cover the casket at the military funeral of members of the military service. It is placed lengthwise on the casket with the union at the head and over the left shoulder of the deceased. The flag is not lowered into the grave and is not allowed to touch the ground.

Display and Use of the Flag. International usage forbids the display of the flag of one nation above the flag of another in time of peace. When the flags of two or more nations are to be displayed, they should be flown from separate staffs, or from separate halyards of equal size and on the same level.

The national flag, when not flown from a staff or mast, should always be hung flat, whether indoors or out. It should not be festooned over doorways, or arches, nor tied in a bowknot, nor fashioned into a rosette. When used on a rostrum, it should be displayed above and behind the speaker's desk. It should never be used to cover the speaker's desk or to drape over the front platform. For this latter purpose, as well as for decoration in general, bunting of the national colors should be used and the colors should be arranged with the blue above, the white in the middle, and the red below. Under no circumstances should the flag be draped over chairs or benches, nor should any object or emblem of any kind be placed above or upon it, nor should it be hung where it can be easily contaminated or soiled. No lettering of any kind should ever be placed upon the flag. It should not be used for advertising purposes. When metal or cloth replicas of the national flag are used as decorations of civilian clothing (such as costume jewelry) or as an identifying symbol, the positioning and treatment should be with the greatest possible respect. When carried with other flags the national flag should always be on the right as color bearers are facing or in front. When a number of flags are grouped and displayed from staffs the national flag should always be in the center or at the highest point of the group.

The display and use of the flag by civilians or civilian groups is prescribed by Public Law 623—77th Congress.

HOW TO DISPLAY THE FLAG.

1. When displayed over the middle of the street, the flag should be suspended vertically with the union to the north in an east and west street, or to the east in a north and south street.

2. When displayed with another flag from crossed staffs, the US flag should be on the right (the flag's own right) and its staff should be in front of the staff of the other flag.

3. When flown at half-mast, the flag should be hoisted to the peak, then lowered to the half-mast position; but before lowering the flag for the day it should again be raised to the peak.

4. When flags of states or cities or pennants of societies are flown on the same halyard with the US flag, the latter should always be at the peak.

5. When the flag is suspended over a sidewalk from a rope extending from house to pole at the edge of the sidewalk, the flag should be hoisted out from the building, toward the pole, union first.

6. When the flag is displayed from a staff projecting horizontally or at any angle from the window sill, balcony, or front of a building, the union of the flag should go to peak of the staff (unless the flag is to be displayed at half-mast).

7. When the flag is used to cover a casket, it should be so placed that the union is at the head and over the left shoulder. The flag should not be lowered into the grave or allowed to touch the ground.

8. When the flag is displayed other than by being flown from a staff, it should be displayed flat whether indoors or out. When displayed either horizontally or vertically against a wall, the union should be uppermost and to the flag's own right, that is, to the observer's left. When displayed in a window it should be displayed in the same way, that is, with the union or blue field to the left of the observer in the street.

9. When carried in a procession with another flag or flags, the US flag should be either on the marching right, or when there is a line of other flags, in front of the center of that line.

10. When a number of flags of states or cities or pennants of societies are grouped and displayed from staffs with our National flag, the latter should be at the center or at the highest point of the group.

11. When the flags of two or more nations are displayed they should be flown from separate staffs of the same height and the flags should be of approximately equal size. International usage forbids the display of the flag of one nation above that of another nation in time of peace.

(Courtesy of Army Information Digest.)
FLAG OF THE UNITED NATIONS.

DISPLAY OF UNITED NATIONS FLAG

When the United States flag and the United Nations flag are displayed together, the United States flag is on the right, best identified as "the marching right." The United States flag will be equal in size or larger, in the position of honor on the right (observer's left), and above the United Nations flag. The United Nations flag will be carried by troops only on occasions when the United Nations or high dignitaries thereof are to be honored. When so carried, the United Nations flag will be carried on the marching left and below the United States flag.

COURTESIES TO THE MILITARY DEAD

General. The military dead are tendered a dignified and ceremonial funeral service in keeping with the military tradition. The flag is displayed at half-mast staff. Prescribed salutes are rendered to the deceased. Briefly, at a military funeral one salutes the caisson or hearse as it passes, the casket as it is being carried, during the firing of the volleys, and the playing of taps. If a unit or detail passes a burial procession or coffin, the officer or NCO in charge will give the command "Eyes Right" (or Left) and salute. When an officer dies, an officer from his unit is detailed to escort the remains to the place of burial.

Officers who are required to plan a military funeral should consult AFR 143-1 which covers in detail several types of funeral ceremonies.

The Flag at Half-Staff. When the national flag is displayed at half staff it is first hoisted to the top of the staff and then lowered to the half-staff position. Before lowering the flag it is again raised to the top of the staff. A flag in any position below the top of the staff (chock-a-block) is technically in the half-staff position. For an unguyed flagstaff of one piece, the middle point of the hoist of the flag should be midway between the top of the staff and the foot thereof. (AFM 50-14.)

Occasions for the Flag at Half-Staff. On Memorial Day, at all Air Force bases and stations, the national flag is displayed at half-staff from reveille to retreat.

On the death of an officer at a base the flag is displayed at half-staff and so remains between reveille and retreat, or until completion of the burial service, after which it is hoisted to the top, or, if the remains are not interred at the base, until they are removed therefrom.

Whenever, due to the death of any person entitled to personal honors, military mourning is prescribed by regulations or specially ordered by proper authority, all flags are displayed at half-staff, and the national and organizational colors are draped.

THE FIRST SPACE WALK.

4

Customs of the Service

Well, the primary custom of the Air Force is to just get the damn job done.
—General Carl Spaatz, First Chief of Staff, USAF

A custom is an established usage. Customs include positive actions—things to do; and taboos—things to avoid doing.

Customs are those reasonable, consistent, universally accepted practices which make for a pleasanter life or more orderly procedures. Continued without interruption for a long period of time they become compulsory. They tend to take the force of law, as indeed they are—the Common Law.

AIR FORCE CUSTOMS

The Air Force has its own customs. Some have been handed down from the past. Others are of comparatively recent origin. For instance, the dining-in, explained in chapter 20, is a relatively recent Air Force custom. Another pleasant practice, which is growing, is for an active Air Force wing to "adopt" a cadet wing at the Air Force Academy or of the AFROTC. Those customs which persist stand on their own merits. They are unwritten in the sense that they are not embodied in official orders or regulations. Many Air Force customs complement procedures which are required by military courtesy, while others merely add to the graciousness of life. The breach of some customs merely brands the offender as ignorant, or careless, or ill-bred; but there are others the violation of which would bring official censure or disciplinary action.

It is an unquestionable fact that most of the present Air Force customs derive from Army customs, as in fact the Air Force itself derives from the Army. Therefore most Air Force customs are almost identical replicas of Army customs. Yet there is a difference. This difference manifests itself in a way so abstract that it defies concrete description. It might be said that the Air Force has adopted practically all of the Army's customs, but with a little more leaven of humor, a little less rigidity, a little more warmth. Perhaps this abstract difference between Army and Air Force customs is pinned down when it is said that the Air Force cares a great deal more for the substance than for the form. For example, an Air Force officer arriving on a new station is likely to be far more impressed with the warmth of your welcome if you take positive action to help him or her unpack, invite the officer and his or her family to dinner, personally introduce him or her to your friends, or give him or her solid tips on how to tackle his or her new job, than he or she will be by a perfunctory, formal social visit executed with clocklike precision at the proper hour and lasting no more, no less, than the prescribed time. The Air Force has no great amount of patience with mere rote. The officers of the Air Force do not, however, ignore custom. It is, in truth, the great "custom" of the Air Force to inject into old customs grown cold through unthinking, routine obedience to them, a new glow—a new warmth—a new humor that says for itself: "I'm not doing this thing because I should, because I have to—I'm doing it because I want to."

The personalities of several of the top leaders have a direct influence upon the conduct and mannerisms of all officers. The simplicity, cordiality and human warmth, combined with personal dignity, displayed on all occasions and with all people of high or low estate by such leaders as Arnold, Doolittle, Vandenberg, and others will serve as a broad pattern to be emulated by many. The day of the austere leader is waning. A "Good morning, Sergeant," with a friendly smile, in returning a salute, is certainly "military" if the practices of some of our greatest leaders are to establish a pattern.

Sanctity of Official Statements. An officer's official word or statement is ordinarily accepted without question. The knowledge that a false official statement is not only a high crime, but is contrary to the ethics of the military profession, has placed personal and official responsibility for an official statement on a high level.

The Officer-Airman Relationship. The officer strives to develop his or her organization to its maximum efficiency, while providing for his or her people an effective leadership, an impartial justice, a wise and fair attitude in every way. Those things which militate against this necessary result must be avoided. It is a psychological fact that undue familiarity breeds contempt. Officers and airmen have not generally associated together in mutual social activities. No officer could violate this ancient custom with one or two people of his command and convince the others of his unswerving impartiality.

The civilian and the inexperienced cannot possibly understand the officer-airman relationship, nor realize that it often develops into something far deeper and more valuable. Only those officers and airmen who have endured together the grueling hardships of bitter campaigns or the ordeals of battle, can understand. Under those conditions there often develops a mutual trust and complete confidence between officer and airman in which each is carried forward to acts of sacrifice, of courage, and leadership beyond themselves to the end that the one shall not be seen as wanting in the eyes of the other. It may be the force which wins wars.

Provide for the Needs of Airmen. The officer must always provide for the needs and requirements of the airman. Officers in command of or responsible for airmen must provide for their housing, feeding, and comfort before they take care of their own personal needs.

Airmen are used as drivers for military vehicles. Officers who are provided with transportation and a driver must provide for the driver's meals and lodging while on extended trips unless suitably-timed stops are made at military installations where the necessities can be obtained.

Expenses of Air Force personnel and other authorized passengers on a normal cross-country training flight are borne entirely by the personnel involved.

Public Breaches of Discipline and Misconduct of Airmen. Officers are responsible for making proper corrections and taking the necessary actions whenever they see airmen conducting themselves in an improper fashion wherever they are encountered. The officer must not fail to take the necessary preventive or corrective steps because the offenders are not from his or her organization or because he or she is "off duty." The officer is never off duty. Generally it is a matter of protecting the airman from his own indiscretions to save him from more serious trouble.

Approach the delinquent offender in a quiet, dignified, unobtrusive manner. Talk in an impersonal officerlike way. Use a tone of voice no louder than necessary to be heard. Avoid threats. Do not scold or argue. Tell the airman the things he has done which bring discredit upon the uniform he is wearing. Obtain his pass or travel orders, and record his name and organization. If necessary, obtain his name and number from his identification card. Generally a contact of this sort will be sufficient. If the airman has been drinking and is unable to care for himself, further measures should be taken. If the situation is aggravated, security police may be summoned. Like commissioned officers, noncommissioned officers are never "off duty" and recalcitrants may be placed in their charge. Mild offenders may be placed in charge of sober and better behaved associates with a caution to keep the man under control. Airmen will look out for one another. Most officers will avoid turning an airman over to civil police except as a last resort, when the offense warrants it clearly. It is better for the military to look out for themselves. Whatever action is taken, there must be a follow-up or the deterring effect will be likely to vanish as soon as the officer is out of sight. No officer worthy of his position of trust will fail or avoid his responsibilities however unpleasant they may seem to be at the time.

Drunkenness in public, foul language and cursing, and acts which discredit the uniform are the things to curb and control. For the most part, the standards of airmen are high and they will give their loyal support to the necessary measures. The officer must distinguish clearly between acts which are merely unwise or on the boisterous side and those which are actually offensive. The former he or she should ignore. A word of caution should suffice for the borderline cases. With force and judgment he or she must "handle" the others.

RANK HAS ITS PRIVILEGES (RHIP)

General. The military system bears some similarity to a hierarchy. Control is exercised by leaders placed in charge of units in the military structure. These leaders are officers and noncommissioned officers. A disciplined obedience combined with loyalty is required of all, from the highest to the lowest in accordance with law and regulations. Subordinates are required to extend an unfailing respect to the authority which issues their orders. Personal admiration is a voluntary tribute to another which the military service does not demand. But the service does demand, and without equivocation, that respect for authority be manifested by unfailing courtesy to individuals who exercise it.

The privileges of rank do not include the privilege of abuse of position. The needs of the organization as a whole come first. An officer who uses official transportation for personal use is abusing the position. The officer who diverts equipment to his or her own use does likewise. In general, the officer who takes the stand that he or she is above the regulations which guide others, especially his or her own subordinates, is in abuse of his or her position, and his or her acts will be so regarded and condemned.

The privileges of rank and position are privileges indeed, well worth striving for and attaining. They exist in about the same ratio, and for the same reasons, as in other walks of life. Position may mean more in the Air Force than in many civil professions because in most instances the civilian leader is provided far greater pay or monetary allowances.

"I Wish" and "I Desire." When the commander states, "I wish," or "I desire" rather than, "I order you to do so-and-so," this wish or desire has all the force of a direct order.

The Place of Honor. The place of honor is on the right. Accordingly, when a junior walks, rides, or sits with a senior he or she takes position abreast and to the left of the senior. The deference which a young officer should pay to elders pertains to this relationship. The junior should walk in step with a senior, step back and allow the senior to be the first to enter a door, and render similar acts of consideration and courtesy. It has been stated aptly that in the relations between seniors and juniors, the senior will never think of the difference in rank; the junior will never forget it.

Addressing a Senior. The word "Sir" is used in military conversation by the junior officer in addressing a senior, and by all airmen in addressing officers. It precedes a report and a query; it follows the answer of a question. For example: "Sir, do you wish to see Sergeant Brown?" "Sir, I report as Officer of the Day." "Sergeant Brown, Sir." "Thank you, Sir." However, a general officer should be addressed as "General."

When addressing a female senior officer, the word "Ma'am" is used.

Departing Before the Commander. Officers should remain at a reception or social gathering until the commander has departed.

Reception of a Newly Joined Officer. It is a custom that newly joined officers shall be cordially received and many acts of courtesy extended to the officer and family to make their arrival more pleasant and convenient. It is taken for granted that a newly joined officer knows his or her professional duties, and that he or she has every intention of performing them ably.

Whenever conditions permit such niceties, the Base Commander or unit commander sends a letter of welcome to an officer under orders to join, informing him or her of local conditions which may be important or interesting for him or her to know before arrival. The Commander will inquire as to the date and hour of arrival, whether traveling by automobile, airplane or train, and the number of persons accompanying him or her. If the arrival is by airplane or train an officer with transportation is at the station or airport to meet the new arrival. A sponsor officer will be detailed to assist him or her.

If quarters are available they should be made ready for immediate occupancy; if not available, steps should be taken to provide temporary accommodations at the BOQ or elsewhere.

The sponsor usually introduces the newly joined officer to the commander, and to the other officers of the unit. The sponsor should also inform the newcomer as to local regulations and customs which he or she will need to know at once. A copy of the air base regulations and a map of the base are especially useful to the stranger.

Military Weddings. Military weddings follow the same procedures as any other except for additional customs which add to their color and tone.

At military weddings all officers should wear the prescribed service dress uniform for the season. Medals to which the individual is entitled may be worn with propriety, or merely the ribbons, whichever is preferred. Whether the wedding is held in the morning, afternoon, or evening has no bearing on the selection of appropriate uniforms. For the wedding party, it would seem that the thoughtful service member would express a preference as to the uniform to be worn.

Birth of a Child. When a child is born to the family of an officer the base commander sends a personal letter of congratulations to the parents on behalf of the command.

The same procedure is appropriate for a child born to the family of an airman except that the letter of congratulation is written by the organization commander, and the gift, if any is presented, would be from the airman's unit.

Death of an Officer or Member of His Family. When an officer dies an officer is immediately designated by the commander to render every possible assistance to the bereaved family. A letter of condolence is written by the base commander on behalf of the base. Flowers are sent in the name of the officers and spouses of the base.

Death of an Airman. When an airman dies a letter of condolence is written to his nearest relative by the immediate commander of the deceased.

Flowers sent in the name of the members of the deceased's unit accompany the body.

The funeral is attended by all officers and members of the deceased airman's unit, his commander, the band, and other members who so desire and whose duties permit.

SUPPORT OF BASE AND ORGANIZATION ACTIVITIES

General. An officer will be expected to support the activities of his or her unit, such as a wing, group, or squadron, as well as the activities of the entire base. The unit to which an officer is assigned consists of a closely-knit group around which are entwined official duties, and athletic, social, religious, and cultural activities for the benefit of all. You are a member of an official family. Your assignment must mean more than the place where required and official duties are performed, important as they may be. You will be expected to support and assist, at least by your presence, many events which form a part of military life. A proper interest and pride in all activities of your unit and base stimulate morale. An officer should be a good military citizen, sharing with other good citizens responsibility for the unofficial life and activities of the base.

Officers' Open Mess. An officers' open mess is a membership association of commissioned officers, established to provide food and beverage services, entertainment, and social and recreation programs for its members and their dependents and bona fide guests. Depending upon space accommodations, membership may be extended to retired officers of all armed services of the United States, officers of the National Guard and Reserve components, and authorized Department of Defense civilian employees working on the installation, as determined by the installation commander. Since all officers are eligible for voluntary membership, they cannot utilize the facilities of the officers' open mess unless they become members. In this sense, the officers' open mess resembles a private civilian club, since funds used to support the open mess services and programs are primarily self-generated by the members themselves in the form of sales, fees, and dues. Traditionally and historically, the officers' open mess is the social center of the Air Force community life for officers and their families. As such, it is customary for every officer to join the officers' open mess on the base of assignment.

The scope of activities and programs offered by the officers' open mess may vary from base to base. However, membership in your installation officers' open mess assures you of reciprocal privileges in all other officers' open messes of the armed forces. Basically, the benefits and services offered include dining and beverage services, private party catering, special events, check cashing, charge privileges, swimming, tennis, and live entertainment attractions. The check-cashing service is an important and convenient privilege of membership, especially during the evenings, weekends, and holidays when banking facilities are closed. The officers' open mess charge system provides you with a 30-day credit policy without incurring any finance charge. Membership dues may vary from $8 to $15 monthly.

Attendance at Athletic Events is Desirable. As a matter of policy to demonstrate an interest in air base affairs, as well as for personal enjoyment, officers should attend athletic events in which their teams participate.

Attendance at Chapel. Similarly, it is good practice for officers to show their support of base religious activities by periodic attendance at services at the base chapel, to the extent that individual religious preferences permit.

Armed Forces Day. The third Saturday in May is usually designated Armed Forces Day. Undoubtedly each base will hold a special ceremony or review and civilians will be invited to attend. This day unifies all the former service "days" and is symbolic of the unity of action and purpose of all our armed services.

RELIEF FROM ASSIGNMENT

Farewell Tendered a Departing Officer. Prior to the departure of an officer from his or her unit or station on change of assignment or upon retirement, a reception or other suitable function is usually given by the officers and spouses in honor of the departing officer and family.

RETIREMENT CEREMONY

Retirement of an officer or airman is an occasion which merits special commemoration. It is a just reward for long, faithful, and honorable service. For most individuals it is a day of serious realization of the value of military friendships accompanied by heart tugs of regret that these fine associations are to be interrupted. It is a day which will be reached by all military people provided they live, serve honorably, and remain on the active list. Therefore it is most fitting that the occasion be marked by suitable ceremony.

X24A LIFTING BODY.

Retirement of an Officer. The officer's unit or the personnel of the base should arrange for a final ceremony such as a parade in his or her honor with a reception for the officer and family.

A custom followed by many organizations is to present a scroll with suitable heading, signed by appropriate members of the base or unit. Such a remembrance is without cost and therefore not in violation of laws or regulations regarding gifts. If suitable for framing, it is a highly cherished memento.

Friends of the officer being retired and family will wish to hold farewell parties of various informal types.

TABOOS

Uniform Must Not Be Defamed. The officer's uniform and his or her official or social position must not be defamed. Conduct which is unbecoming an officer is punishable under the Uniform Code of Military Justice. The confidence of the nation in the integrity and high standards of conduct of officers is an asset which no individual may be permitted to lower.

Members of Other Services. Never belittle members of the Army, Navy, or Marines. They are part of our team. For that matter, don't downgrade your own service.

Interview Commander Through the Aide or Executive. It is the custom of officers to ask the aide, or the executive, for an appointment with the commander. Often this may be done informally by stating the desire with an inquiry as to when the commander can see you.

Proffer No Excuses. Never volunteer excuses or explain a shortcoming unless an explanation is required. More damage than good is done by proffering unsought excuses.

Servility Is Scorned. Servility, "bootlicking," and deliberate courting of favor are beneath the standard of conduct expected of officers, and any who practice such things earn the scorn of their associates.

"Old Man" to be Spoken with Care. The commander acquires the accolade, "The Old Man," by virtue of his position and without regard to his age. When the term is used it is more often in approbation and admiration than otherwise. However, it is never used in the presence of the commander, and if so used would be considered disrespectful.

Avoid "Going Over an Officer's Head." The jumping of an echelon of command is called "Going over an officer's head." For example, a squadron commander may make a request of the group commander concerning a matter which should first have been presented to the group operations officer. The act is contrary to military procedure and decidedly disrespectful.

Harsh Remarks Are to be Avoided. The conveying of gossip, slander, harsh criticism, and fault finding are unofficerlike practices. In casual conversation it is wiser to follow this guide: "All the brothers are valiant, and all the sisters virtuous." Don't criticize or correct fellow officers in front of airmen.

Avoid Vulgarity and Profanity. Foul and vulgar language larded with profanity is repulsive to most self-respecting people. Its use by officers is reprehensible. They must avoid foulness, repulsiveness, and vulgarity if respect is to be obtained and no officer can lead others unless he or she holds their respect.

Excessive Indebtedness to be Avoided. There are few offenses which injure the standing of an officer more than earning the reputation of being a poor credit risk. Officers are people, and they are subject to the same temptations and the same hazards of life as any other adults. There are times when assumption of debt is unavoidable and necessary. Debts incurred must be liquidated. When circumstances intervene which prohibit the payment which is due, the officer should write or visit the creditor and make an arrangement which is mutually satisfactory. Some payment, however small, should be made at the time a payment is due; this protects the legal standing of the obligation and shows the intention to pay. The practice of permitting bills to accumulate, with no attempt to pay or to arrange a method of payment, is reprehensible and subject to official censure.

Officers enjoy an exceptional individual and group credit standing. It has been earned and deserved because of the scrupulous care officers have taken through the years to meet obligations when due. An individual who violates this custom brings discredit upon the entire officer corps.

Never Lean on a Senior Officer's Desk. Avoid leaning or lolling against a senior officer's desk. It is resented by most officers and is unmilitary. Stand erect unless invited to be seated. Don't lean.

Never Keep Anyone Waiting. Report on time when notified to do so. Never keep any one waiting unnecessarily.

Avoid Having People Guess Your Name. Do not assume that an officer whom you have not seen nor heard from for a considerable period will know your name when a contact is renewed. Say at once who you are, and then renew the acquaintance. If this act of courtesy is unnecessary it will be received only as an act of thoughtfulness, while if it happens to be necessary it will save embarrassment. *At official receptions always announce your name to the aide.*

Don't Trade on Your Commission. An officer is prohibited from using, or permitting others to use, his military title in connection with commercial enterprises of any kind. (See Chapter 8.)

Stay Out of Politics. Don't become embroiled in politics. Political activity is contrary to American military tradition. As a citizen you have a right to your opinions, and a *duty* to vote, but keep your opinions to yourself, within your home, or within your own circle of friends. You can do this without being an intellectual eunuch. Also remember that criticism of the President is particularly improper—he *is* the Commander-in-Chief of the armed forces.

Appearance in Public. Avoid smoking or chewing gum when on the streets in uniform. Keep your hands out of your pockets, your uniform coat (blouse) buttoned, your tie tight, and your cap squared away properly.

What Not to Discuss. Avoid talking shop at social occasions, and it *should* go without saying that classified matters are never discussed, no matter how obliquely. At the mess, use discretion in discussing politics and religion.

CUSTOMS CONCERNING AIRCRAFT

The base Airdrome Officer (AO) meets all transient aircraft and determines crew and passenger transportation requirements. General and flag officers will usually be met by the base commander if practicable.

Pilots of aircraft carrying classified material are responsible for safeguarding it unless it can be removed from the aircraft and stored in an adequately guarded area.

Passenger vehicles are allowed on the ramp or flight line only with the permission of the AO.

The aircraft commander, regardless of rank, is the final authority on operation of his aircraft. If he decides to alter his flight plan, passengers of whatever rank or service will not question his decision.

Aircraft passengers must be prompt, obey safety regulations, and avoid moving around unnecessarily. Don't be one of those that pilots castigate as "waltzing mice"—it requires frequent trimming of the aircraft.

The pilot is the last to leave his aircraft if it is abandoned.

Aircraft flyovers are often used in parades and other ceremonies honoring dignitaries. When airplanes participate in the funeral of an aviator, they will fly in tactical formation, less one aircraft over the graveside service, but not so low as to drown out the service with noise.

SOME ARMY CUSTOMS

Since the Army is the parent of the Air Force, most of our military observances are similar. There are a few differences, however, and the Air Force officer should know them. For instance, the Air Force no longer employs the cannon salute, but it is fired at reveille and retreat at Army posts, and at ceremonies honoring distinguished visitors entitled to gun salutes. On Independence Day a 50-gun salute to the Union is fired, one shot for each state. At this salute one comes to attention, but does not salute.

Organization Day. Army regiments and separate battalions celebrate the anniversary of their units' founding as Organization Day. This festive occasion often includes a parade and review with the reading of the unit's history and feats of arms, an athletic field day, holiday dinner, and ball.

Regimental Traditions. Some of the older Army units are rightly proud of their traditions, and if you are placed with Army units on joint service you should ascertain them. Such units include the oldest Regular Army unit, Battery D, 5th Field Artillery Battalion, commanded by Captain Alexander Hamilton in the Revolutionary War; the 3d Infantry "Old Guard," oldest Regular Army regiment, stationed at Fort Myer, Va., for ceremonial duties in the Capital; the "Cottonbalers" of the 7th Infantry with 60 battle streamers dating back to the Battle of New Orleans; the 9th Infantry with their Liscum Bowl from Peking; and the "Garry Owens" of the 7th Cavalry, Custer's regiment. Some National Guard regiments are even older: Massachusetts' 182d Infantry, formed in 1636 as the Puritans' "Train Bands"; and Mississippi's 155th Infantry, commanded by Colonel Jefferson Davis at Buena Vista in the Mexican War. Unit traditions are upheld in the other services, too, as witness the Navy ships named Wasp, Hornet, Essex, and Bonhomme Richard; the "China Fourth" Marines, and the Fifth and Sixth Marines of Belleau-Wood fame. Despite the organizational changes imposed by modern military technology, we of the Air Force have units with prideful histories, too: Capt. Eddie Rickenbacker's "Hat-in-the-Ring" 94th Fighter Squadron, part of ADC's 1st Fighter Wing at Selfridge AFB; SAC's 2d Bomb Wing, which flew in Billy Mitchell's great Meuse-Argonne bomber offensive; TAC's 4th Fighter Wing, formed from the RAF Eagle Squadrons; TAC's 23d Fighter Wing, formed from Chennault's Flying Tigers of the A.V.G.; and SAC's 19th Bomb Wing, which went on from Clark Field in December 1941 to win 9 U.S., one Philippine, and one Korean Presidential Unit Citations.

"How." The traditional Old Army toast, "How," in drinking one's health is the equivalent of "Cheers," *"Prosit,"* or *"Skoal."* There are several stories as to its origin, and it apparently results from the old Indian-fighting days. One story is that it began in Florida in 1841 during the Seminole Wars, when Chief Coacoochee tried to imitate the officers' toasts by shouting "Hough!" The officers of the 8th Infantry and the 2d Dragoons picked it up and spread it through the Army.

Massing the Colors. Some Army posts, in conjunction with local chapters of the Military Order of the World Wars, have adopted the British ceremony of the Massing of the Colors and the Searchlight Tattoo.

Dancing Old Year Out and New Year In. It is customary for officers and their partners to dance the old year out and the new year in, and formerly a bugler sounded tattoo at about 11:50 p.m., at 12 o'clock taps, and immediately afterwards the orchestra would play the reveille of the New Year.

NAVY CUSTOMS AIR FORCE OFFICERS SHOULD KNOW

Courtesies. Air Force officers will sometimes visit Navy ships. On ships having 180 or more men of the seaman branch, visiting officers of our Armed Services, except when in civilian clothes, and officers of the foreign service are attended by "side boys" when the officer comes on board and departs. This courtesy is also extended to commissioned officers of the armed services of foreign nations. Officers of the rank of lieutenant to major inclusive are given two side boys, from lieutenant colonel to colonel four side boys, from brigadier general to major general six side boys, and lieutenant general and above, eight side boys. Full guard and band are given to general officers; for a colonel, the guard of the day but no music.

During the hours of darkness or low visibility an approaching boat is usually hailed, "Boat ahoy?" which corresponds to the sentry's challenge, "Who is there?" Some of the answers are as follows:

Answer:	*Meaning:* Senior in boat is:
"Aye aye"	Commissioned officer
"No no"	Warrant officer
"Hello"	Enlisted man
"Enterprise"	CO of Enterprise
"Third Fleet"	Admiral Commanding Third Fleet

Similarly, if the commander of the 18th Fighter Wing is embarked or the commander of Langley Air Force Base, the answers would be "18th Wing" or "Langley Air Force Base."

On arrival, at the order "Tend the side," the side boys fall in fore and aft of the approach to the gangway, facing each other. The boatswain's mate-of-the-watch takes station forward of them and faces aft. When the boat comes alongside the boatswain's mate pipes and again when the visiting officer's head reaches the level of the deck. At this latter instant the side boys salute.

On departure, the ceremony is repeated in reverse, the bos'n's mate begins to pipe and side boys to salute as soon as the departing officer steps toward the gangway between the side boys. As the boat casts off the bos'n's mate pipes again. (Shore boats and automobiles are not piped.)

You uncover when entering a space where men are at mess and in Sick Bay (Quarters) if sick personnel are present. You uncover in the wardroom at all times if you are junior. All hands except when under arms uncover in the captain's cabin.

Admirals and captains when in uniform fly colors astern when embarked in boats. When on official visits they also display their personal flags (pennants for commanding officers) in the bow. Flag officers' barges are distinguished by the appropriate number of stars on each side of the barge's hull. Captains' gigs are distinguished by the name or abbreviation of their ships surcharged by an arrow.

Visiting. Where gangways are rigged on both sides, the starboard gangway is reserved for officers and the port for enlisted. Stress of weather or expedience, in the discretion of the officer of the deck (OOD), may make either gangway available to both officers and enlisted.

Seniors come on board ship first. When reaching the deck you face toward the ensign (or aft if no ensign is hoisted) and salute the ensign (quarterdeck). Immediately thereafter you salute the OOD and request permission to come on board. The usual form is, "Request permission to come on board, Sir." The OOD is required to return both salutes.

On leaving the ship the inverse order is observed. You salute the OOD and request permission to leave the ship. The OOD will indicate when the boat is ready (if a boat is used). Each person, juniors first, salute the OOD; then faces toward the ensign (quarterdeck), salutes, and embarks.

The OOD on board ship represents the captain and as such has unquestioned authority. Only the executive and commanding officer may order him relieved. The authority of the OOD extends to the accommodation ladders or gangways. He is perfectly within his rights to order any approaching boat to "lie off" and keep clear until in his judgment he can receive her alongside.

The OOD normally conveys orders to the embarked troops via the troop commander but in emergencies he may issue orders direct to you or any person on board.

The *bridge* is the "Command Post" of the ship when underway, as the quarterdeck is at anchor. The officer-of-the-deck is in charge of the ship as the representative of the captain. Admittance to the bridge when underway should be at the captain's invitation or with his permission. You may usually obtain permission through the executive officer.

The *quarterdeck* is the seat of authority; as such it is respected. The starboard side of the quarterdeck is reserved for the captain (and admiral if a flagship). No person trespasses upon it except when necessary in the course of work or official business. All persons salute the quarterdeck when entering upon it. When pacing the deck with another officer the place of honor is outboard, and when reversing direction each turns toward the other. The port side of the quarterdeck is reserved for commissioned officers, and the crew has all the rest of the weather decks of the ship. However, every part of the deck (and the ship) is assigned to a particular division so that the crew has ample space. Not unnaturally every division considers it has a prior though unwritten right to its own part of the ship. For gatherings, such as smokers and movies, all divisions have equal privileges at the scene of assemblage. Space and chairs are reserved for officers and for CPO's, where available, and mess benches are brought up for the enlisted. The seniors have the place of honor. When the captain (and admiral) arrive those present are called to attention. The captain customarily gives "carry on" at once through the executive officer or master-at-arms who accompanies him to his seat.

Messes. If you take passage on board a naval vessel you will be assigned to one of several messes on board ship, the wardroom, or junior officers' mess. In off-hours, particularly in the evenings, you can foregather there for cards, yarns, or reading. Generally a percolator is available with hot coffee.

The executive officer is ex-officio the president of the wardroom mess. The wardroom officers are the division officers and the heads of departments. All officers await the arrival of the executive officer before being seated at lunch and dinner. If it is necessary for you to leave early ask the head of your table for permission to be excused as you would at home. The seating arrangement in the messes is in order of seniority.

Naval officers are required to pay their mess bills in advance. The mess treasurer takes care of the receipts and expenditures and the management of the mess. The mess chooses him by election every month. When assigned to a mess you are an honorary member. Consult the mess treasurer as to when he will receive payment for mess bills. Your meals are served by stewards who in addition, clean your room, make up your room, make up your bunk, shine your shoes. This is their regular work, hence they are not tipped.

The Cigar Mess is the successor of the old Wine Mess. You may make purchases from this mess, for example, of cigarettes, cigars, pipe tobacco and candies. The cigar mess treasurer will make out your bill at the end of the month or before your detachment. Before you are detached be sure that the mess treasurer and the cigar mess treasurer have sufficient warning to make out your bills before you leave. Once a ship has sailed long delays usually occur before your remittances can overtake it. The unpaid mess bill on board is a more serious breach of propriety than the unpaid club bill ashore because of the greater inconvenience and delay in settlement.

Calls. Passenger officers should call on the captain of the ship. If there are many of you, you should choose a calling committee and consult with the executive officer as to a convenient time to call. The latter will make arrangements with the captain.

Ceremonies. Gun salutes in the Navy are the same as in the Army except that flag officers below the rank of fleet admiral or general of the Army or Air Force are, by Navy regulations, given a gun salute upon departure only.

Other Naval Customs. The commander of any ship, regardless of his grade, is addressed as "Captain." Navy officers below the rank of commander are addressed as "Mister," chief petty officers as "Chief," and other ratings usually by their last names. Marine lieutenants are also often addressed as "Mister" (this was also the practice in the Old Army), and Marine NCOs by their rank titles.

The naval services do not salute when uncovered.

If overtaking a senior, draw abreast of him on the left, slow down, salute and say, "By your leave, Sir." The senior normally will reply "Granted." A senior will overtake a junior on the right and the junior initiates the salute when he sees the senior out of the corner of his eye.

In sending verbal messages, the junior uses the formula: "My respects to Captain Hornblower, and ...." The senior uses the phrase: "My compliments to Mr. Roberts, and ...." In correspondence, a senior "calls attention" to something; a junior only "invites" it. A junior closes a letter with "Very respectfully"; a senior, with "Respectfully."

ORIGIN OF SOME MILITARY CUSTOMS

The newly-commissioned officer may find some military customs more meaningful if he or she understands how they began. The origins of many are shrouded in antiquity, and some of the stories may be apocyrphal, but they should be of interest. Here they are:

The dress parade was originally intended to impress visiting emissaries with the strength of the monarch's troops, rather than to honor the visitor.

Inspecting the guard of honor began with the restoration of Charles II to the throne of England. When one of Cromwell's regiments offered its allegiance, the King carefully scrutinized the face of each soldier in ranks looking for signs of treachery. Convinced of their sincerity, he accepted the escort.

The "Sound off" in which the band plays the "Three Cheers" and marches down the front of assembled troops stems from the Crusades. Those selected as Crusaders were stationed at the right of the line of troops, and the band marched past them in dedication, while the populace gave three cheers. "The Right of the Line" was the critical side in ancient battle formations and is the unit place of honor in ceremonies.

The saber salute, still employed on occasion of ceremony in the other services, demonstrated the officer's dedication to Christ in his bringing the hilt (symbolic of the Cross) to his lips, while dipping the saber tip signified submission to his liege lord commanding him.

Precedence among units is determined by age, and for that reason Air Force units usually follow the sister services in parades.

THE MACH 3 SR 71

Raising the right hand in taking the oath stems from ancient days when the taker called upon God as his witness to the truth and pledged with his sword hand. If gloved, the right glove is removed, supposedly because criminals were once branded on the right palm, and it could thus be determined if one were a reputable witness.

Wearing decorations and medals on the left breast also goes back to the Crusaders, who wore the badge of honor of their order over their hearts and protected by their shields which were carried on the left arm.

The origin of the aiguillette and the fourragere is much disputed. One story has it that squires used to carry metal-tipped thongs (aiguillette is French for little spike, or lace-tag) in a roll over their shoulder to lace their knights into their armor, and it thus became the badge of an aide-de-camp. Another yarn concerns the fourragere, which states that Marlborough's foragers carried fodder sacks for their mounts attached to their shoulders by a looped-up cord hooked to the jacket.

The white flag of truce may derive from the Truce of God arranged on certain days by Pope Urban V in 1095 between warring medieval barons.

The use of the arch of sabers in military weddings recalls the days when the groom's men pledged to protect the wedded couple, particularly elopers.

The term "chaplain" supposedly derives from St. Martin of Tours, who gave half his military cloak or "cappa" to a beggar. The kings of France made of it a relic and war talisman, which was guarded by special clerical custodians, who celebrated services in the field.

Saluting a ship's quarterdeck is believed to derive from ancient times when a pagan altar was carried in the ship's stern to propitiate the gods. Later the Christian crucifix was carried there and it is still the seat of authority.

At a military funeral the national flag is laid on the coffin to indicate that the dead died in service of the state, which acknowledges its responsibility.

The firing of three volleys over the grave may derive from the Romans who honored their dead by casting earth thrice on the grave, calling the name of the dead, and saying "Vale" (farewell) three times.

Taps is played at a military funeral to mark the beginning of the serviceman's last sleep and to express hope in the ultimate reveille.

STRATEGIC AIR COMMAND HEADQUARTERS LOCATED AT OFFUTT AFB, EIGHT MILES SOUTH OF OMAHA. A MINUTEMAN MISSILE IS ON DISPLAY IN FRONT OF THE BUILDING.

5

Pay and Personal Allowances

The nation which forgets its defenders will itself be forgotten.

—Calvin Coolidge, 1920

An officer's pay and allowances are composed in the main of four fundamental elements (Department of Defense Military Pay and Allowances Entitlement Manual—DODPM)

1. Basic Pay.
2. Basic Allowance for Subsistence.
3. Basic Allowance for Quarters.
4. Incentive and Special Pay.

FUNDAMENTAL ELEMENTS

Basic Pay. Pay table information is usually readily available for reference by the Serviceman at the accounting and finance office and the personal affairs office serving the station or area in which he or she is assigned for duty. It also appears from time to time in various journals and periodicals circulated primarily to the military and therefore is not included in this text. The service creditable for basic pay will be computed as prescribed in DODPM.

Basic Allowances. The Subsistence Allowance is the same for all officers of all grades, with or without dependents. ($98.17 per month.)

The amount of pay provided as *Basic Allowance for Quarters* for officers authorized but not occupying Government quarters differs between officers having dependents and those without dependents. A detailed definition of what constitutes *"dependents"* is found later in this chapter.

The monthly quarters allowance for officers varies with the pay grade and there are two schedules, with dependents and without dependents. The dependents schedule is the same regardless of the number of dependents you have.

BASIC ALLOWANCE FOR QUARTERS (BAQ)

Pay Grade	Without Dependents Full Rate	Without Dependents Partial Rate	With Dependents
O-10	508.50	50.70	636.30
O-9	508.50	50.70	636.30
O-8	508.50	50.70	636.30
O-7	508.50	50.70	636.30
O-6	456.60	39.60	556.80
O-5	420.90	33.00	506.70
O-4	374.70	26.70	452.10
O-3	329.40	22.20	406.50
O-2	286.20	17.70	361.80
O-1	223.50	13.20	290.70

Variable Housing Allowance. The Nunn-Warner Bill of 1980 authorizes additional quarters allowances up to the difference between 115 percent of the Basic Allowance for Quarters and the average cost of housing (off-base) in each area.

Hostile Fire Pay. As prescribed in regulations of the Department of Defense, except in time of war declared by Congress a service member may receive hostile fire pay of $65 for any month in which he or she draws basic pay *and* was:

(1) Subject to hostile fire or explosion of hostile ordnance;

(2) On duty in an area in which he or she was in imminent danger of being exposed to hostile fire or explosion of hostile ordnance; or

(3) On duty in a Hostile Fire Area as designated by DODPM.

Special Pay for Medical, Dental, and Veterinary Officers. In addition to regular pay, the Uniformed Services Health Professionals Special Pay Act of 1980 provides special incentive pay for medical and dental officers.

Pay and Allowances. With the foregoing facts in mind, an officer may determine from the tables the four fundamental elements of his pay and allowances.

Incentive pay for duty as a crew member as determined by the Secretary of the Air Force, involving frequent and regular participation in aerial flights, will be at the rates shown in the table.

Officers in pay grade O-4 and above are authorized to elect to live off base and receive the BAQ for their grade without dependents even if government quarters are available.

Personal money allowances are authorized as follows: Lieutenant General, $500 a year; General, $2,200.00 a year; The Chief of Staff, USAF, $4,000 a year.

INTERMITTENT AND NON-RECURRING ELEMENTS OF PAY AND ALLOWANCES

Family Separation Allowance. This allowance is payable only to service members with dependents. The Family Separation Allowance is of two types, both of which are payable to the same individual if he or she meets the qualifications for both listed below. Each type of allowance may be paid to a member who qualifies therefor in addition to any other allowance or per diem to which he or she may be entitled.

(1) One allowance, equal to the basic monthly *allowance for quarters payable to a member in the same pay grade without dependents,* is payable to a member with dependents, who is on permanent duty in Alaska or anywhere outside the United States, when:

(a) Movement of dependents to the permanent station or place nearby is not authorized at government expense, and dependents are not residing at or near the station; and

(b) No government quarters are available for assignment to him or her.

(2) The other type of Family Separation Allowance, equal to $30 per month, is payable (except in war-time or national emergency declared by Congress) to a member with dependents who is authorized a basic quarters allowance, when:

(a) Movement of dependents to the permanent station or place nearby is not authorized at government expense and dependents are not residing at or near the station; and

(b) He or she is on duty aboard ship away from the ship's home port continuously for more than 30 days *or*

(c) He or she is on temporary duty away from his permanent station continuously for more than 30 days and dependents do not reside at or near the temporary duty station.

Uniform Allowances. *Reserve Officers. Upon Initial Appointment.* A newly appointed Reserve officer is entitled to $200 to purchase required uniforms. This accrues upon first reporting for active duty for a period in excess of 90 days; upon completion as a member of a Reserve component of not less than 14 days active duty or active duty for training; or completion of 14 periods of not less than two hours each in the Ready Reserve where wearing of the uniform is required. An additional allowance of $100 may be authorized after 90 days of duty.

The Allowance for Re-entry upon Active Duty. A Reserve officer who has been on inactive duty for a period of two years, and is then recalled to active duty of more than 90 days duration, may qualify for $100 as reimbursement for additional uniforms and equipment required on such duty.

Dislocation and Trailer Allowances. *Dislocation Allowance.* The allowance is to partially reimburse a service member for the expense incurred in closing out his household at one location and setting it up again at a new place when he is transferred on permanent change of station. The allowance is an amount equal to the service member's monthly basic allowance for quarters. It may be paid only once in connection with any one permanent change of station. It can be paid only when dependents have completed travel in connection with a permanent change of station where transportation of dependents is authorized to be furnished or travel allowances are authorized to be paid. It cannot be paid for the movement from home to first duty station, nor from last duty station to home. Under normal conditions service members are not entitled to payment of the dislocation allowance more than once in any fiscal year except where the Secretary of the Air Force finds that the exigencies of the service require the additional moves.

The following conditions govern the application of the statutory limitation of one dislocation allowance payment per fiscal year:

a. For the purpose of determining the fiscal year in which entitlement to a dislocation allowance occurs, the date of departure (detachment) from the old permanent duty station in compliance with permanent change of station orders shall govern.

b. Prior permanent changes of station in the same fiscal year for which a dislocation allowance was not authorized shall be excluded from computation.

c. Permanent changes of station to, from, or between courses of instruction conducted at an installation of the uniformed services shall be excluded from computation.

Trailer Allowance. The trailer allowance is authorized in lieu of transportation of baggage and household effects when a member transports a house trailer or mobile dwelling within the United States for use as a residence, provided the member would otherwise be entitled to transportation of baggage and household effects. A member receiving a trailer allowance would not be authorized to receive a dislocation allowance for the same move.

The allowance is payable in advance of the movement of the trailer.

Travel Allowance. An allowance of $50 per diem and 13 cents a mile for the official distance, up to $300 per diem, is authorized for officers moving under PCS orders.

Station Allowances, Overseas. Station allowances to equalize the cost-of-living and housing are authorized for uniformed personnel stationed in a number of foreign countries. The allowances are planned to help offset the total of selected essential expenditures of the foreign station with those in the United States. It is to be noted that there must be an officially computed difference in the cost of living factors which are considered, as to the United States and oversea stations, and the cost greater in the oversea station, for the allowance to be authorized.

Advance Pay. Advance pay of not more than three months' net basic pay may be drawn when an officer is ordered to change permanent stations.

DEPENDENTS

Definition. The term "dependents" shall include at all times and in all places the lawful spouse and unmarried children, under twenty-one years of age, of any member of the uniformed services, except as hereinafter limited. Such term shall include the father or mother of such member, *provided he or she is in fact dependent on such member for over half of his or her support. It shall also include unmarried children, over twenty-one years of age, of such member who are incapable of self-support because of being mentally defective or physically incapacitated,* and who are in fact dependent on such member for over half of his or her support: *Provided,* That the term "children" shall be held to include stepchildren and adopted children when such stepchildren or adopted children are in fact dependent upon such member: *Provided further,* That in the case of female members of the uniformed services, the term "dependent" shall include a husband in addition to those persons otherwise defined as dependents in this subsection.

The term "father" or "mother" shall include a stepparent, or parent by adoption, and any person, including a former stepparent, who has stood in loco parentis to the person concerned at any time for a continuous period of not less than five years during the minority of such member: *Provided,* That a stepparent-stepchild relationship shall be deemed to be terminated by the stepparent's divorce from the blood parent: *Provided further,* That no member claiming a dependent may be paid increased allowances on account of such dependent for any period during which such dependent is entitled to receive basic pay for active or training duty.

Credit for Dependents. Credit for basic allowances for quarters because of a dependent parent will be made only where the dependent parent actually is residing in the household of the officer claiming the increased allowance.

DEDUCTIONS FOR SOCIAL SECURITY TAX

The Air Force is obliged by law to deduct Social Security taxes from individual pay. The percentage deducted and the amount on which it is computed has edged upward gradually. For details, check your finance office.

DEDUCTIONS FOR FEDERAL INCOME TAX

Members of the armed services are subject to the federal law providing for payroll deductions for income tax. Taxable income includes all pay, but does not include subsistence and quarters allowances.

DEDUCTIONS FOR STATE INCOME TAX

Members of the armed services are subject to the laws of their state of official residence providing for payroll deductions for state income taxes.

INCOME TAX EXEMPTION

Taxation of Pay, Retired Service Personnel. Members of the armed services who are retired for length of service or age pay the same Federal income taxes as other citizens. The Accounting and Finance Officer makes the same tax withholdings as for active duty officers.

Members retired for physical disability receive an important exemption in the computation of federal income tax. The amount of their exemption depends upon the degree of disability and is applied against their active duty base pay at the time of retirement. An example will clarify the benefit. Assume an officer with active duty base pay of $1,085, retired with 30% disability. By tax rulings, apply the 30% to the active duty base pay, which is $325.50, the amount exempt from income tax. The retired pay is $813.75. Deduct the non-taxable $325.50, and the remainder of $488.25 is taxable. Moreover, this officer could deduct $5,200 as sick pay ($100 weekly) until reaching statutory retirement age, under an Internal Revenue ruling. He or she would therefore pay no tax at all if retired pay is his or her sole income.

ALLOTMENTS AND STOPPAGES

Allotments. The Secretary of the Air Force is authorized to permit all personnel to make allotments from their pay, under such regulations as he or she may prescribe, for the support of their families or relatives or for other proper purposes which in his or her discretion warrant such action.

The word "allotment" as used in this chapter refers to a definite portion of the pay of an officer which is authorized to be paid to another person or institution in a manner prescribed by the Secretary.

The "allotter" is the person who makes the allotment.

The "allottee" is the person or institution to whom the allotment is made.

Classification. Allotments are divided into classes, some of which follow:

Class B. Purchase of Government Bonds.

Class C. Charity Drive Donation (Combined Federal Campaign).

Class D. Allotments made to dependents.

Class H. Repayment of home loans.

Class I. Commercial life insurance.

Class N. Allotments made for the payment of premiums on National Service Life Insurance.

Term of allotment. Generally allotments are made for an indefinite period beginning on the first of the month for which deduction is to be made from the allotter's pay, but not prior to date of commission or appointment.

While Prisoner of War or Reported Missing. Any officer who is interned or a prisoner of war or reported missing, and who has made an allotment of pay for the support of dependents or for the payment of insurance premiums, is entitled to have such allotments or insurance deductions continued for a period of 12 months from date of commencement of absence. Allotments of officers under any of the above conditions may not continue beyond a 12 months' period following the officially reported date of commencement of absence, except that when the 12 months' period from date of commencement of absence is about to expire and no official report of death or of being a prisoner of war or of being interned has been received, the head of the department concerned will cause a full review of the case to be made. Following such review and when the 12 months' absence has expired, or following any subsequent review of the case, the head of the department concerned is authorized to direct the continuance of the officer's missing status, if he may reasonably be presumed to be living. Such missing officers continue to be entitled to have pay and allowances credited, and payment of allotments as authorized to be continued, increased, or initiated.

In the absence of a previously executed allotment, or where the allotment made is not sufficient for reasonable support of a dependent and for the payment of insurance premiums, the head of the department concerned may direct that an allotment, not to exceed the total basic pay of the person concerned, be paid by the appropriate accounting and finance officer to the insurer or such dependents as has been designated in official records or, in the absence of such designation, to such person as may be determined by the head of the department concerned, or by such person as he may designate, to be a bona fide dependent.

Authorized Stoppages. The pay of officers may be withheld on account of an indebtedness to the United States admitted or shown by the judgment of a court, but not otherwise unless specifically ordered by the Secretary of the Air Force.

CREDITABLE SERVICE FOR BASIC PAY

DODPM establishes the rules under which the various conditions of active duty may be counted in computing cumulative years of service for credit toward basic pay. This manual is of such length and detail that any officer having an unusual background of service is urged to peruse it carefully. The following are general statements of certain provisions of the manual.

Active Duty. Full time credit is given for all periods of active duty served in *any*

Regular or Reserve component of *any* uniformed service in *any* military status, commissioned or enlisted.

Retired Status. Additional credit is given for any period of time while the individual is on any retired list. This provision does *not* operate to increase retired pay, but should the individual be returned to active duty from a retired list he or she benefits from inclusion of his or her time on the retired list in computing basic pay.

Academies. Service as a cadet or midshipman in the Service Academies or ROTC is *not* creditable.

Fraudulent Enlistment. Service in a fraudulent enlistment *not voided by the Government* is credited.

Beginning Dates of Service. Service as an officer is counted from the date of acceptance of appointment; as an enlisted man from the date of enlistment.

REIMBURSEMENT FOR TDY TRAVEL EXPENSES

Travel at Personal Expense. When authorized travel is performed at personal expense the member will be reimbursed a monetary allowance in lieu of transportation at the rate of 16 cents per mile for the official distance in addition to authorized per diem.

Travel by Privately-owned Conveyance. For travel actually performed by privately-owned conveyance under orders authorizing such mode of transportation as more advantageous to the Government, an officer will be paid a monetary allowance in lieu of transportation at the rate of 16 cents per mile for the official distance in addition to authorized per diem. Military passengers on TDY may also claim 16 cents per mile.

Per Diem. The normal per diem allowance within the United States is $50.00. This allowance covers the cost of quarters, subsistence, and other necessary expenses. In rare cases, a per diem allowance up to $75.00 may be authorized.

In instances requiring submission of certificates relative to use or non-use of Government quarters and messing facilities, you must adhere to the provisions of paragraph M4451 of the Joint Travel Regulations if you wish to receive proper per diem reimbursement.

Interrupted Leave. When an officer is called back to his permanent station within twenty-four hours after his departure on a leave of five or more days, his travel will be at government expense. The government will also pay the cost of returning the officer to his point of leave if resumption of leave is feasible.

Reimbursable Travel Expenses. When traveling under orders, keep a detailed record of all expenses, however minor. Many are reimbursable. Note the following list of reimbursable items:

Taxi Fares. Reimbursement is authorized for taxi fares between places of abode or business and stations, wharves, airports, other carrier terminals or local terminus of the mode of transportation used, between carrier terminals, while en route when free transfer is not included in the price of the ticket or when necessitated by change in mode of travel, and from carrier terminals to lodgings and return in connection with unavoidable delays en route incident to the mode of travel. Itemization is required.

Allowed Tips. Tips incident to transportation expenses are reimbursable as follows: Tips to Pullman porters, not to exceed $1 per day. Tips of 15 percent to taxi drivers. Tips to baggage porters, red caps, etc., at not to exceed customary local rates, but not including tips for baggage handling at hotels; the number of pieces of baggage handled will be shown on the claim. Itemization is required.

Checking and Transfer of Baggage. Expenses incident to checking and transfer of baggage are reimbursable. The number of pieces of baggage checked will be shown on the claim. Itemization is required.

Excess Baggage. When excess baggage is authorized, actual costs for such excess baggage in addition to that carried free by the carrier are reimbursable.

Government Aircraft. Cost of gasoline, oil, repairs, nonpersonal services, guards, and storage are reimbursable when such expenses are necessary by reason of landing at other than a Government field. Receipts are required.

Government Auto. Cost of storage of Government automobiles when necessary is reimbursable if Government storage facilities are not available. Receipt is required.

Telephone, Telegraph, Cable, etc. Cost of official telephone, telegraph, cable, and similar communication services is reimbursable when incident to the duty enjoined or in connection with items of transportation. Such services when solely in connection with reserving hotel room, etc., are not considered official. Copies of messages sent are required for all mechanical transmissions unless the message is classified in which case a full explanation and a receipt will suffice. Local verbal transmissions are allowable when itemized. Long distance verbal transmissions will be fully explained in the claim.

Local Public Carrier Fares. Expenses incident to travel on streetcar, bus, or other usual means of transportation may be allowed in lieu of taxi fares under the conditions and limitations stated. Itemization is required.

Toll Fares. Ferry fares, and road, bridge, and tunnel tolls are reimbursable when travel is performed by special conveyance or Government highway transportation. Itemization is required.

Local Transportation. Reimbursement is authorized for transportation obtained at personal expense in the conduct of official business within and around either the permanent or temporary duty station. See JTR, Vol. I, Part K.

Non-Reimbursable Travel Expenses. The examples of travel expenses listed below are NOT payable from Government funds:

(1) Expenses incurred during period of travel which are incident to other duties (such as traveling aboard a vessel in performance of temporary duty on such vessel).

(2) Travel under permissive orders to travel in contrast to orders directing travel requires the individual to pay the costs.

(3) Travel under orders but not on public business. Example: travel as a participant in an athletic contest. Such travel may be paid from unit or command welfare funds which are generated through operation of exchanges and motion picture theaters. Reimbursement for such travel is not authorized from appropriated funds.

(4) Return from leave to duty abroad. Unless Government transportation is available, such as space on a transport, the individual on leave in the United States from an oversea command must defray his own return expenses.

(5) Attendance at public ceremonies or demonstrations whose expenses are borne by the sponsoring agency.

Travel Outside the United States. Travel expenses for travel outside the United States are furnished in advance or are reimbursable on essentially the same basis as temporary duty travel performed in the United States. That is to say, the traveler is entitled to the costs of transportation, per diem allowance, and cost of incidental necessary expenses.

A deduction of 54 percent of the travel per diem allowance will be made when Government quarters are available to the traveler on that day.

An officer will certify that Government quarters were available even though, in connection with the occupancy of such quarters, he was required to pay incidental room or service charges.

Certificates are required from the commander, or his designated representative, of an installation at which a traveler performs temporary duty, as shown below:

> "I certify that Government quarters as defined in pars. 1150-5, JTR, were not available to .... on the following dates: .... and that a Government mess, as defined in pars. 1150-4, JTR, was not available on the following dates for the number of meals indicated: (Feb. 22 (1); 23 (3); 24-26 (9); 27 (0) ...."

The traveler will certify that quarters and mess were not utilized for the same periods.

Amount of Oversea Travel Allowances. The allowances for TDY travel costs overseas are variable. The traveler may expect that necessary expenses will be borne without use of personal funds provided he or she uses reasonable care.

Before undertaking travel outside the United States the traveler should consult local transportation or accounting and finance officers to determine the exact amount of the expected allowances which are stated in Appendix A, Joint Travel Regulations.

Preparation of Vouchers for Payment. Consult the local accounting and finance officer.

Advance Payment of Travel and Transportation Allowances. Travel and transportation allowances are authorized to be paid in advance, except in connection with retirement.

Advance payment is not authorized for movement of household goods.

WEIGHT ALLOWANCES, CHANGE OF STATION

The weight allowances authorized for shipment of household goods at government expense on change of station are as shown in the accompanying table.

TABLE OF WEIGHT ALLOWANCES (Pounds)

Grade*	Temporary Change of Station	Permanent Change of Station
General of the Air Force and General	**2,000	***24,000
Lieutenant General	1,500	***18,000
Major General	1,000	***14,500
Brigadier General	1,000	13,500
Colonel	800	13,500
Lieutenant Colonel	800	13,000
Major (and warrant officer in pay grade W-4)	800	12,000
Captain (and warrant officer in pay grade W-3)	600	11,000
1st Lieutenant (and warrant officer in pay grade W-2)	600	10,000
2d Lieutenant (and warrant officer in pay grade W-1)	600	9,500

*Members of reserve components and officers holding temporary commissions in the Air Force of the United States are entitled to weight allowances for corresponding relative grades listed. The weight allowance of an individual is based upon his grade or rating at the time of his detachment from the last duty station.

**Exception to this limitation may be authorized by the respective Secretaries for the Chiefs of Staff, U.S. Air Force and Army, and Chief of Naval Operations in such additional amounts, not exceeding 2,000 pounds, as they may consider appropriate.

***See footnote 3 on page 8-4 of JTR.

6

Uniforms and Insignia

A soldier must learn to love his profession, must look to it to satisfy all his tastes and his sense of honor. That is why handsome uniforms are useful. —Napoleon

The Air Force is a uniformed service and, as such, requires all members to maintain a high standard of dress and personal appearance. Pride in each member's daily personal appearance and uniform wear greatly enhances the esprit de corps essential to an effective military force. Officers who violate the specific prohibitions of AFR 35-10 are subject to appropriate administrative action and/or prosecution under the Uniform Code of Military Justice.

Officers are required to provide, at their own expense, many uniforms and uniform items. For information concerning uniform allowances for newly commissioned officers, see Chapter 5.

WEARING THE UNIFORM

When to Wear the Uniform. Air Force personnel will wear the prescribed service uniform at all places of duty during duty hours, except as specifically authorized by AFR 35-10. The appropriate uniform combinations will be designated by the installation commander in accordance with local climatic and mission requirements. Those occasions requiring wear of the dress uniform are described later in this chapter.

Standards of Dress and Appearance. As previously stated, all Air Force personnel are required to maintain a high standard of dress and appearance. This is especially true for officers, whose manner of dress and personal appearance provide a visual example for enlisted personnel. The uniform should be worn in a manner which emphasizes pride, and an individual's personal appearance should be above reproach.

The elements of the Air Force standard are neatness, cleanliness, safety, and military image. The first three are absolute, objective criteria required for the efficiency and well-being of the Air Force. The fourth standard, military image, is subjective in that the American public and its elected representatives draw certain conclusions based on the image presented by Air Force members. The military image, therefore, must instill public confidence and leave no doubt that Air Force members live by a common standard and respond to military order and discipline.

Appearance in uniform is an important part of the military image. Because judgment as to what constitutes the proper image differs in and out of the military, the Air Force must spell out what is and what is not acceptable. The image of a disciplined and reliable service member excludes the extreme, the unusual, or the fad.

Uniform Standards. Uniforms must be clean, neat, correct in design and specification, properly fitted, pressed and in good condition (i.e., not frayed, worn out, torn, faded, patched, or stained). Uniform items, including pockets, are to be buttoned, snapped or zippered as appropriate. Shoes must be shined and in good repair. Badges, insignia, belt buckles, and other metallic devices are to be kept clean and free from scratches or corrosion. Ribbons must be clean and not frayed.

Wristwatches and identification bracelets may be worn, but they should be of conservative design. The identification bracelet must be no wider than one inch, and must not subject the wearer to potential injury. Rings may also be worn. However, only a total of three rings on both hands at any one time is permitted.

Female members with pierced ears may wear, on a daily basis, small, conservative, spherical earrings in gold, silver or white pearl. When worn, earrings will fit flat against the ear and will not extend below the earlobe. Earrings will not be worn when safety considerations dictate otherwise.

Personal Appearance and Grooming Standards. One of the most important elements of an officer's personal appearance is his or her hair. Hair must be clean, well-groomed, and neat. Extreme or fad hairstyles, such as mohawks, ducktails, or corn rows, must not be worn. If hair is dyed, it must look natural. When groomed, hair must not touch the eyebrows or protrude in front below the band of properly worn headgear. Female members, however, must have their hair visible when wearing the beret.

For males, hair must not touch the ears, and only the closely shaved hair on the back of the neck may touch the collar. The bulk of a male's hair may not exceed 1¼ inches. Sideburns can be no longer than the lowest part of the exterior ear opening. Females may not wear their hair longer than the back bottom edge of their shirt collar. Hair bulk may not exceed 3 inches.

Hair ornamentation is not permitted. However, to keep their hair in place, females may wear plain pins, combs or barrettes similar in color to their own hair.

Males who choose to wear mustaches must insure that they do not extend downward beyond the lip line of the upper lip or sideways beyond a vertical line drawn upward from the corner of the mouth. Handlebar mustaches are prohibited. Beards are also prohibited. However, on the advice of a medical officer, the commander may authorize otherwise.

CLASSIFICATION OF UNIFORMS

Uniforms worn by USAF personnel are categorized into three groups: (1) service uniforms, (2) utility uniforms, and (3) dress uniforms.

Service Uniforms. The service uniform is worn during regular duty hours while performing assigned work. All service uniform combinations are authorized for year-round wear. However, installation commanders may prescribe when members will wear certain combinations. If nothing has been prescribed, good judgment, based on weather conditions and duties, should be used when selecting the particular service uniform combination to be worn.

Personnel assigned to a non-Air Force military installation will wear the service uniform which most closely matches the host service order of dress.

The blue maternity uniform is mandatory for wear by all pregnant members.

Utility Uniforms. The utility uniform is worn whenever mission requirements render wear of the service uniform inappropriate. The utility uniform is not to be worn to off-base businesses if extended shopping, dining, or socializing is anticipated or intended. When wearing the utility uniform, especially if the wearer will be making an off-base stop, members must present the proper standards of cleanliness, neatness, and military image.

A maternity utility uniform is currently under development. However, until it becomes available, commanders may authorize the wearing of civilian clothes when duties require wearing of the utility uniform.

Dress Uniforms. The formal evening mess dress and the winter and summer mess dress uniforms are authorized for year-round wear at the discretion of the installation commander. The black mess dress uniform is normally worn during the winter months, while the white mess dress is worn in the summer.

REQUIRED COMBINATIONS

Service Uniforms. *Men's Combination 1.* Dark blue service coat and trousers of matching shade and fabric; dark blue belt with silver buckle and tip; light blue long-sleeved shirt with epaulets; undergarments; black socks and shoes; dark blue necktie; service cap or flight cap; name tag; ribbons, badges, and insignia.

Men's Combination 4. Same as Combination 1 above, except the dark blue ser-

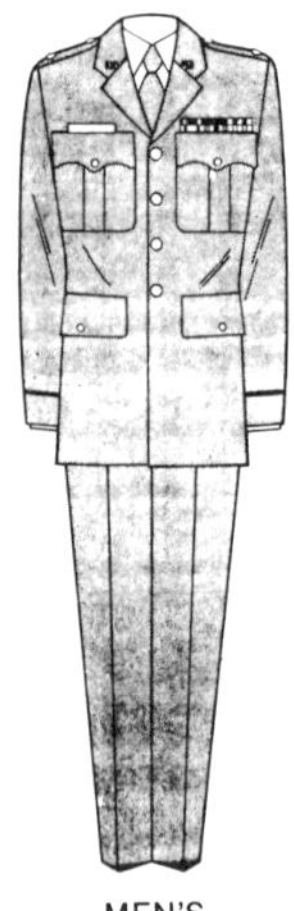

MEN'S COMBINATION 1 (Required).

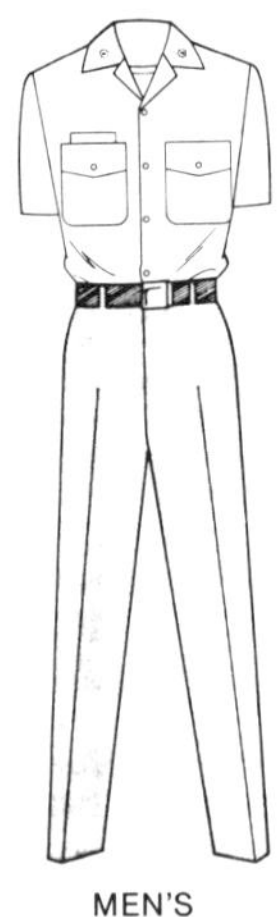

MEN'S COMBINATION 4 (Required).

WOMEN'S COMBINATION 1 (Required).

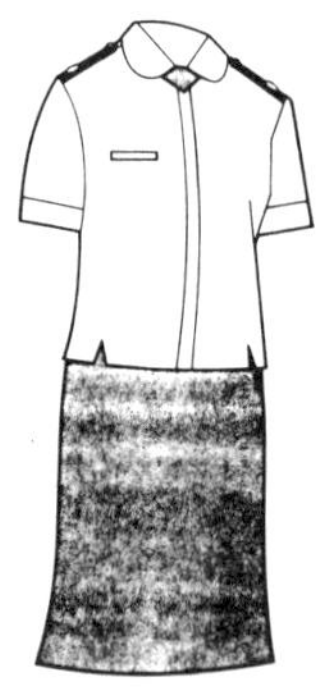

WOMEN'S COMBINATION 2 (Required).

vice coat and necktie are not worn, the light blue long-sleeved shirt is replaced by a light blue short-sleeved shirt without epaulets, and ribbons are not required.

Women's Combination 1. Dark blue semi-box skirt and coat of matching shade and fabric; light blue overblouse, standard length with epaulets and dark blue tie tab; undergarments; nylon hose of a neutral or dark shade; black pumps, oxfords, or dress boots; beret or service hat; name tag; ribbons, badges, and insignia. (The dark blue blazer from Combination 1A may be substituted for the semi-box coat.)

Women's Combination 2. Same as the Combination 1 uniform described above, except the dark blue semi-box coat is not worn, and ribbons are not required.

Utility Uniforms. *Men's and Women's Subdued Work Utility Uniforms.* Olive drab green trousers and shirt; dark blue belt with black buckle and tip; undergarments; black oxfords, low-quarters or combat boots; black socks; olive green utility cap; name tape and USAF tape in subdued colors; badges, organizational patches and insignia in subdued colors.

MEN'S WORK UTILITY UNIFORM (Required).

WOMEN'S WORK UTILITY UNIFORM (Required).

Dress Uniforms. *Men's Mess Dress Uniform.* Black or white jacket (officers must have both); black trousers with grosgrain side-stripes (when black jacket is worn, the trousers must be of matching shade and fabric; the white jacket must be of the same fabric as the trousers); plain white shirt; undergarments; black socks and low-quarter shoes; shoulder board grade insignia; miniature medals and badges; black cummerbund (tucks facing up); black bow tie with square edges; white suspenders; plain, silver satin-finish studs and cufflinks; silver chain button; black or white dress cap with silver bullion chin strap (optional).

Women's Mess Dress Uniform. Black or white jacket (officers must have both); black street- or ankle-length skirt (when black jacket and skirt are worn, they must be of matching shade and fabric; white jacket must be of fabric matching the skirt); white mess dress shirt with black tie tab; undergarments; nylon hose of a neutral shade; black pumps; shoulder board grade insignia; silver or black

WINTER MESS DRESS UNIFORMS (Required).

Miniature badges (i.e., aviation, chaplain, missileman, airborne, etc.) and medals must be worn with all dress uniform combinations.

The cummerbund will be worn with the tucks facing up for men and down for women.

cummerbund (silver worn only with ankle-length skirt); miniature medals and badges.

SUMMER MESS DRESS UNIFORMS (Required).

Sleeve braid on the winter and summer dress uniforms is required. Men will wear the braid 3 inches from the end of the sleeve; women will wear the braid 2½ inches from the end of the sleeve.

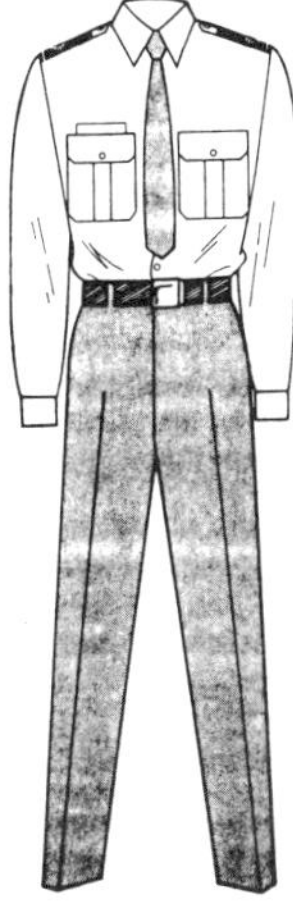
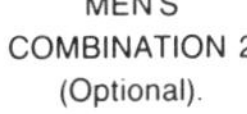

MEN'S COMBINATION 2 (Optional).

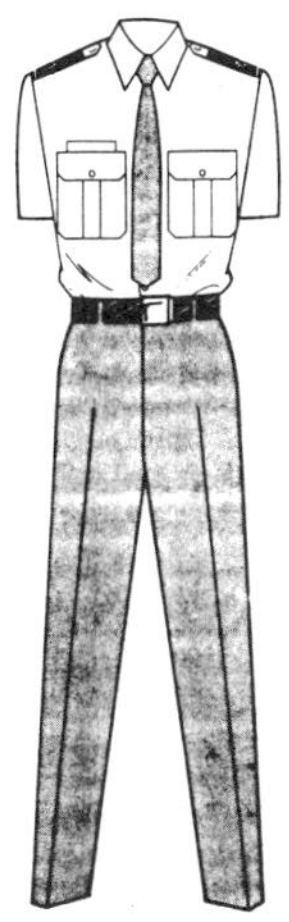

MEN'S COMBINATION 2A (Optional).

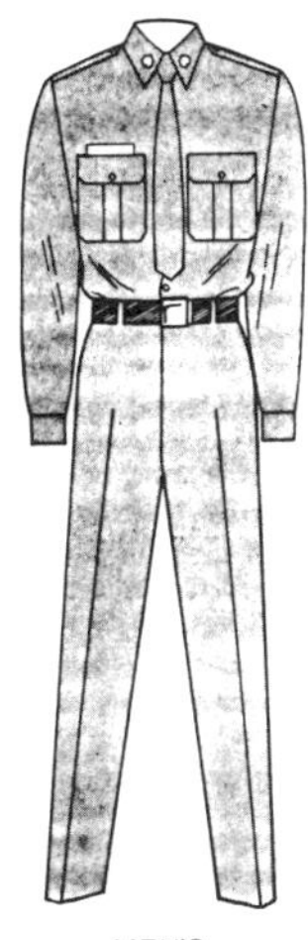

MEN'S COMBINATION 3 (Optional).

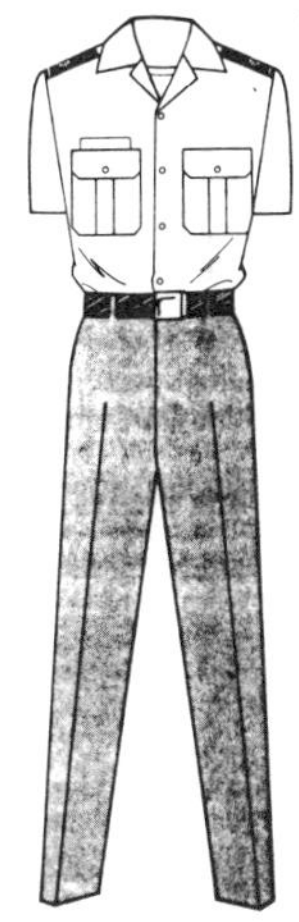

MEN'S COMBINATION 4A (Optional).

OPTIONAL COMBINATIONS

Service Uniforms. *Men's Combination 2.* Dark blue trousers; dark blue belt with silver buckle and tip; light blue long-sleeved shirt with epaulets; dark blue necktie; undergarments; black shoes and socks; service cap or flight cap; name tag; badges and shoulder mark grade insignia.

Men's Combination 2A. Same as Combination 2, except a light blue short-sleeved shirt with epaulets and shoulder mark grade insignia is worn in place of the light blue long-sleeved shirt.

Men's Combination 3. Same as Combination 2, except a dark blue long-sleeved shirt with epaulets is worn in place of the light blue long-sleeved shirt. The dark blue shirt in this combination must be of a shade and fabric matching the trousers. Miniature grade insignia are worn on the collar or epaulet. Epaulet stitching may be removed to permit wearing of shoulder mark insignia.

Men's Combination 4A. Same as Combination 2A, except the necktie is not worn.

Women's Combination 1A. Dark blue blazer and slacks of matching shade and fabric; light blue standard-length overblouse with epaulets and dark blue tie tab; undergarments; nylon hose of neutral or dark shade or black socks (socks worn only with oxfords or boots); black pumps, oxfords, or dress boots; service hat or beret; name tag; ribbons, badges, and insignia.

Women's Combination 2A. Same as Combination 1A except the dark blue blazer is not worn and ribbons are not required.

Women's Combination 5. (Mandatory for all pregnant members.) Dark blue tunic and skirt of matching shade and fabric; light blue standard-length overblouse with epaulets looped over tunic shoulder seams; shoulder mark grade in-

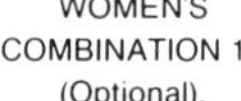
WOMEN'S COMBINATION 1A (Optional).

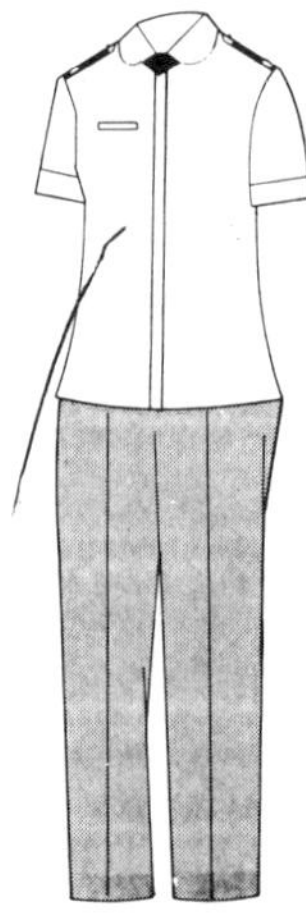
WOMEN'S COMBINATION 2A (Optional)

WOMEN'S COMBINATION 5 (Required for pregnant members).

WOMEN'S COMBINATION 5A (Required for pregnant members).

signia; undergarments; nylon hose of a neutral or dark shade; black pumps, oxfords, or dress boots; service hat or beret; name tag; badges and insignia.

Women's Combination 5A. Same as Combination 5 except the dark blue skirt is replaced with dark blue slacks.

Utility Uniforms. *Men's and Women's Subdued Field Formation Utility Uniform.* Same as the subdued work utility uniform except that a camouflage bib scarf is worn and trousers are bloused; only combat boots are authorized for footwear.

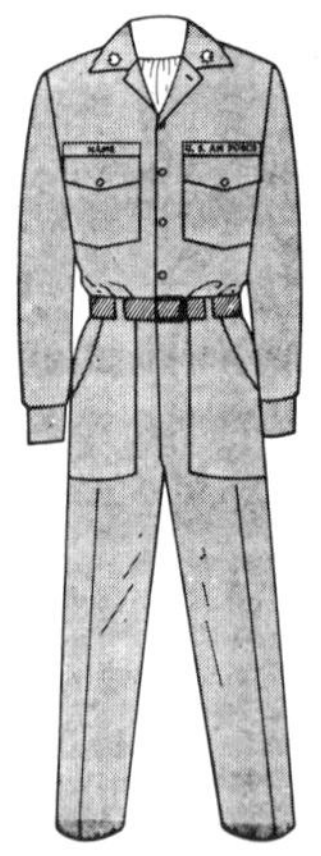
MEN'S FIELD FORMATION UTILITY UNIFORM (Optional).

WOMEN'S FIELD FORMATION UTILITY UNIFORM (Optional).

FORMAL EVENING MESS DRESS UNIFORMS (Optional).

Dress Uniforms. *Men's Formal Evening Dress Uniform.* Black jacket and trousers in matching shades and materials; white full dress shirt; undergarments; black socks and low-quarter shoes; spearhead sleeve ornament grade insignia; miniature medals; badges and insignia; white bow tie, square edged; pearl studs and cufflinks; black mess dress cap; white suspenders; silver chain button; white gloves; white single-breasted, low-cut vest.

Women's Formal Evening Mess Dress Uniform. Black jacket and ankle-length skirt; white mess dress shirt with silver tie tab; undergarments; nylon hose of a neutral or dark shade; black pumps; spearhead sleeve ornament grade insignia; miniature medals; badges and insignia; silver cummerbund.

OPTIONAL UNIFORM ITEMS

Certain uniform items not mentioned previously are available for wear as noted below.

Outergarments. The raincoat or the lightweight blue jacket may be worn.

Sweaters. The V-necked, front-buttoned sweater is authorized for indoor wear only, but the pullover sweater with epaulet grade insignia can be worn both indoors and outdoors.

Umbrella. A black umbrella may be carried by all personnel.

Gloves. Black or grey leather gloves can be worn with all outergarments or with the men's Combination 1 uniform.

Light Blue Long-Sleeved Shirt. Women are authorized to wear the light blue long-sleeved shirt (must be tucked in) as follows: Combination 1—with skirt with waistband and blazer jacket combination only; Combination 2—with skirt with waistband only. There are no restrictions for wear of this shirt with the other uniform combinations.

OTHER UNIFORM REGULATIONS

Reserve Officers. Except as otherwise prescribed, a Reserve officer on active duty will wear the uniform, including insignia, prescribed for officers of the Regular Air Force.

Reserve officers not on active duty, when within the limits of the United States or its possessions, may wear the uniform when participating in inactive duty training, on occasions of military ceremony, at social functions and informal gatherings of a military character, when engaged in military instruction, or when responsible for the military discipline at an educational institution.

Reserve officers will wear the uniform when participating in an inactive duty training period, when performing equivalent or appropriate duties, or when engaged in equivalent training or instruction, when such duty, training, or instruction is performed within the confines of a military installation. For the purpose of this paragraph, any installation or facility normally utilized for military purposes will be considered a military installation.

Reserve officers will wear the appropriate uniform when engaged in military flying activities.

Reserve officers not on active duty and outside the United States or its possessions, except when granted authority by the Secretary of the Air Force, will *not* wear the uniform. These officers on occasions of military ceremony or other military functions, upon reporting to the nearest military attache and having their status accredited, may be granted authority to appear in uniform. In a country to which no military attache is accredited, authority to wear the uniform for a specific occasion will be obtained from the proper civil or military authorities of the country concerned.

Warrant officers and airmen of the Regular Air Force who hold commissions in the Air Force Reserve may wear the uniform of their grade in the Air Force Reserve as follows:

(1) When participating in inactive duty training designed for Reserve officers in which they have been authorized to participate by proper authority, and in going to and returning from such training.

(2) When attending meetings or functions of associations formed for military purposes, the membership of which is composed largely or entirely of officers or former officers of the service.

Retired Officers. Retired officers and warrant officers on active duty wear the uniform prescribed for officers and warrant officers on the active list.

The uniform of retired officers and warrant officers not on active duty will be, at their option, either that for officers and warrant officers of corresponding grade at date of retirement, or that for officers and warrant officers on the active list, but the two uniforms will not be mixed.

Retired officers and warrant officers, not on active duty status, when attending ceremonies and social functions of an official character, or when calling at or visiting the White House, may wear the appropriate civilian dress for the occasion, except that when attending New Year's Day receptions in formation with officers and warrant officers on the active list they will wear the uniform.

Retired officers and warrant officers, not on active duty, are prohibited from wearing the uniform in connection with nonmilitary, civilian or personal enterprises or activities of a business nature.

Medal of Honor winners may wear the uniform any time wear is not prohibited by AFR 35-10.

Responsibilities. Individuals are responsible for insuring that their uniforms are neat, correct, and in good condition and that their appearance reflects credit upon themselves and the Air Force.

Commanders are responsible for insuring that personnel either singly or collectively present an excellent appearance, reflecting credit on the Air Force. This responsibility will not be construed to permit commanders to prescribe the purchase of optional items.

Special Occasions. The wearing of the proper uniform for special occasions such as ceremonies, weddings, funerals, and White House social functions, and similar duties will be largely dictated by circumstances in each case. Gloves either in white silk, cotton, or nylon-type; or gray suede or double-weave cotton may be worn on such occasions.

In Foreign Countries. Air Force personnel prior to departing for foreign countries in an official capacity will be responsible for obtaining proper information regarding uniform matters. Questions regarding uniforms which arise while in a foreign country should be referred to the U.S. Air Attache or the Senior Military Attache.

The uniform will *not* be worn by members of the Air Force, active or retired, visiting or residing in a foreign country in an unofficial capacity except when attending, by formal invitation, ceremonies or social functions at which the wearing of the uniform is required by the terms of the invitation or by the regulation or customs of the service.

The wearing of civilian clothes when traveling through foreign countries is prescribed in AFR 5-30.

Uniforms for Civilians. The Secretary of the Air Force may authorize the wearing of the uniform by civilians when their duties are deemed to require identification of this type.

Former Members of the Armed Forces. All persons honorably separated and who served in the Armed Services of the United States may wear, on occasions of ceremony, the uniform of the highest grade held during that service. (Sec. 125 of the National Defense Act, as amended.) The uniform may be either that authorized at time of separation or that authorized by current regulations. Occasions of ceremony will be those with military significance. Persons honorably discharged from the service may wear their uniform from the place of discharge to their homes, provided such wear is within three months of the date of discharge.

Illegal Wearing. Any person within the jurisdiction of the United States who wears a uniform, or a distinctive part of a uniform of the Armed Services without authority is subject to the penalties prescribed by the U.S. Code, Crimes and Criminal Procedure, as amended (AFR 35-10 and AFR 900-48).

Illegal Manufacture, Sale, and Possession. The protection of law extends to the wearing, manufacture, sale, possession, and reproduction in regular size of any United States decoration, medal, badge, and insignia which require the approval of the Secretary of the Air Force (AFM 35-10 and AFM 900-48).

CARE AND PRESERVATION OF UNIFORMS

Good uniforms and the accessories which are worn with them deserve the treatment which will assure maximum durability and appearance to the owner. Rumpled and soiled uniforms, and frayed service ribbons, present an unsightly appearance, regardless of the quality of the articles.

The care which should be given to uniforms and equipment is considerable but need not be burdensome. Regulations require the uniform to be neat, clean, and well pressed. It should be quite obvious that an old uniform of good quality which is well-fitting, clean, neat and unfaded, will look far better than a new and costly one which is slightly soiled or out of press. Economy is obtained and appearance enhanced by the exercise of care in preservation.

Coats should be placed on a good hanger whenever removed. If there is any moisture in the garment it should be placed where it may be dried. The hanger should have sufficient width at the shoulders to hold the garment in shape; wire hangers which are commonly used do not satisfy this requirement. Trousers should be placed on hangers so that they may hang full length. Dust is injurious to clothing, and uniforms which are worn infrequently should be placed in containers which provide protection and which may be closed tightly to exclude dust. Preparations destructive of moths may be placed within such containers to avoid this hazard. A clothing brush with stiff bristles should be used each time a garment is worn; it removes loose dust and freshens the nap.

Dry Cleaning Precautions. A competent cleaner should be entrusted with removal of spots or stains, especially those from unusual causes. Underarm sweating is especially injurious to the color and wearing qualities of fabrics, and protection should be provided by attaching an impervious lining at the armpit. The nature of unusual stains should be reported to the cleaner so that he may select the correct solvent without experimentation.

ITEMS OF INSIGNIA

Purpose of Air Force Insignia. Air Force insignia identifies the wearer as a member of the United States Air Force. There are five different types of insignia used in conjunction with the wear of Air Force uniforms. They are: (1) lapel insignia, (2) grade insignia, (3) cap and hat insignia, (4) specialty insignia, and (5) aides' insignia.

Lapel Insignia. Lapel insignia is different for officers and enlisted personnel, but both identify the wearer as a member of the United States Air Force. Officers wear the letters "U.S." in oxidized silver-color metal, 7/16 inch high. Enlisted personnel wear the same "U.S." letters. However, they are contained within a one-inch circle.

All Officers and Warrant Officers.

All Enlisted Personnel.

LAPEL INSIGNIA.

Grade Insignia. Officer grade insignia is available in two different styles. Metal grade insignia is used on all outergarments, on men's uniform Combinations 1, 3, and 4, and on women's uniform Combinations 1, 1A, and the maternity uniform. Embroidered epaulets may be used on all other uniform combinations.

Cap and Hat Insignia. The insignia worn on the men's service cap, women's service hat, and the women's beret is the United States coat of arms. On the women's beret, the miniature coat of arms is worn, while the larger-sized device is worn on the service cap or hat. On the flight cap, metal grade insignia is worn, in either regular or miniature size, except general officers wear a 5/8-inch star or connecting bar. On the utility cap, subdued cloth grade insignia is sewn onto the front of the cap. For lieutenant colonels and above, a design of lightning, clouds, and darts is worn on the visor of the service cap.

General Officers.

Colonel and Lieutenant Colonel.

Other Officers and Airmen.

SERVICE CAP VISORS.

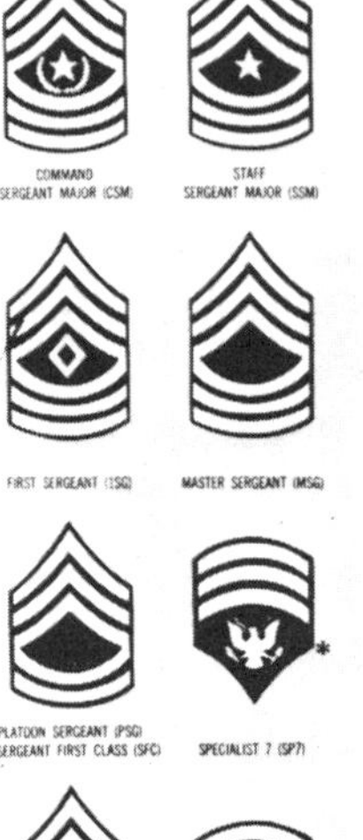

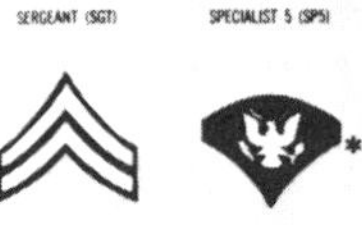
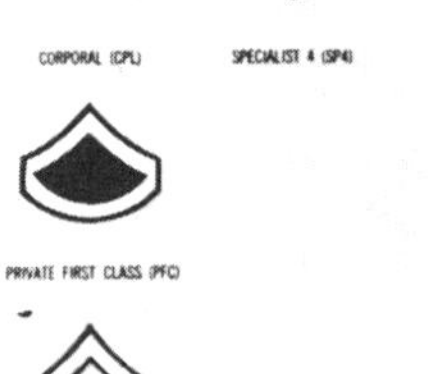

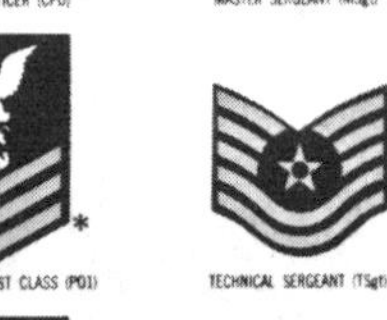

INSIGNIA OF GRADE, ENLISTED.

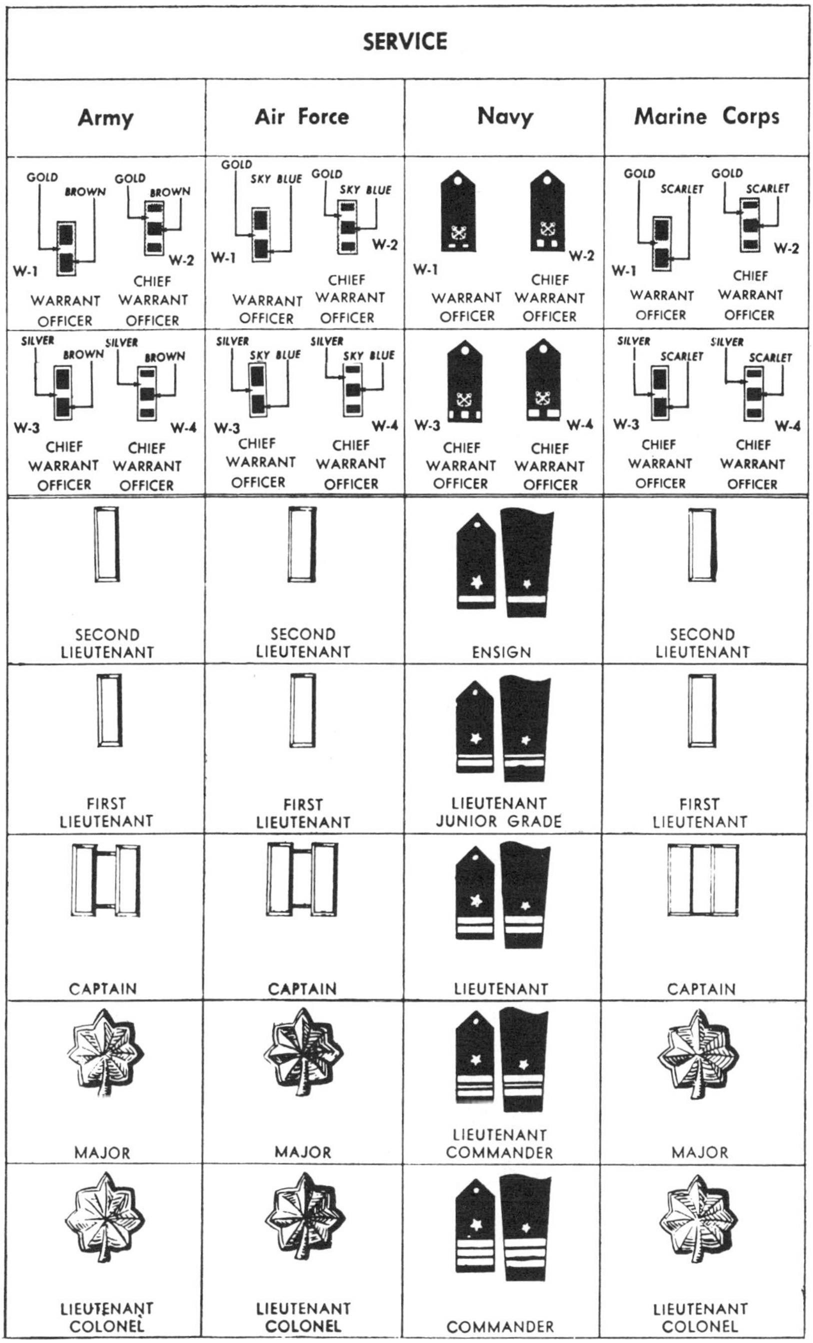

SERVICE			
Army	**Air Force**	**Navy**	**Marine Corps**
GOLD BROWN W-1 WARRANT OFFICER; GOLD BROWN W-2 CHIEF WARRANT OFFICER	GOLD SKY BLUE W-1 WARRANT OFFICER; GOLD SKY BLUE W-2 CHIEF WARRANT OFFICER	W-1 WARRANT OFFICER; W-2 CHIEF WARRANT OFFICER	GOLD SCARLET W-1 WARRANT OFFICER; GOLD SCARLET W-2 CHIEF WARRANT OFFICER
SILVER BROWN W-3 CHIEF WARRANT OFFICER; SILVER BROWN W-4 CHIEF WARRANT OFFICER	SILVER SKY BLUE W-3 CHIEF WARRANT OFFICER; SILVER SKY BLUE W-4 CHIEF WARRANT OFFICER	W-3 CHIEF WARRANT OFFICER; W-4 CHIEF WARRANT OFFICER	SILVER SCARLET W-3 CHIEF WARRANT OFFICER; SILVER SCARLET W-4 CHIEF WARRANT OFFICER
SECOND LIEUTENANT	SECOND LIEUTENANT	ENSIGN	SECOND LIEUTENANT
FIRST LIEUTENANT	FIRST LIEUTENANT	LIEUTENANT JUNIOR GRADE	FIRST LIEUTENANT
CAPTAIN	CAPTAIN	LIEUTENANT	CAPTAIN
MAJOR	MAJOR	LIEUTENANT COMMANDER	MAJOR
LIEUTENANT COLONEL	LIEUTENANT COLONEL	COMMANDER	LIEUTENANT COLONEL

INSIGNIA OF GRADE, WARRANT OFFICERS AND OFFICERS.

Note: Grade insignia of second lieutenant and major are gold; of other officer grades, silver. Navy officers' shirt collar insignia same as Marine Corps, except those for warrant officers, where navy-blue enamel is substituted for scarlet.

SERVICE			
Army	**Air Force**	**Navy**	**Marine Corps**
COLONEL	COLONEL	CAPTAIN	COLONEL
BRIGADIER GENERAL	BRIGADIER GENERAL	COMMODORE	BRIGADIER GENERAL
MAJOR GENERAL	MAJOR GENERAL	REAR ADMIRAL	MAJOR GENERAL
LIEUTENANT GENERAL	LIEUTENANT GENERAL	VICE ADMIRAL	LIEUTENANT GENERAL
GENERAL	GENERAL	ADMIRAL	GENERAL
GENERAL OF THE ARMY	GENERAL OF THE AIR FORCE	FLEET ADMIRAL	(NONE)

INSIGNIA OF GRADE, OFFICERS—Continued.

Note: Navy officers' shirt collar insignia same as other services. In the Army, General of the Armies Pershing, and in the Navy, Admiral of the Navy Dewey outranked the five-star grades authorized in World War II, but neither had prescribed distinctive grade insignia for themselves.

SPEARHEAD SLEEVE ORNAMENTATION.

General Officers (top); Field Officers (bottom left); Company Grade (bottom right).

Description: On a blue-black broadcloth background with regular insignia of grade embroidered 1/4 inch above the horizontal braid. Insignia of rank will be placed horizontally and colonel's eagle will face to the front of the wearer. Insignia and sleeve ornamentation are wire embroidery.

Service Hat (Colonel and Lieutenant Colonel).

Beret (Officer Insignia).

HAT INSIGNIA (WOMEN).

Large U.S. Coat of Arms is worn on all service hats except women's beret.

Specialty Insignia. Chaplains and medical personnel have distinctive specialty insignia worn on all uniform combinations.

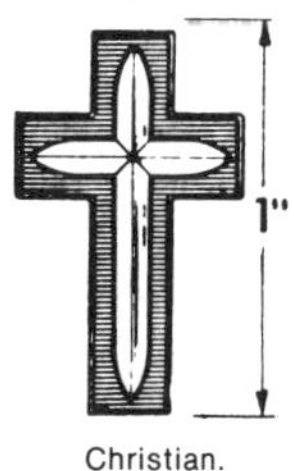

Christian.

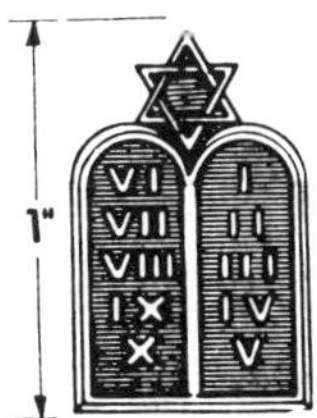

Jewish.

CHAPLAIN'S INSIGNIA.

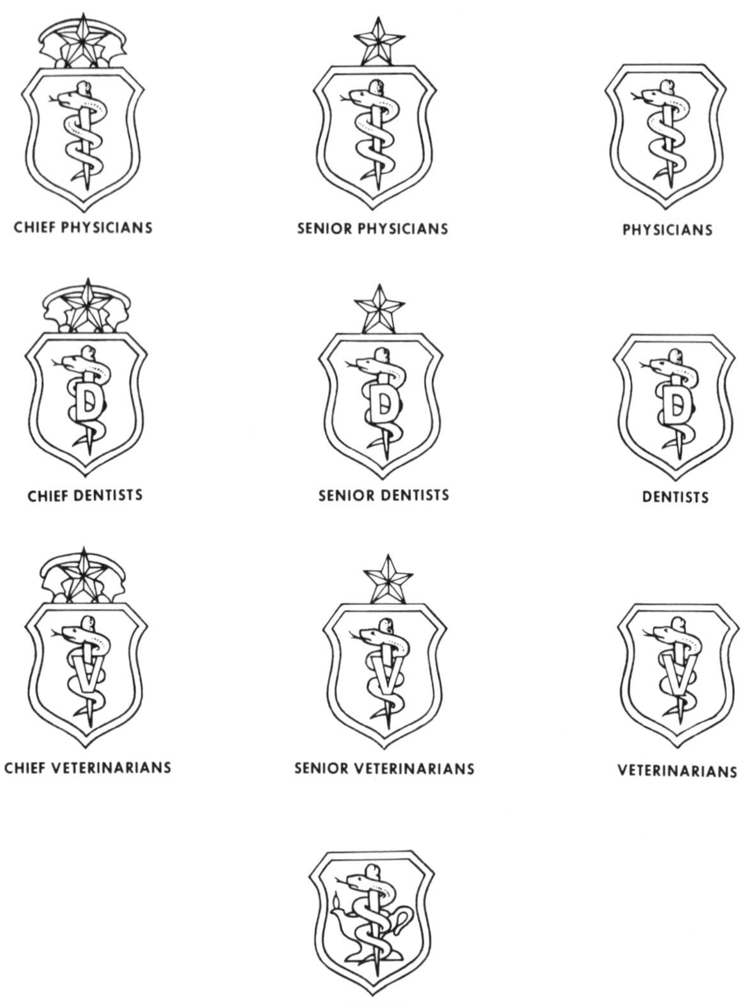

MEDICAL INSIGNIA.

Aides' Insignia. Aides and attaches are distinguished by aiguillettes of silver-colored rayon or metallic cord worn with the service and dress uniforms. Presidential aides wear their aiguillettes on the right shoulder, while all others use the left. Aides are distinguished by metal insignia designating the office or the commander to which they are assigned.

DRESS AIGUILETTES.

Loop-type ends should go over the shoulder board strap approximately midway. The button on the lapel should be sewn to the body of the jacket under the lapel. Although the button is hidden, it should be the same color as the jacket, and of the smallest size that will adequately accommodate the loop on the aiguillette. Aiguillettes may be shortened when worn by women. Description: Silver color rayon or metallic cord.

NOTE: Aides to the President, White House social aides, and aides to foreign heads of state will wear aiguillettes on the right side.

AIDE'S INSIGNIA.

Buttons. Oxidized silver-colored metal of suitable composition and weight, circular and slightly convex with raised rim, the Great Seal of the Department of the Air Force in clear relief against a horizontally-lined background.

Brassards. Armbands are authorized to be made of blue shade 1083 felt background and gray, shade 1155, letters. An exception to this will be for noncombatant personnel who come under the provisions of the Geneva Convention of August 12, 1949. For these personnel, the brassards will be white with a red cross superimposed thereon.

WEARING OF INSIGNIA, DEVICES, BADGES, AND DECORATIONS

Appurtenances. Appurtenances identify and distinguish the grade, service, and specialty of persons wearing the uniform. The manner in which insignia, devices, badges, and decorations are worn is prescribed herewith and illustrated by drawings.

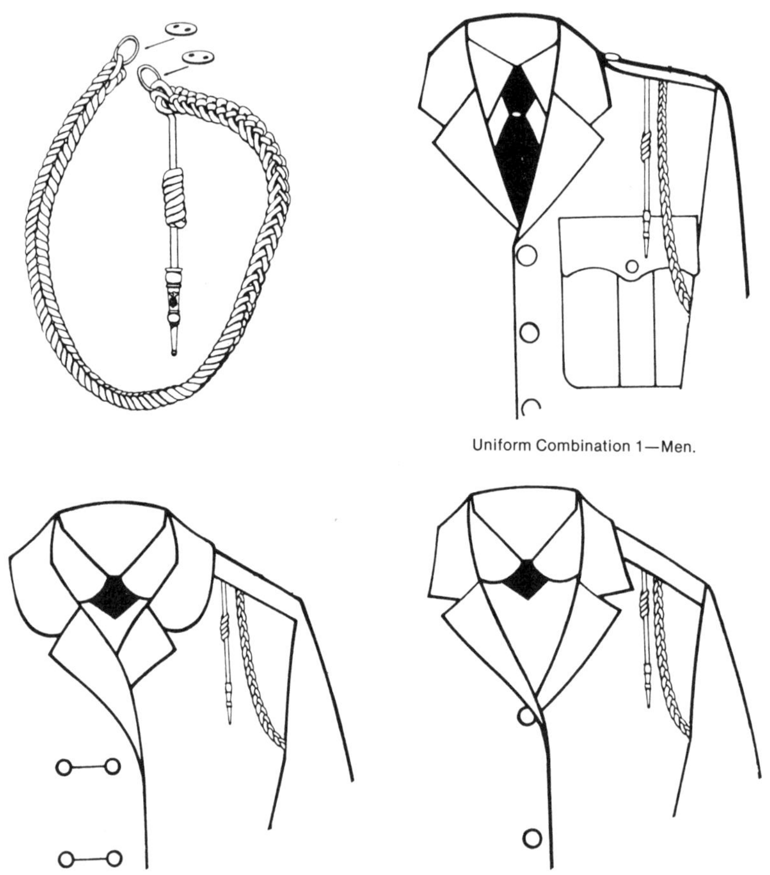

Uniform Combination 1—Men.

Uniform Combination 1—Women (Semibox coat).

Pantsuit—Women.

SERVICE AIGUILLETTES (Men and Women).

Loop-type ends should be secured by sewing button or hook fastener to the underside of the epaulet approximately midway between the epaulet button and the shoulder seam to avoid interfering with the clutch fasteners on the officer grade insignia.

Collar Insignia. Insignia is worn on the collar when the shirt is worn as an outergarment. Officers wear grade insignia and airmen wear the regular-size "U.S.," insignia. Miniature-size grade insignia is optional. Nurses and women medical specialists will wear miniature type grade insignia on both sides of collar of hospital-duty uniform.

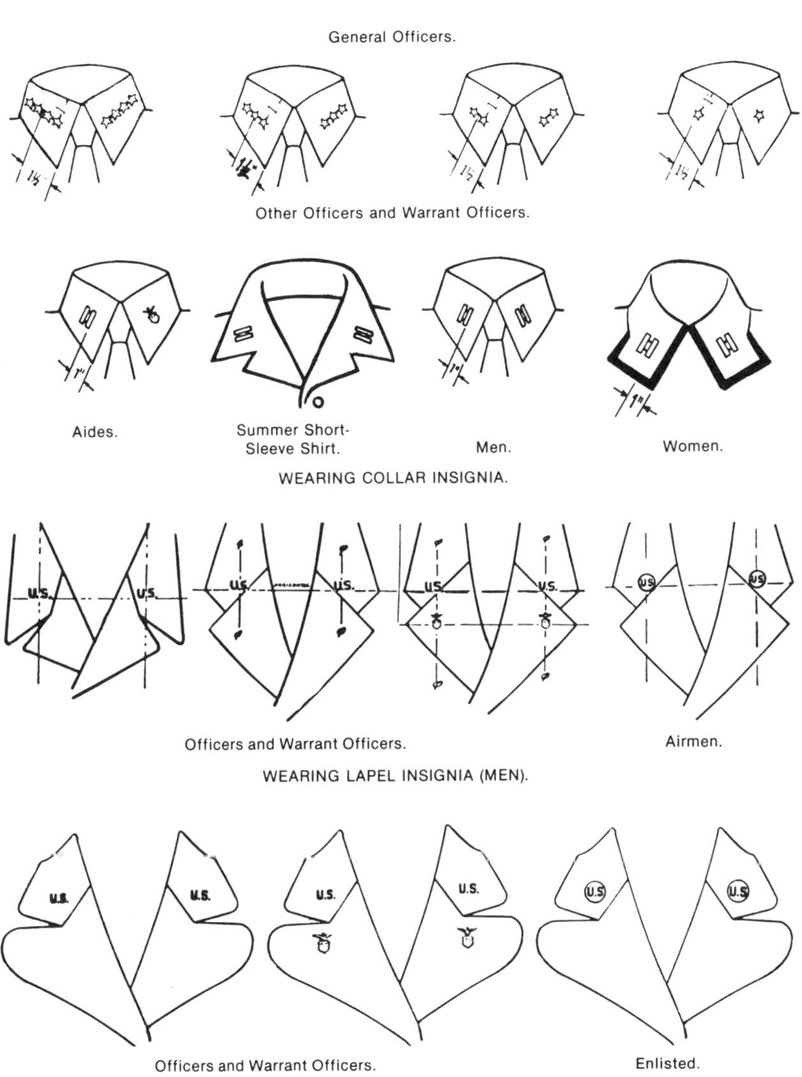

WEARING LAPEL INSIGNIA (WOMEN).

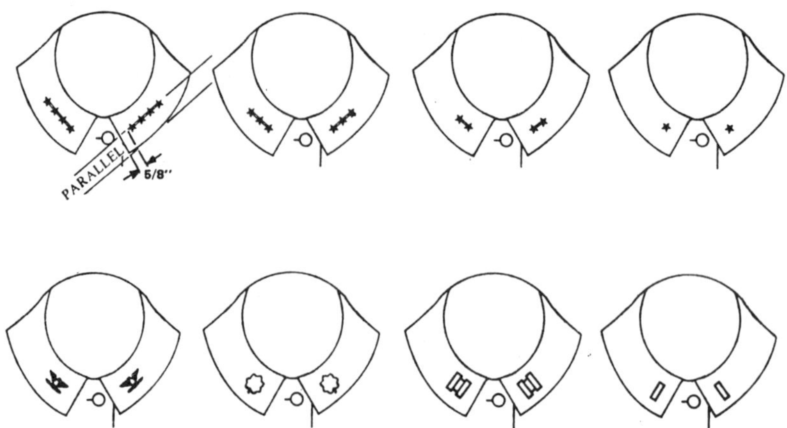

Wear of the grade insignia on the women's raincoat without epaulets; on the raincoat with epaulets (men's and women's), grade insignia is worn ⅝ inch from the shoulder seam.

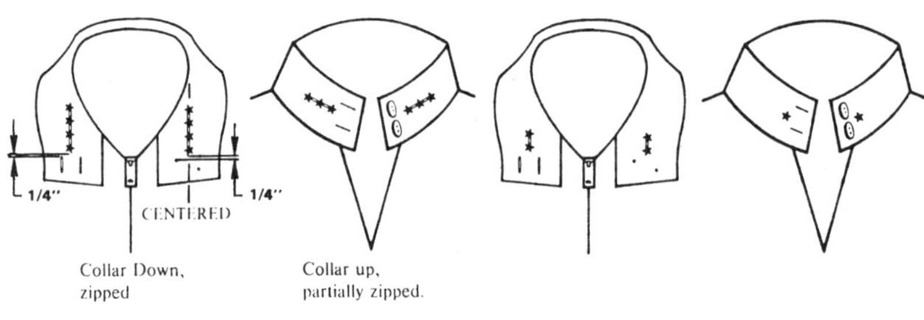

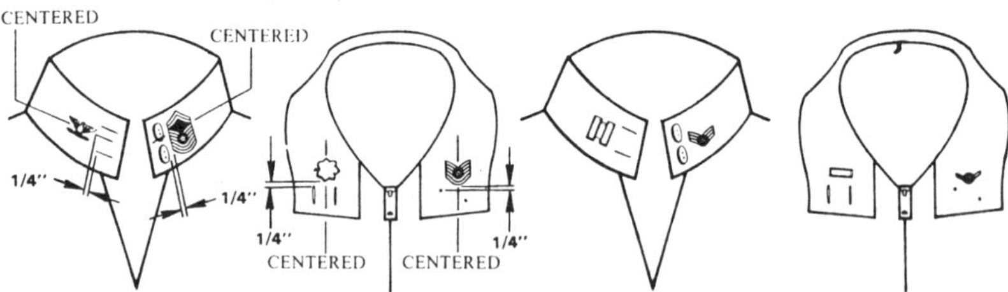

Wear of grade insignia on the lightweight blue jacket (women's); officer grade insignia on the men's lightweight blue jacket is worn on the shoulder epaulets, ⅝ inch from the shoulder seams. Insignia are centered ¼ inch away from buttons and button holes.

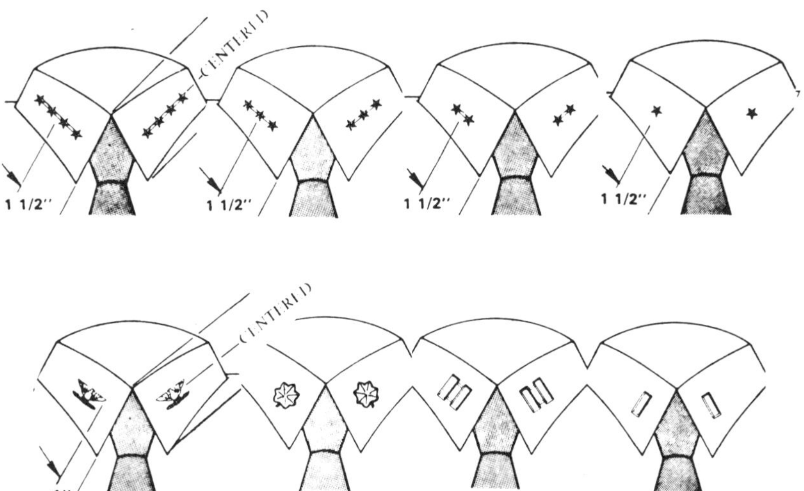

Wear of miniature grade insignia on collar of Men's Combination 3.

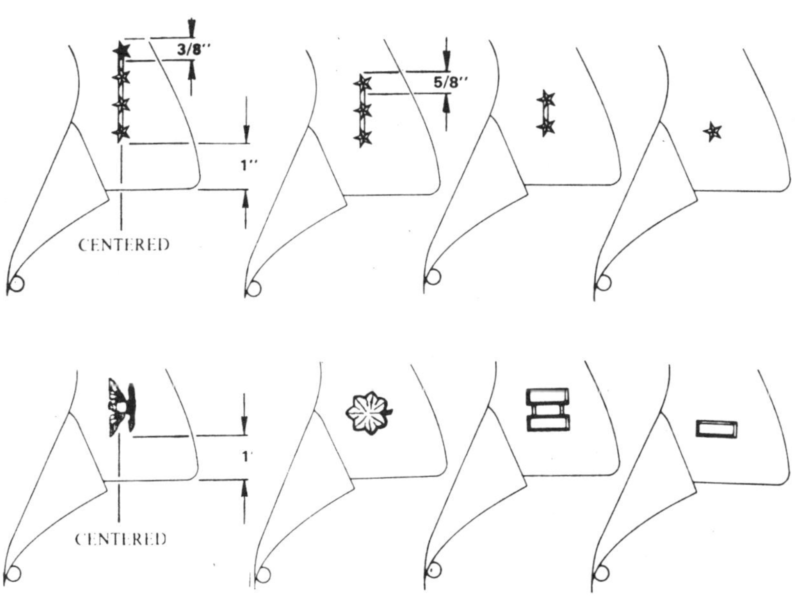

Wear of miniature grade insignia on collar of Men's Combination 4.

Lapel Insignia (All Personnel). The lower edge of "U.S." insignia is centered on each upper part of lapel (collar) measured on a horizontal line from upper lapel peak.

Aides' insignia is centered on each lower part of lapel on a horizontal line from lower lapel peak.

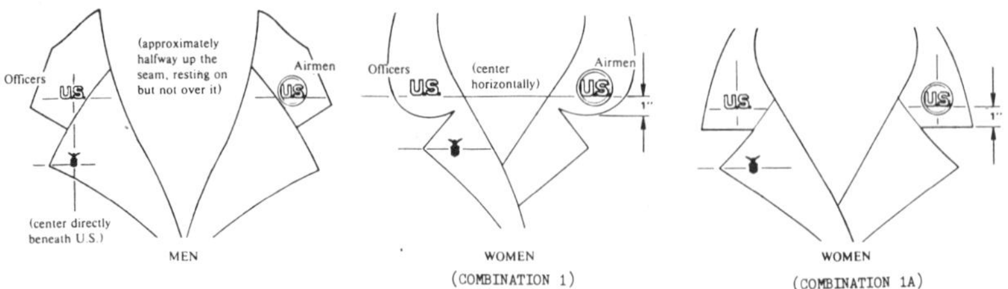

Wear of U.S. and aide insignia on service uniforms.

Cap and Hat Insignia. *Service Cap and Sun Helmet.* Insignia for the service cap is worn centered on the front rise of the cap and for the sun helmet is worn in the guide hole.

Flight Cap. Insignia of grade (officers only) is worn on the curtain as illustrated.

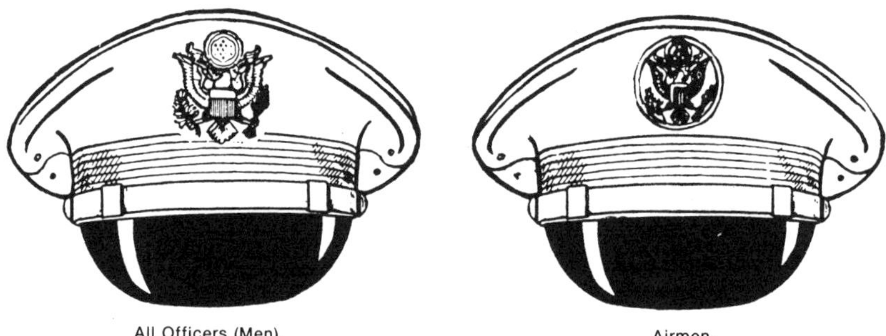

All Officers (Men).

Airmen.

SERVICE CAP INSIGNIA (MEN).

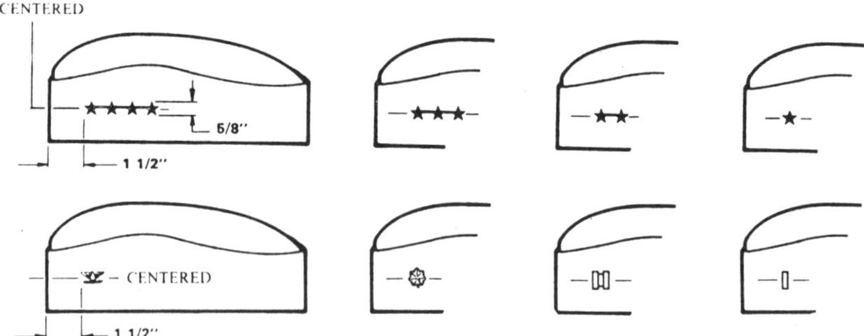

Wear of metal grade insignia on the flight cap. Either regular or miniature insignia is authorized.

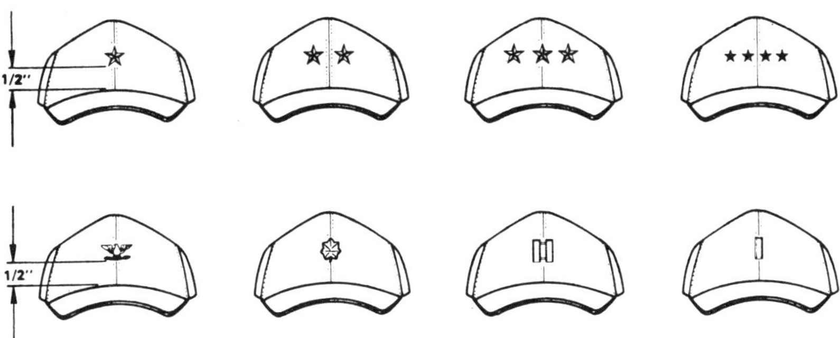

On the utility hat subdued cloth grade insignia is sewn onto the front of the hat.

Shoulder Loop Insignia. Insignia of grade is worn by all officers and warrant officers on the shoulder loops of overcoat, topcoat, and coat.

Insignia Denoting Specialty. The wearing of special insignia is described as follows:

Physicians, Dentists and Nurses. These insignia are worn ½" above the left breast pocket of coat, or shirt or ½" above the top row of ribbons.

THE C-5A GALAXY

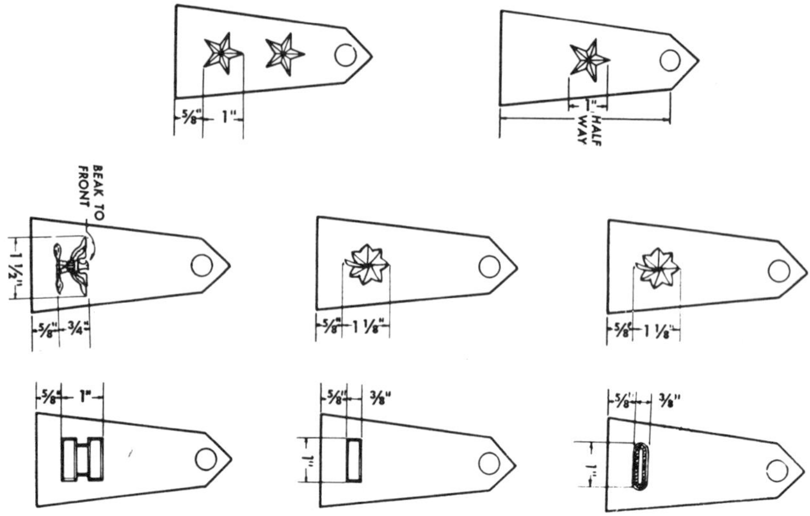

WEARING SHOULDER LOOP INSIGNIA.

Chaplain. Chaplains' insignia are worn centered ½″ above the left breast pocket of the coat or shirt when worn as an outer garment. When ribbons are worn, the insignia will be centered ½″ above the top row of ribbons, except that in no instance will the insignia be placed closer than ¼″ horizontally to the lapel of the coat.

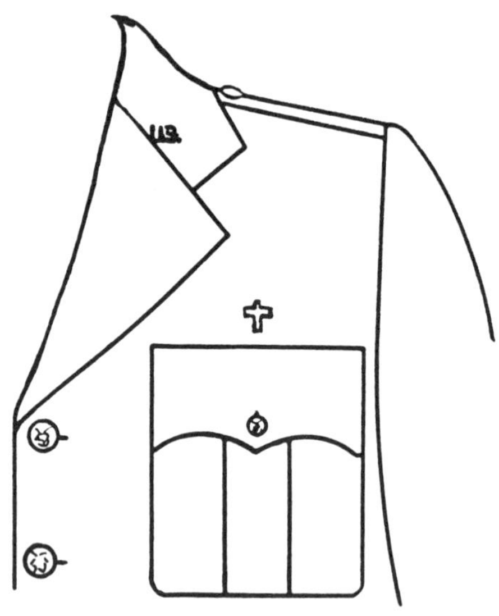

CHAPLAIN'S INSIGNIA.

PLEASANT TEAMWORK—AIR FORCE UNIFORMS AS SEEN IN MANY EVERYDAY DUTY SETTINGS.

7

Decorations, Achievement Awards, Service Medals, and Badges

Decorations awarded to members of the armed services are a symbol of acknowledgment by the Government of our nation for a job well done. They consist of awards for heroism, the highest of which is the Medal of Honor, and awards for achievement and service, the highest of which is the Distinguished Service Medal. Just as there is a variation in degree of heroism or achievement above and beyond the call of duty, so is there also a variation in the importance or rank of the several awards given for these two purposes.

HISTORY

The granting of awards by governments stems at least as far back in history as the Roman era. It was during that period that Roman rulers took the laurel wreath unto themselves. The Greeks crowned citizens who were outstanding in war, athletics, literature, and oratory with the laurel wreath and therefore it is natural that it forms a part of our nation's highest award for valor. During the Age of Feudalism there grew up a system of "rewards" in the form of titles and prerogatives. The "strong men," or barons, would honor the "knights" who performed valiant feats of arms not only by giving them increased titles, lands, and pensions, but also by encouraging them in the growing habit of decorating their shields and other armor with insignia which were, in effect, a pictorial history of their deeds. As additional deeds increased the number and complexity of insignia, there developed the whole field of heraldry and heraldic art, and the production of "coats of arms" for the families which had evolved them.

Prior to the Napoleonic period such rewards were for the aristocracy and for the few. The rewards given the drafted peasantry, the yeomen, the lowly fighting men, were infrequent indeed.

Napoleon started the modern concept. He originated a decoration which could be worn by *anyone,* regardless of rank or social background. This was the *Legion D'Honneur.* It was the spearhead of the system he so shrewdly utilized to instill loyalty in the nation he led—because *anybody could win it.* Thereafter there arose the Russian Order of St. George, the Iron Cross of Germany, and the English decoration of the Victoria Cross. They were products of the era of nationalism, created to reward the citizen for his loyalty to his state.

The Purple Heart was the first military decoration established in the United States. It was authorized by General George Washington in 1782 as a decoration for "singularly meritorious action." Three men had been awarded it in 1783 but the records show no others. It was reestablished in 1932. There were six other medals awarded in the early years of the Republic. The first was given to General Washington by a resolution of the Continental Congress and approved on 25 March 1776. The second was awarded to General Horatio Gates on 25 November 1777 for the defeat of the British at Saratoga. The third was given to Henry Lee, father of Robert E. Lee, who won the nickname "Light Horse Harry" on 24 September 1779 in recognition of his successful attack on the British at Paulus Hook, N.J., in July of that year, during which he captured 160 of the enemy without loss to his own forces. The fourth, fifth, and sixth were awarded to enlisted men and are known as the "André" medals, given to the three American militiamen who captured the British intelligence major, John André, while he was en route to New York from West Point, wearing civilian clothes, after having plotted with Benedict Arnold for the betrayal of the American cause.

In the first winter of the Civil War, a feeling developed that a way must be found to recognize and honor the heroism and gallantry of American soldiers who had distinguished themselves in the fighting. There was great argument and discussion at the time that the award of decorations was unbecoming a republic. However, the award of a Naval medal was approved by Mr. Lincoln on 21 December 1861, and on 12 July 1862 a Medal of Honor was provided for award to enlisted men of the Army "who shall most distinguish themselves by their gallantry in action, and other soldier-like qualities." It was amended by an act approved 3 March 1863 which extended its provisions to include officers as well as enlisted men. This legislation stood as the basis upon which the Army Medal of Honor could be awarded until 9 July 1918 when it was superseded by a completely revised statute.

All other present decorations have been authorized in 1918 or later years.

PRINCIPLES GOVERNING THE AWARD OF DECORATIONS

The award of decorations is a powerful stimulus to pride of service and to the encouragement of heroism or achievement. But if the greatest benefit is to be obtained, the process of making awards must be most carefully administered.

Awards must be made only to those who have truly earned them. The making of unmerited awards is cheapening and destructive of the purpose for which authorized. But to fail to recognize true valor or merit promptly, and make the awards, also defeats the purpose. Both a proper evaluation of facts plus promptness in making awards are required.

This requirement presents a serious problem of decision to all commanders of combat organizations. Evidence is often sketchy. The witnesses may be dead as a consequence of combat. The weighing of facts as to which of the several awards is most appropriate is also a delicate choice. To justify any award the act must be above and beyond the call of duty. Final consideration of the facts before decision by a senior commander will rest with a board of officers who can establish a reasonable yardstick against which to measure and compare these events. The individual commander, and especially the junior commander, must recognize or identify those acts which are truly above and beyond the call of duty; he or she must make prompt recommendation for the award which he or she considers most appropriate; and he or she must document it with the statements of witnesses, all to the end that the reviewers may arrive at an independent and accurate decision. Unless a careful process is followed by all there will be many serious errors, anomalies, variations between organizations, the sum of which may be to reduce greatly or even to nullify the powerful stimulus to pride of individual service which may be attained through awards.

AFR 900-3 describes the policy on award of decorations.

Decorations, Service Medals, and Achievement Awards. *Decorations* are awarded in recognition of and as a reward for extraordinary, unusual, or outstanding acts or services. They are the visible evidence of such acts or services. Properly utilized, they are potent incentives to greater effort and instrumental in building and maintaining morale. It will be noted that decorations are of two categories: those awarded for *heroism* and those awarded for *achievement.* While most decorations are awarded for one or the other distinct purpose, the Distinguished Flying Cross and the Bronze Star Medal may be won for either reason.

Service medals are awarded to members of the active military service of the United States for performance of specified duty, usually during periods of war or national emergency.

Achievement Awards are presented to recognize specific types of achievements for individuals serving on active duty in the Air Force or as members of the Air Reserve Forces.

Oak Leaf Clusters. The Oak Leaf Cluster is awarded in lieu of a second award of the same decoration. It consists of a bronze twig of four oak leaves with three acorns on the stem. It is 13/32 of an inch in length and is affixed to the ribbon. One Silver Oak Leaf Cluster is worn in lieu of five bronze ones.

Award by Foreign Governments. No person holding any office of profit or trust under the United States shall, without the consent of Congress, accept any present, emolument, office, or title, of any kind whatever, from any King, Prince, or foreign State (AFR 900-48).

In this connection special authority was granted by the Congress with respect to foreign decorations awarded during World War II, the Berlin Airlift, the Korean War, and the Vietnam War. Officers to whom such awards may be tendered are advised to obtain the advice of their headquarters personnel division or judge advocate.

Penalty for Unauthorized Wearing of Decorations, Service Medals, or Badges. A federal statute provides for a fine of not more than $250 or imprisonment for not more than six months for individuals convicted of unauthorized wearing of any decoration, service medal, ribbon or rosette, or badge.

SUMMARY OF UNITED STATES AIR FORCE DECORATIONS

Award or Decoration	Awarded for	Rank of awards for valor	Rank of awards for achievement	May be awarded to civilians	May be awarded in peacetime
Medal of Honor	Gallantry and intrepidity at the risk of life above and beyond the call of duty.	1	. . .	No	No
Air Force Cross	Extraordinary heroism in military operations against an armed enemy.	2	. . .	Yes	No
Distinguished Service Medal	Exceptionally meritorious service in a duty of great responsibility.	. . .	1	Yes	Yes
Silver Star	Gallantry in action.	3	. . .	Yes	No
Legion of Merit	Exceptionally meritorious conduct in the performance of outstanding services.	. . .	2	No	Yes
Distinguished Flying Cross	Heroism, extraordinary achievement while participating in aerial flight.	4	3	No	Yes
Airman's Medal	Heroism not involving actual conflict with an enemy.	5	. . .	No	Yes
Bronze Star Medal	Heroic or meritorious achievement or service against an ememy not involving aerial flight.	6	4	Yes	No
Meritorious Service Medal	Outstanding non-combat meritorious achievement or service.	. . .	5	No	Yes
Air Medal	Meritorious achievement while participating in aerial flight.	. . .	6	Yes	Yes
Air Force Commendation Medal	Meritorious achievement not in operations against enemy.	. . .	7	No	Yes
Air Force Achievement Medal	Awarded for meritorious achievement.	. . .	8	No	Yes
Purple Heart	Wounds received in action against an enemy of the United States.	7	. . .	Yes	No

The same penalty is provided as to any person who wears the uniform or decorations of a foreign nation for the purpose of deception.

USAF DECORATIONS FOR HEROISM (AFR 900-48)

The decorations for heroism are stated below. The order of precedence with decorations for achievement is illustrated in the accompanying chart.

Medal of Honor (MH). The Medal of Honor is awarded for conspicuous gallantry and intrepidity at the risk of life above and beyond the call of duty, while a member of the Air Force was engaged in armed conflict against an enemy of the United States. Each recommendation for the Medal of Honor must incontestably prove that the self-sacrifice or personal bravery involved conspicuous risk of life, the omission of which could not justly cause censure.

The Medal of Honor is a gold star with the head of the Statue of Liberty centered upon it and surrounded with green enamel laurel leaves suspended by rings from a trophy consisting of a bar inscribed with the word VALOR above an adaptation of the thunderbolt from the U.S. Air Force Coat of Arms. The bar is suspended from a light blue moire silk neckband behind a square pad in the center with corners turned in and charged with 13 white stars in the form of a triple chevron.

Air Force Cross (AFC). On 6 July 1960, Congress established the Air Force Cross to parallel the U.S. Army Distinguished Service Cross and the U.S. Navy Cross. Prior to this date the Air Force awarded the Distinguished Service Cross.

The Air Force Cross is our Nation's second highest military decoration and is awarded to U.S. Air Force airmen for extraordinary heroism in military operations against an enemy of the United States. The Air Force Cross may be awarded to members of foreign military forces and to American and foreign civilians serving with the Armed Forces of the United States.

The Air Force Cross is a bronze cross with an oxidized satin finish. Centered on the cross is a gold-plated American bald eagle with wings spread against a cloud formation. The eagle is encircled by a laurel wreath finished in green enamel. The cross is suspended from a ribbon of brittany blue, edged with old glory red and bears a narrow white vertical stripe inside the red edges.

MEDAL OF HONOR. AIR FORCE CROSS.

Silver Star (SS). The Silver Star was instituted by Congress in 1918. It is granted to persons serving in any capacity with the Air Force cited for gallantry in action which does not warrant the award of a Medal of Honor or the Air Force Cross.

It is a small silver star within a wreath centered on a larger star of gold colored metal. The ribbon has a center band of red, flanked by equal bands of white; the white bands are flanked by equal blue bands having borders of white lines with blue edgings.

Distinguished Flying Cross (DFC). *(As an Award for Heroism.)* The Distinguished Flying Cross has been awarded to airmen since 1917 for heroism while participating in an aerial flight. The heroism must be evidenced by voluntary action in the face of great danger and beyond the line of duty.

The Distinguished Flying Cross is a bronze cross with rays on which is displayed a propeller. The ribbon is predominantly blue, with a narrow band of red bordered by white lines in the center. The edges of the ribbon are outlined with equal bands of white inside blue.

SILVER STAR. DISTINGUISHED FLYING CROSS. AIRMAN'S MEDAL.

Airman's Medal (AmnM). The Airman's Medal is awarded for heroism involving voluntary risk of life under conditions other than conflict with an armed enemy of the United States.

The Airman's Medal is a bronze metal disk with an oxidized satin finish. The pendant bears a representation of Hermes, son of Zeus, releasing an American bald eagle. The ribbon is brittany blue displaying alternately in the center, 13 vertical stripes of the Air Force colors, golden yellow and ultra-marine blue.

Bronze Star Medal (BSM). *(Awarded for Valor.)* The Bronze Star Medal is awarded for heroism while engaged in military action against an enemy of the United States.

The medal consists of a bronze star bearing in the center a small star of the same color. The ribbon, on which is a small bronze letter "V", is predominantly red with a white edged narrow blue band in the center, and white lines at each edge.

Purple Heart (PH). The Purple Heart is our Nation's oldest medal. It was first established by General George Washington 7 August 1782. The Purple Heart is awarded for wounds received, or death after being wounded, in action against an enemy of the United States or as a direct result of an act of such enemy.

The Purple Heart is a heart-shaped pendant of purple enamel bearing a gold replica of the head of General George Washington, in relief, and the Washington Shield. The shield is in colors. The ribbon is dark purple with white edges.

BRONZE STAR MEDAL. PURPLE HEART. DISTINGUISHED SERVICE MEDAL.

USAF DECORATIONS FOR ACHIEVEMENT (AFR 900-48)

The decorations for achievements in order of precedence with decorations for heroism were shown earlier in the chapter.

Distinguished Service Medal (DSM). The Distinguished Service Medal is awarded to members of the Armed Forces who, while serving in any capacity with the Air

Force, distinguish themselves by exceptionally meritorious service to the Government in a duty of great responsibility. (It is awarded to a foreign national only rarely.)

The term "duty of great responsibility" means duty of such a character that exceptionally meritorious service therein has contributed in high degree to the success of a major command, installation, or project. The performance of the duty must be such as to merit recognition of the service as clearly exceptional. A superior performance of the normal duties of the position will not alone justify the award. The accomplishment of the duty for which the award is recommended should have been completed, or it should have progressed to an exceptional degree if the person rendering the service has been transferred to other duties prior to its full accomplishment.

The Air Force Distinguished Service Medal features a blue-stone representing the firmament at the center of a sunburst of 13 gold rays separated by 13 white enamel stars. The center motif represents the vault of the heavens; the stars symbolize the 13 original colonies. The stylized wings on the ribbon bar are symbolic of the USAF.

LEGION OF MERIT.

The Legion of Merit (LM). *As to United States Armed Forces.* The Legion of Merit, without reference to degree, is awarded to members of the Armed Forces of the United States who, while serving in any capacity, distinguish themselves by exceptionally meritorious conduct in the performance of outstanding services. In peacetime, awards by the Air Force are generally limited to recognizing services of marked national or international significance, services which aided the United States in furthering national policy or national security.

This decoration, like the Purple Heart, stems from the Badge for Military Merit, America's oldest decoration, established by George Washington in 1782. As was the case with the Badge for Military Merit, it will be awarded for "extraordinary fidelity and essential service." It will constitute a reward for service in a position of responsibility, honorably and well performed.

The design of the Legion of Merit has been developed from the Great Seal of the United States, also approved by Congress in 1782.

The obverse or front of the badge of the Legion of Merit is a five-pointed American star of heraldic form, bordered in purplish red enamel, centered with a constellation of the 13 original stars on a blue enameled field breaking through a circle of clouds. The star is backed by a laurel wreath, the symbolic award for achievement, which is interlaced with crossed war arrows in gold pointing outward, representing the protection afforded by the armed forces to the Nation.

On the reverse are the words, "United States of America," inscribed on a circling ribbon. In the center is space left for inscription of the name and grade of the individual to whom the award is made. Surrounding this is a band which carries the words (taken from the reverse of the Great Seal) "Annuit Coeptis" (He [God] has favored our undertakings), and the date MDCCLXXXII, the year of the founding of the decoration. The ribbon is a purple-red color, edged with white.

Legion of Merit to Armed Forces of Foreign Nations. The Legion of Merit, in four degrees, is awarded to *personnel of the armed forces* of friendly foreign nations who distinguish themselves by exceptionally meritorious conduct in the performance of outstanding service. The degrees are: Chief Commander; Commander; Officer; and Legionnaire.

The criteria for award of the various degrees are:

Chief Commander: Chief of state or head of government.

Commander: Equivalent of U.S. military chief of staff.

Officer: Other general or flag rank, equivalent assignments, and foreign attaches.

Legionnaire: All other eligibles.

Distinguished Flying Cross (DFC). (As an Award for Achievement.) The Distinguished Flying Cross is awarded to members of the Armed Forces who, while serving in any capacity with the Air Force, distinguish themselves by heroism or extraordinary achievement while participating in aerial flight.

To warrant an award of the Distinguished Flying Cross for extraordinary achievement while participating in aerial flight, the results accomplished must be so exceptional and outstanding as clearly to set the individual apart from his comrades who have not been so recognized.

The requirements for the award for heroism are stated earlier in this chapter.

Bronze Star Medal (BSM). (For Achievement.) The Bronze Star Medal is awarded to members of the Armed Forces who, while serving in any capacity, distinguished themselves by meritorious achievement or meritorious service not involving participation in aerial flight, in connection with military operations against an enemy of the United States. The Bronze Star Medal awarded for valor is worn with a "V" as previously discussed.

The required meritorious achievement or meritorious service for award of the Bronze Star Medal is less than that required for award of the Legion of Merit, but must nevertheless be accomplished with distinction. The Bronze Star Medal may be awarded to recognize meritorious service or single acts of merit.

Meritorious Service Medal (MSM). The Meritorious Service Medal is awarded for outstanding non-combat meritorious achievement or service to the United States, although the required achievement or service is less than that required for the award of the Legion of Merit. It must, nevertheless, be accomplished with distinction and above and beyond that for the award of the Air Force Commendation Medal. The Meritorious Service Medal ranks with, but after, the Bronze Star.

The Meritorious Service Medal is bronze consisting of six rays issuant from the upper three points of a five-pointed star, beveled edges, and containing two small stars defined by an incised outline. In front of the lower part of the star appears an eagle with wings upraised, standing upon two upward curving branches of laurel tied with a ribbon beneath the feet of the eagle. The ribbon is predominantly of ruby color with white vertical stripes and ruby lines at each edge.

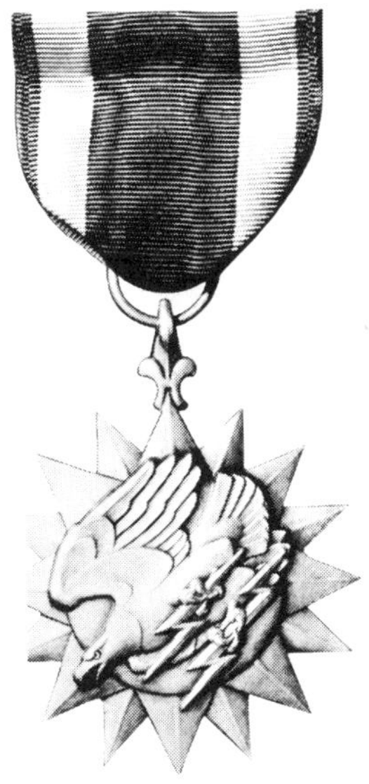

AIR MEDAL.

Air Medal (AM). Authorized by Executive Order of the President in 1942, the Air Medal is awarded to any person who, while serving in any capacity with the Army or Air Force of the United States subsequent to September 8, 1939, distinguishes himself by meritorious achievement while participating in an aerial flight. This decoration is awarded in those cases where the act of meritorious service does not warrant the award of the Distinguished Flying Cross.

Pendant from a ribbon striped with the Air Force colors of blue and gold, is a fleur-de-lis which surmounts a compass rose. In relief on the rose is a swooping eagle with lightning bolts clutched in his talons.

Air Force Achievement Medal (AFAM). May be awarded for outstanding achievement, meritorious service, or acts of courage not meriting award of the AFCM. Intended primarily for outstanding junior officers and airmen. Pendant is a silver-colored nebular disc bearing the winged thunderbolt from the USAF seal. Ribbon is predominantly silver-gray with 12 vertical ultra-marine stripes.

AIR FORCE ACHIEVEMENT MEDAL.

DOD DECORATIONS FOR ACHIEVEMENT (AFR 900-48)

Besides the USAF decorations listed above, the Department of Defense awards four military decorations which take precedence with their service counterparts, but which are worn before them.

Defense Distinguished Service Medal (DDSM). May be awarded to any U.S. Armed Forces officer assigned to a joint staff or other DOD joint activity for exceptionally meritorious service in a position of unique and great responsibility. Pendant is basically a blue enamel pentagon superimposed by a gold eagle bearing the United States Shield and grasping three crossed arrows. Ribbon has blue, yellow, and red stripes. Worn before the service DSM.

Defense Superior Service Medal (DSSM). May be awarded to United States military personnel who render superior meritorious service in a position of significant responsibility on a joint staff or in a joint activity. Pendant design is that of the DDSM, except that eagle, etc., are in silver. Ribbon has yellow, blue, white, and red stripes. Worn before the LM.

Defense Meritorious Service Medal (DMSM). May be awarded to United States military personnel who render incontestably exceptional meritorious service on a joint staff or in a joint activity. Pendant consists of a bronze laurel wreath and overlapping pentagon surmounted by an eagle. Ribbon is light blue broken by a gold band that is broken by a smaller red band. Worn before the MSM.

Joint Service Commendation Medal (JSCM). The Joint Service Commendation Medal is awarded to personnel on duty in the Office of the Secretary of Defense, the Joint Staff, the Defense Supply Agency, the National Security Agency, other joint agencies reporting to the Joint Chiefs of Staff, joint task forces, or NATO organizations. The award is for meritorious achievement. The associated ribbon is composed of blue, white, and green stripes. It takes precedence with the AFCM, but before it when worn. A bronze "V" is authorized for a combat award after 25 June 1963.

JOINT SERVICE
COMMENDATION MEDAL.

UNIT AWARDS (AFM 900-48)

Unit awards are authorized as recognition of certain types of service, and as a means of promoting esprit de corps. They are of the following categories:

Unit decorations
Campaign and expeditionary streamers
War service streamers

United States Unit Decorations. United States unit decorations have been established to recognize outstanding heroism or exceptionally meritorious conduct in the performance of outstanding services:

Presidential Unit Citation
Air Force Outstanding Unit Award
Air Force Organizational Excellence Award

PRESIDENTIAL UNIT CITATION (with Oak Leaf Cluster).

AIR FORCE OUTSTANDING UNIT AWARD.

Presidential Unit Citation (PUC). The Presidential Unit Citation (formerly Distinguished Unit Citation [DUC]) is awarded to units of the armed forces of the United States and co-belligerent nations for extraordinary heroism in action against the armed enemy. The unit must display such gallantry, determination, and esprit de corps in accomplishing its mission under extremely difficult and hazardous conditions as to set it apart and above other units participating in the same campaign. The degree of heroism required is the same as that which would warrant award of an Air Force Cross to an individual. Extended periods of combat duty or participation in a large number of operational missions, either ground or air, is not sufficient. Only on rare occasions will a unit larger than a battalion or air group qualify for award of this decoration. It is a blue ribbon set in a gold colored metal frame of laurel leaves; it is worn above the pocket of the left breast after decorations and preceding service ribbons.

Air Force Outstanding Unit Award (AFOUA). Established in 1954, this consists of a predominantly blue streamer with a narrow red band center bordered by white lines and red bands. Theater or area of operations is embroidered in white on the streamer. The individual emblem is a ribbon of the streamer color. It is awarded to units not larger than a wing for meritorious achievement or service in support of military operations or of great significance in accomplishment not involving combat operations. The ribbon bar is worn on the left breast pocket.

Air Force Organizational Excellence Award (AFOEA). Established in August 1969, this ribbon is predominantly red with a narrow blue band center bordered by white lines and blue bands at each edge separated by white lines. It is awarded to Air Force internal organizations that are organizational entities within larger organizations. The organizations are unique, unnumbered organizations or activities that perform staff functions as well as functions normally performed by numbered wings, groups, squadrons, etc. The ribbon is worn on the left breast pocket.

War Service Streamers and Campaign and Expeditionary Streamers. War service streamers are awarded to organizations for service in a theater or area of operations. They represent the unit's service in the same manner that service medals represent the individual's service in a theater or area of combat operations. Campaign and Expeditionary Streamers represent the unit's participation in a campaign in the same manner that battle stars and arrowheads on the service medal represent the individual's participation in a campaign or in an airborne landing or amphibious assault. The streamers are carried on the organizational flag or guidon on ceremonial occasions.

ACHIEVEMENT AWARDS (AFR 900-48)

Air Force achievement awards recognize specific types of achievement by members serving on active duty in the Air Force or in the Air Reserve Forces.

Combat Readiness Medal (CRM). The Combat Readiness Medal is authorized for members of combat crews of manned weapons delivery systems for sustained meritorious performance. The medal is a circle marked with arrowheads and the points of overlapping triangles to indicate the hours of round-the-clock duty and performance. The ribbon is predominantly Old Glory red and banded in blue, with a narrow dark blue stripe separated by two wider stripes of light blue.

Air Force Good Conduct Medal (AFGCM). This is a medal awarded only to enlisted persons in recognition of exemplary behavior, efficiency, and fidelity, under prescribed conditions as to time and ratings. A distinctive clasp is awarded for each successive period of three years' service which meets the requirements. The ribbon is light blue with red, white, and blue vertical stripes to the right and left of the center.

COMBAT READINESS MEDAL.

AIR FORCE GOOD CONDUCT MEDAL.

Outstanding Airman of the Year Ribbon (OAYR). Awarded to airmen nominated by major air commands and separate operating agencies for competition in the 12 Outstanding Airmen of the Year Program sponsored by the Air Force Association. Ribbon is oriental-blue with a white center bordered by equal strips of ultramarine-blue and flaming red.

Air Force Recognition Ribbon (AFRR). Awarded to named individual Air Force recipients of special trophies and awards listed in AFR 900-29, except 12 Outstanding Airmen of the Year nominees. Ribbon is predominantly turquoise-blue centered with a wide red stripe and with red and white stripes at either end.

Air Force Overseas Ribbon (AFOSR). Awarded for completion of an overseas tour of duty after 1 September 1980, provided the tour is not recognized by another service award. Ribbon is predominantly brittany-blue with three vertical narrow white stripes on either side.

Air Force Longevity Service Award (AFLSA). All members of the Air Force on active duty and all Reservists not on active duty are eligible for the Longevity Service Award. Requirements for the basic award are four years of honorable active federal military service with any branch of the United States Armed Forces. A bronze oak leaf cluster is worn on the ribbon for each additional four years of service. A silver oak leaf cluster is worn in lieu of five bronze clusters.

The Air Force Longevity Service Award is an ultramarine blue service ribbon divided by four equal stripes of turquoise. There is no medal authorized.

AIR FORCE LONGEVITY
SERVICE AWARD RIBBON.

Air Reserve Forces Meritorious Service Medal. This medal was approved in April 1964 to be effective 1 April 1965 as an award equivalent to the Air Force Good Conduct Medal for issuance to Air Reserve personnel. Qualifications are as follows: Exemplary behavior, efficiency, and fidelity for four continuous years; attendance at ninety percent of all scheduled training periods each year of four continuous years; completion of active duty requirements in the four-year period. The ribbon is predominantly light blue with white, ultramarine and yellow stripes at the edges.

NCO Professional Military Education Graduate Ribbon (NCOPMEGR). Awarded to graduates of certified Noncommissioned Officer Professional Military Education Schools, Phase III, IV, and V. The ribbon is red with a narrow white stripe near each edge and two blue stripes near the center.

USAF Basic Military Training Honor Graduate Ribbon (BMTHGR). Awarded to honor graduates of BMT who have demonstrated excellence in all phases of academic and military training and limited to the top 10 percent of the training flight. Ribbon has an ultramarine-blue center with yellow, brittany blue and white bands on either side.

Small Arms Expert Marksmanship Ribbon (SAEMR). Awarded after 1 January 1963 to persons qualifying as "expert" on weapons specified in AFR 50-57. The ribbon has a green center with a wide blue stripe at each edge and two narrow yellow stripes separating the green and blue.

Air Force Training Ribbon (AFTR). Awarded to Air Force members on completion of initial military accession training (BMT, OTS, ROTC, Academy, Medical Services, Judge Advocate, Chaplain orientation, etc.). Ribbon is predominantly Air Force blue centered with a red stripe with one gold stripe near each end.

SERVICE MEDALS (AFR 900-48)

Service medals are awarded to members of the Armed Forces of the United States to denote the honorable performance of duty. Most service medals pertain to federal duty in a time of war or national emergency. A person's entire service during the period for which the award is made must be honorable.

Service Medals, World War I Period. There are but two service medals covering the entire World War I period. Attachments to the ribbon of the battle clasps, service clasps, and service stars follow.

World War I Victory Medal. These are requirements for the award of this service medal:

Service, 6 April 1917—11 November 1918.

American Expeditionary Forces in European Russia, 12 November 1918—5 August 1919.

American Expeditionary Forces in Siberia, 12 November 1918—1 April 1920.

Battle Clasps. The award of a battle clasp required specific combat service.

Service Clasps. The award of a service clasp required service in France, Italy, Siberia, European Russia, or England, as a member of a crew of a transport sailing between the United States and those countries.

Service Stars. Possession of a battle clasp or defensive sector clasp is denoted on the service ribbon of the medal, one bronze star for each clasp.

Army of Occupation of Germany Medal. Requirement is honorable service in Germany or Austria-Hungary, 12 November 1918—11 July 1923.

Service Medals, World War II. For the period of World War II, and the national emergency preceding the declaration of war, Executive Orders were issued for a series of separate service medals, in contrast to the World War I method of one service medal with a number of service or battle clasps.

American Defense Service Medal (ADSM). Required is service between 8 September 1939, the initiation of the period of national emergency, to 7 December 1941, date of declaration of war, for a period of 12 months or longer. A foreign service clasp was provided for wear on the suspension ribbon of the

AMERICAN DEFENSE SERVICE MEDAL.

WOMEN'S ARMY CORPS SERVICE MEDAL.

AMERICAN CAMPAIGN MEDAL.

medal, denoted by a service star worn on the service ribbon. Description: Stripes of yellow, blue, white, red, golden yellow band, stripes of red, white, blue, yellow.

Women's Army Corps Service Medal. Required is service in the Women's Army Corps between 20 July 1942 and 30 August 1943 or between 1 September 1943 and 2 September 1945. The ribbon is moss green with gold borders.

American Campaign Medal (ACM). Required is service within the continental limits of the United States for an aggregate period of 1 year, or specified service within the American Theater of shorter duration. The service ribbon consists of stripes of blue, white, black, red, white, blue, dark blue, white, red, blue, white, red, black, white, blue. A person who served in a unit accorded battle credit for the Anti-submarine Campaign is entitled to wear a bronze service star on this medal.

Asiatic-Pacific Campaign Medal (APCM). Required is service within the areas indicated in the map illustration, 7 December 1941—2 March 1946, on permanent assignment, or in passenger status or temporary duty for 30 consecutive days or 60 days not consecutive, or in active combat which latter requires proof. This ribbon is of stripes of orange, white, red, white, orange, blue, white, red, orange, white, red, white, orange. Service stars are provided for wear with this medal and

ASIATIC-PACIFIC CAMPAIGN MEDAL.

EUROPEAN-AFRICAN-MIDDLE EASTERN CAMPAIGN MEDAL

WORLD WAR II VICTORY MEDAL.

ribbon for combat campaigns. An arrowhead is also provided for participation in a combat parachute jump, combat glider landing, or amphibious assault.

European-African-Middle Eastern Campaign Medal (EAMECM). Required is service within the areas indicated in the map illustration, 7 December 1941—8 November 1945, under the same conditions, and with service star and arrowhead as stated for the Asiatic-Pacific Campaign Medal. This ribbon is of stripes of brown, green, white, red, green, blue, white, red, green, white, black, white, brown.

World War II Victory Medal (WWIIVM). Required is honorable service, 7 December 1941—31 December 1946. No stars or arrowhead. This ribbon is a double rainbow in juxtaposition, white stripe, red band, white stripe, and double rainbow in juxtaposition.

Army of Occupation Medal (AOM). This medal requires service for 30 consecutive days (as contrasted to inspector, visitor, courier, escort, passenger status, temporary duty, or detached service), while assigned to any of the following armies of occupation:

Germany (exclusive of Berlin) between 9 May 1945 and 5 May 1955.
Austria, 9 May 1945—27 July 1955.
Italy, 9 May 1945—15 September 1947.
Japan, 3 September 1945—27 April 1952.
Korea, 3 September 1945—29 June 1949.

A clasp appropriately inscribed is issued with each award to denote the area in

ARMY OF OCCUPATION MEDAL.

MEDAL FOR HUMANE ACTION.

NATIONAL DEFENSE SERVICE MEDAL.

which occupation duty was performed.

The ribbon is composed of white stripe, black band, red band, white stripe.

Medal for Humane Action (MHA). Required is service for at least 120 days during the period 26 June 1948—30 September 1949, within the boundaries of the Berlin airlift operations while participating in or in direct support of the Berlin airlift. Persons other than members of the armed forces may receive the award when recommended for meritorious participation. Posthumous award may be made without regard to length of the prescribed service. This ribbon consists of a black band, white stripe, blue band, white stripe, red stripe, white stripe, blue band, white stripe, black band.

National Defense Service Medal (NDSM). This medal has been authorized for honorable active service for any period between 27 June 1950 and 27 July 1954. Persons on active duty for purposes other than extended active duty are not eligible for this award. The medal is of bronze as pictured. The ribbon is red with a yellow center bordered with white, blue, white, and red. The NDSM has also been authorized for cold war service from 1 January 1961 to 14 August 1974. A bronze service star is authorized for persons who have served during both periods.

Korean Service Medal (KSM). Required is service between 27 June 1950 and 27 July 1954 under any of the following conditions:

ANTARCTICA
SERVICE MEDAL.

KOREAN
SERVICE MEDAL.

Service must have been on permanent assignment, or on temporary duty for 30 consecutive days or 60 days not consecutive, or in active combat under specific conditions of combat award or command certificate. Service is required within Korea or the waters immediately adjacent thereto; or within a unit under operational control of CINCFE supporting directly the effort in Korea; or substantiated by individual certificate by the Commander-in-Chief, Far East, testifying to material contribution made in direct support of the military effort in Korea.

The ribbon is composed of a white stripe, United Nations blue band, white stripe, United Nations blue band, white stripe.

Service star. Combat service within the Korean Theater between 27 June 1950 and 27 July 1954, one bronze star for each campaign.

Arrowhead. Awarded for participation in combat parachute jump, combat glider landing, or amphibious assault within the Korean Theater.

Antarctica Service Medal (ASM). Awarded to members of any of the armed forces who as members of a United States expedition participated in scientific, direct support, or exploratory operations on the Antarctic Continent, or in a foreign expedition cooperating with US expeditions, or who participated in flights to and from Antarctica in support of such operations, or who have served in a US ship operating south of the 60th parallel in support of such US operations. The medal includes a "Wintered Over" clasp for those who stayed in Antarctica during the winter months. There is a corresponding circular device for wear on the ribbon bar, depicting the map of the Antarctic Continent.

ARMED FORCES EXPEDITION MEDAL.

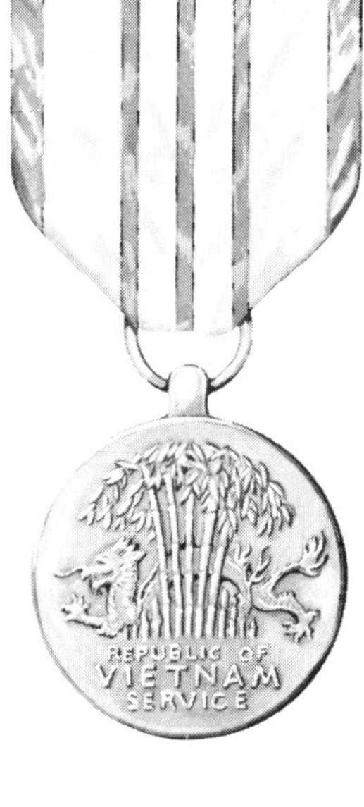

VIETNAM
SERVICE MEDAL.

Armed Forces Expeditionary Medal (AFEM). Awarded to members of the US armed forces who have served during the cold war in places where there was fighting or the threat of it, such as Berlin, Quemoy, Matsu, Lebanon in 1958, the Dominican Republic, Laos, Vietnam, or Korea after 1 October 1966. (Those who served in Vietnam between 1 July 1958 and 3 July 1965 may wear either the AFEM or VSM for such service, but not both.)

Vietnam Service Medal (VNSM). The Vietnam Service Medal has been authorized by the President to be awarded to personnel who served in Vietnam, Cambodia, Thailand, Laos, or contiguous waters or airspace, from 4 July 1965 through 28 March 1973. The ribbon is yellow, edged in green, with three red stripes in the center.

Humanitarian Service Medal (HSM). Awarded to those who after 1 April 1975 distinguished themselves by meritorious direct participation in approved military acts of a humanitarian nature. Bronze pendant displays an opened right hand on obverse and a sprig of oak on reverse. Ribbon has vertical stripes of purple, white, light and dark blue.

Armed Forces Reserve Medal. Required is honorable and satisfactory service in one or more of the Reserve components of the Armed Forces for a period of 10 years, not necessarily consecutive, provided such service was performed within a period of 12 consecutive years. Periods of service as a member of a Regular component are excluded from consideration.

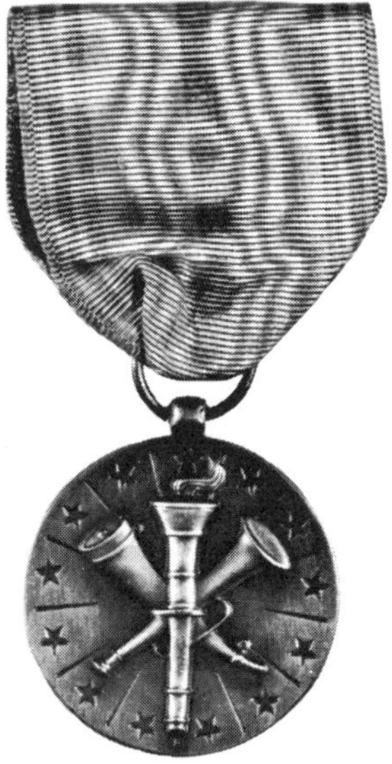

ARMED FORCES
RESERVE MEDAL.

The ribbon is composed of stripes of blue, buff, blue, buff, blue, buff band, continuing with stripes of blue, buff, blue, buff, blue.

10-Year Device. One 10-year device is authorized to be worn on the suspension and service ribbon to denote service for each 10-year period in addition to and under the same conditions as prescribed for the award of the medal. It is an hour glass, bronze, with a Roman numeral "X" superimposed thereon 5/16 inch in height.

AUTHORIZED FOREIGN SERVICE AWARDS

Philippine Defense Ribbon (PDR). *Description.* A silk moire ribbon composed of a red stripe, white stripe, red band, white stripe, and a red stripe. In the center of the red band are three white stars.

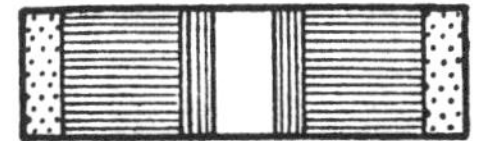

PHILIPPINE SERVICE RIBBONS.

Philippine Liberation Ribbon (PLR). *Description.* A silk moire ribbon composed of a red band, blue stripe, white stripe, and a red band.

Philippine Independence Ribbon (PIR). *Description.* A silk moire ribbon composed of a yellow stripe, blue stripe, red stripe, white stripe, red stripe, blue stripe, and yellow stripe.

Republic of Vietnam Campaign Medal (RVNCM). *Description.* A silk moire ribbon of vertical green and yellow stripes with a metal device reading "1960- ."

Requirements. Republic of Vietnam Armed Forces Order No. 48, dated 24 March 1966, authorized award of this RVN service medal to US military personnel serving six months or wounded in Vietnam during the recent conflict. US military personnel outside Vietnam who contributed direct combat support to the RVNAF for six months were also eligible.

UNITED NATIONS SERVICE MEDALS

UNITED NATIONS SERVICE MEDAL. UNITED NATIONS MEDAL.

United Nations Service Medal (UNSM). The President on 15 December 1951 authorized the wearing by eligible personnel of the United Nations Service Medal. It is given for service in defense of the principles of the charter of the United Nations to those who have participated in the Korean campaign from the start of the conflict to 27 July 1954.

United Nations Medal. The United Nations Medal is awarded by the United

States to specific individuals for service with United Nations Forces, specifically with the UN Observer Group in Lebanon, UN Truce Supervisory Organization in Palestine, or the UN Military Observer Group in India and Pakistan. The medal is a round disc with the UN emblem and the letters "UN" on the obverse, and on the reverse the inscription "In the Service of Peace." The ribbon is of UN blue with two single narrow white stripes ¼-inch from each edge.

UNITED STATES AIR FORCE MEDAL OF HONOR WINNERS—1918-1977

NAMES, ALPHABETICALLY BY WARS AND RANK AT TIME OF ACTION	HOME TOWN	DATE AND PLACE OF ACTION	PRESENT ADDRESS OR DATE OF DEATH
		WORLD WAR I	
Bleckley, 2d Lt. Erwin R.	Wichita, Kan.	Oct. 6, 1918, Binarville, France	KIA, Oct. 6, 1918
Goettler, 2d Lt. Harold E.	Chicago, Ill.	Oct. 6, 1918, Binarville, France	KIA, Oct. 6, 1918
Luke, 2d Lt. Frank, Jr.	Phoenix, Ariz.	Sept. 29, 1918, Murvaux, France	KIA, Sept. 29, 1918
Rickenbacker, Capt. Edward V.	Columbus, Ohio	Sept. 25, 1918, Billy, France	Deceased, July 23, 1973
		WORLD WAR II	
Baker, Lt. Col. Addison E.	Chicago, Ill.	Aug. 1, 1943, Ploesti, Romania	KIA, Aug. 1, 1943
Bong, Maj. Richard I.	Superior, Wis.	Oct. 10-Nov. 15, 1944, Southwest Pacific	Killed, Aug. 6, 1945, Burbank, Calif.
Carswell, Maj. Horace S., Jr.	Fort Worth, Tex.	Oct. 26. 1944, South China Sea	KIA, Oct. 26, 1944
Castle, Brig. Gen. Frederick W.	Manila, P.I.	Dec. 24, 1944, Liège, Belgium	KIA, Dec. 24, 1944
Cheli, Maj. Ralph	San Francisco, Calif.	Aug. 18, 1943, Wewak, New Guinea	Died as POW, Mar. 6, 1944
Craw, Col. Demas T.	Traverse City, Mich.	Nov. 8, 1942, Port Lyautey, French Morocco	KIA, Nov. 8, 1942
Doolittle, Lt. Col. James H.	Alameda, Calif.	Apr. 18, 1942, Tokyo, Japan	Los Angeles, Calif. (Ret. Lt. Gen.)
Erwin, SSgt. Henry E.	Adamsville, Ala.	Apr. 12, 1945, Koriyama, Japan	Birmingham, Ala.
Femoyer, 2d Lt. Robert E.	Huntington, W. Va.	Nov. 2, 1944, Merseburg, Germany	KIA, Nov. 2, 1944
Gott, 1st Lt. Donald J.	Arnett, Okla.	Nov. 9, 1944, Saarbrücken, Germany	KIA, Nov. 9, 1944
Hamilton, Maj. Pierpont M.	Tuxedo Park, N.Y.	Nov. 8, 1942, Port Lyautey, French Morocco	Santa Barbara, Calif. (Ret. Maj. Gen.)
Howard, Lt. Col. James H.	Canton, China	Jan. 11, 1944, Oschersleben, Germany	Washington, D.C. (Ret. Brig. Gen.)
Hughes, 2d Lt. Lloyd H.	Alexandria, La.	Aug. 1, 1943, Ploesti, Romania	KIA, Aug. 1, 1943
Jerstad, Maj. John L.	Racine, Wis.	Aug. 1, 1943, Ploesti, Romania	KIA, Aug. 1, 1943
Johnson, Col. Leon W.	Columbia, Mo.	Aug. 1, 1943, Ploesti, Romania	McLean, Va. (Ret. Gen.)
Kane, Col. John R.	McGregor, Tex.	Aug. 1, 1943, Ploesti, Romania	Barber, Ark. (Ret. Col.)
Kearby, Col. Neel E.	Wichita Falls, Tex.	Oct. 11, 1943, Wewak, New Guinea	KIA, Mar. 5, 1944, Wewak, New Guinea
Kingsley, 2d Lt. David R.	Portland, Ore.	June 23, 1944, Ploesti, Romania	KIA, June 23, 1944
Knight, 1st Lt. Raymond L.	Houston, Tex.	Apr. 25, 1945, Po Valley, Italy	KIA, Apr. 25, 1945
Lawley, 1st Lt. William R., Jr.	Leeds, Ala.	Feb. 20, 1944, Leipzig, Germany	Montgomery, Ala. (Ret. Col.)
Lindsey, Capt. Darrell R.	Jefferson, Iowa	Aug. 9, 1944, Pontoise, France	KIA, Aug. 9, 1944
Mathies, SSgt. Archibald	Scotland	Feb. 20, 1944, Leipzig, Germany	KIA, Feb. 20, 1944
Mathis, 1st Lt. Jack W.	San Angelo, Tex.	Mar. 18, 1943, Vegesack, Germany	KIA, Mar. 18, 1943
McGuire, Maj. Thomas B., Jr.	Ridgewood, N.J.	Dec. 25-26, 1944, Luzon, P.I.	KIA, Jan. 7, 1945, Negros, P.I.
Metzger, 2d Lt. William E., Jr.	Lima, Ohio	Nov. 9, 1944, Saarbrücken, Germany	KIA, Nov. 9, 1944
Michael, 1st Lt. Edward S.	Chicago, Ill.	Apr. 11, 1944, Brunswick, Germany	Fairfield, Calif. (Ret. Col.)
Morgan, 2d Lt. John C.	Vernon, Tex.	July 28, 1943, Kiel, Germany	Greenwich, Conn. (Ret. Col.)
Pease, Capt. Harl, Jr.	Plymouth, N.H.	Aug. 7, 1942, Rabaul, New Britain	KIA, Aug. 7, 1942
Pucket, 1st Lt. Donald D.	Longmont, Colo.	July 9, 1944, Ploesti, Romania	KIA, July 9, 1944
Sarnoski, 2d Lt. Joseph R.	Simpson, Pa.	June 16, 1943, Buka, Solomon Is.	KIA, June 16, 1943
Shomo, Maj. William A.	Jeannette, Pa.	Jan. 11, 1945, Luzon, P.I.	Pittsburgh, Pa. (Ret. Lt. Col.)
Smith, SSgt. Maynard H.	Caro, Mich.	May 1, 1943, St. Nazaire, France	Long Island City, N.Y.
Truemper, 2d Lt. Walter E.	Aurora, Ill.	Feb. 20, 1944, Leipzig, Germany	KIA, Feb. 20, 1944
Vance, Lt. Col. Leon R., Jr.	Enid, Okla.	June 5, 1944, Wimereaux, France	Killed, July 26, 1944, near Iceland
Vosler, TSgt. Forrest L.	Lyndonville, N.Y.	Dec. 20, 1943, Bremen, Germany	Poland, N.Y.
Walker, Brig. Gen. Kenneth N.	Cerrillos, N.M.	Jan. 5, 1943, Rabaul, New Britain	KIA, Jan. 5, 1943
Wilkins, Maj. Raymond H.	Portsmouth, Va.	Nov. 2, 1943, Rabaul, New Britain	KIA, Nov. 2, 1943
Zeamer, Maj. Jay, Jr.	Carlisle, Pa.	June 16, 1943, Buka, Solomon Is.	Hyannis, Mass. (Ret. Lt. Col.)
		KOREA	
Davis, Maj. George A., Jr.	Dublin, Tex.	Feb. 10, 1952, Sinuiju-Yalu River, No. Korea	KIA, Feb. 10, 1952
Loring, Maj. Charles J., Jr.	Portland, Me.	Nov. 22, 1952, Sniper Ridge, No. Korea	KIA, Nov. 22, 1952

Sebille, Maj. Louis J.	Harbor Beach, Mich.	Aug. 5, 1950, Hamch'ang, So. Korea	KIA, Aug. 5, 1950
Walmsley, Capt. John S., Jr.	Baltimore, Md.	Sept. 14, 1951, Yangdok, No. Korea	KIA, Sept. 14, 1951

VIETNAM

Bennett, Capt. Steven L.	Palestine, Tex.	June 29, 1972, Quang Tri, So. Vietnam	KIA, June 29, 1972
Day, Col. George E.	Sioux City, Iowa	Conspicuous gallantry while POW	Shalimar, Fla. (Ret. Col.)
Dethlefsen, Maj. Merlyn H.	Greenville, Iowa	Mar. 10, 1967, Thai Nguyen, No. Vietnam	Fort Worth, Tex. (Ret. Col.)
Fisher, Maj. Bernard F.	San Bernardino, Calif.	Mar. 10, 1966, A Shau Valley, So. Vietnam	Kuna, Idaho (Ret. Col.)
Fleming, 1st Lt. James P.	Sedalia, Mo.	Nov. 26, 1968, Duc Co, So. Vietnam	Active duty, Lt. Col., Randolph AFB, Tex
Jackson, Lt. Col. Joe M.	Newnan, Ga.	May 12, 1968, Kham Duc, So. Vietnam	Kent, Wash. (Ret. Col.)
Jones, Lt. Col. William A. III	Norfolk, Va.	Sept. 1, 1968, Dong Hoi, No. Vietnam	Killed. Nov. 15, 1969, Woodbridge, Va.
Levitow, A1C John L.	Hartford, Conn.	Feb. 24, 1969, Long Binh, So. Vietnam	Vienna, Va.
Sijan, Capt. Lance P.	Milwaukee, Wis.	Conspicuous gallantry while POW	Died while POW, Jan. 1968
Thorsness, Lt. Col. Leo K.	Walnut Grove, Minn.	Apr. 19, 1967, No. Vietnam	Santa Monica, Calif. (Ret. Col.)
Wilbanks, Capt. Hilliard A.	Cornelia, Ga.	Feb. 24, 1967, Dalat, So. Vietnam	KIA, Feb. 24, 1967
Young, Capt. Gerald O.	Anacortes, Wash.	Nov. 9, 1967, Da Nang area, So. Vietnam	Anacortes, Wash. (Ret. Maj.)

BADGES (AFM 35-10)

Badges are appurtenances of the uniform. In the eyes of their wearers several badges have a significance equal to or greater than all but the highest decorations. There is no established precedence between badges except that aviation badges are worn above any others.

Eligibility to Wear Badges. Eligibility to wear aviation badges will be as prescribed by the Chief of Staff of the Air Force (AFM 35-10).

The illustration shows current aviation and medical badges. Those individuals granted aeronautical ratings no longer current are authorized to wear the aviation badge which was in effect when the rating was granted.

Also illustrated are the Astronaut, Marksmanship, Missileman, Security Police, Combat Crew, Recruiting, ATC Instructor Parachutist and Badges.

Air Force Combat Crew Badge. This badge may be worn only by aircraft and missile launch crew members who: (1) are assigned to a USAF operational unit having a weapons delivery mission and are currently maintaining a war mission alert posture; (2) have been designated and certified as combat ready according to USAF and major air command qualification criteria; and (3) are currently serving in rated or missile launch crew positions as combat crew members.

The badge is a rectangular oxidized silver badge bearing the Air Force coat of arms and the words "COMBAT CREW." It is worn over the right breast pocket immediately above the name tag. The Combat Crew badge is *not* worn on the mess dress or formal evening dress.

Astronaut Wings. To distinguish Air Force space pilots, the USAF has established badges for pilot astronauts, the Command Pilot Astronaut and the Senior Pilot Astronaut. In the new insignia a shooting star, symbolizing the astronaut's spatial environment, has been superimposed over the traditional pilot's badge. To be eligible for the Senior Pilot Astronaut wings, a pilot must fly at least 50 miles above the earth in a powered space vehicle.

Air Force Distinguished and Excellence-in-Competition Rifleman and Pistol Shot Badges. *Distinguished Rifleman and Pistol Shot Badges.* Each "Distinguished" badge consists of a gold bar and a gold pendant designed as a shield, fastened together with rings. The bar is lettered "U.S. Air Force." An enameled

COMMAND PILOT

SENIOR PILOT

PILOT

MASTER NAVIGATOR OR MASTER AIRCRAFT OBSERVER

SENIOR NAVIGATOR OR SENIOR AIRCRAFT OBSERVER

NAVIGATOR OR AIRCRAFT OBSERVER

CHIEF FLIGHT SURGEON

SENIOR FLIGHT SURGEON

FLIGHT SURGEON

CHIEF FLIGHT NURSE

SENIOR FLIGHT NURSE

FLIGHT NURSE

CHIEF NONRATED OFFICER AIRCREW MEMBER

SENIOR NONRATED OFFICER AIRCREW MEMBER

NONRATED OFFICER AIRCREW MEMBER

CHIEF ENLISTED AIRCREW MEMBER

SENIOR ENLISTED AIRCREW MEMBER

ENLISTED AIRCREW MEMBER

ASTRONAUT DESIGNATOR
(This design is embossed on top of badge member is awarded.)

AIR FORCE ASTRONAUT AND AVIATION BADGES. (AFR 35-13, OPR: HQ AFMPC/MPCAJD)

target is at the center of the pendant below the word "Distinguished" and above either the word "Rifleman" or "Pistol Shot."

Excellence in Competition Rifleman or Pistol Shot Badges. These are gold, silver, or bronze, and consist of a circular pendant suspended from a bar by rings.

How Worn. These badges are worn centered on the flap of the left breast pocket of the coat or shirt, the pendant covering the button and button hole. These badges are *not* worn on the mess dress uniform, fatigue, or flying clothing.

WEARING OF DECORATIONS, SERVICE MEDALS, RIBBONS AND BADGES

Individuals entitled to wear decorations, service medals, ribbons and badges should be certain that they are placed on the uniform in the prescribed location and in the prescribed order.

United States Military Service Medals. *Definition.* United States military and naval service medals are:

Service medals awarded by the Department of the Air Force and Department of the Army as listed in this chapter.

Service medals awarded by the Navy Department for performance of duty as a member of the United States Naval Forces, as listed in Chapter 1, Part A, of Bureau of Naval Personnel Manual.

Service medals awarded by the Philippine Government for performance of active military service during the period the Commonwealth of the Philippines was a possession of the United States and while the Philippine Army was in the service of the United States.

The Republic of Vietnam Campaign Medal (RCVM) awarded by that government for service in Vietnam between 1 March 1961 and 18 March 1973.

When Worn. Commanders may prescribe the wearing of decorations and service medals on the following occasions:

Parades, reviews, inspections, and funerals; and

Ceremonial and social occasions of a formal nature.

Decorations and service medals may be worn at the option of the wearer on the following occasions:

Holidays when not on duty with troops; and

Social occasions of a private nature.

Decorations and service medals will *not* be worn on the following occasions:

When equipped for combat or simulated combat;

By officers while suspended from either rank or command; or

By airmen while serving sentence of confinement.

United States Nonmilitary Decorations. These may be worn on the uniform only if military decorations or service medals are worn. These include the Medal for Merit, National Security Medal, Presidential Medal for Freedom, and decorations awarded by NASA, the Treasury Department, Public Health, and Maritime Services. Nonmilitary service *awards* may *not* be worn on the Air Force uniform.

State decorations and service awards may be worn only by Air National Guardsmen in non-active duty status.

Foreign Service Medals. The acceptance or wearing of foreign decorations and service medals for service performed while a member of the armed forces of the United States is prohibited, except as provided in AFM 900–48.

Decorations and service medals awarded by the national government of a friendly country may be worn provided such service medals were earned while a bona fide member of the armed forces of that friendly foreign nation. At least one United States decoration or service medal must be worn at the same time that a foreign service medal is worn.

The wearing of foreign service medals (except those indicated above), including civilian service medals awarded by a foreign national government and all service medals awarded by an inferior foreign jurisdiction, is prohibited.

DEVICES

Service Ribbons. *Description.* The service ribbon is a strip of ribbon identical with that from which the service medal is suspended. (Not all ribbons reflect the award of a medal, e.g. service longevity ribbons.) Service ribbons will not be impregnated with unnatural preservatives nor worn with artificial protective coverings.

Wearing. Only service ribbons representing possession of corresponding authorized military decorations and service awards may be worn. Service ribbons will be worn on the service coat or shirt, and in the same order and position as prescribed for decorations and service awards. They may be sewed in place or attached by means of a bar. When more than one line is necessary, the bottom line is placed immediately above the pocket on the left breast, additional lines being parallel to and above the first line.

Clasps. Clasps are authorized for the Good Conduct Medal, World War I Victory Medal, American Defense Service Medal, Army of Occupation Medal, and Antarctica Service Medal.

They are placed on the suspension ribbon of the appropriate service medal. Where more than one clasp is to be worn on the suspension ribbon of a service medal, the clasps will be worn in the order of date of service performed beginning at the top, with a space of 1/8 inch between clasps. The clasp for the Good Conduct Medal may be worn on the service ribbon.

Service Stars. *Description.* The service star is a bronze or silver five-pointed star 3/16 inch in diameter. A silver service star is worn in lieu of five bronze service stars.

Wearing On Suspension Ribbons. Service stars are authorized to be worn with one point of each star up in a vertical position and all stars arranged in a horizontal row on the suspension ribbons of the American Campaign Medal, the Asiatic-Pacific Campaign Medal, the European-African-Middle Eastern Campaign Medal, the Korean Service Medal, and the Vietnam Service Medal, to represent combat service.

Wearing On Service Ribbons. Service stars are authorized to be worn on the service ribbons of the following medals only:

World War I Victory Medal, to represent possession of battle clasps.

American Defense Service Medal, to represent possession of the foreign service clasps.

American Campaign Medal, Asiatic-Pacific Campaign Medal, European-African-Middle Eastern Campaign Medal, Korean Service Medal, and Vietnam Service Medal, to represent possession of service stars on the suspension ribbons.

National Defense Service Medal to represent a second award.

Silver service stars are worn to the wearer's right of bronze service stars, but to the left of the arrowhead.

Arrowheads. *Description.* The arrowhead is a bronze replica of an Indian arrowhead 1/4 inch in height and 1/8 inch in width.

Wearing. The arrowhead is authorized for wear on the suspension ribbons and service ribbons of the Asiatic-Pacific Campaign Medal, European-African-Middle Eastern Campaign Medal and the Korean Service Medal to represent a combat parachute jump, combat glider landing, or amphibious assault landing. The arrowhead will be worn point up in a vertical position and to the wearer's right of all service stars. Only one bronze arrowhead will be worn on any suspension ribbon or service ribbon.

Berlin Airlift Device. This is a gold colored metal miniature of a C–54 plane to be worn on the service ribbon or the suspension ribbon of the occupation medal. This device is awarded for 90 days' consecutive service assigned or attached to a unit in the Army of Occupation of Germany which has been designated in general orders of the Department of the Air Force as participating in the Berlin Airlift between 26 June 1948 and 30 September 1949, inclusive.

Miniature Service Medals and Appurtenances. *Description.* Miniature service medals and appurtenances are replicas of the corresponding service medals and decorations on the scale of 1/2. The Medal of Honor is not worn in miniature, but its rosette is authorized for civilian wear.

Wearing. Miniature service medals with miniature appurtenances are worn attached to a bar on the left breast of the mess dress, informal black, and the formal evening dress uniforms.

Miniature Service Ribbons. *Description.* Miniature service ribbons are replicas of corresponding service ribbons on a scale of about 1/2.

Wearing. Miniature service ribbons with miniature appurtenances may be worn attached to a bar on service and semi-formal uniforms.

LAPEL BUTTONS AND LAPEL RIBBONS FOR WEAR WITH CIVILIAN CLOTHING.

Lapel Buttons. *For all decorations and service medals except Victory Medals.*

The lapel button is 21/32 inch in length and 1/8 inch in width, in colored enamel, being a reproduction of the service ribbon. Miniature appurtenances may be placed on lapel buttons.

For World War I Victory Medal. The lapel button is a five-pointed star 5/8 inch in diameter on a wreath with the letters "U.S." in the center. For persons wounded in action, the lapel button is of silver, for all others, of bronze.

For World War II Victory Medal. No lapel button is authorized. The honorable service lapel button is worn in lieu of a lapel button for the World War II Victory Medal.

Description of Honorable Service Lapel Button. The gold-color metal lapel button consists of a dexter eagle with wings displayed perched within a ring composed of a chief and thirteen vertical stripes; the dexter wing of the eagle is behind the ring, the sinister wing is in front of the ring.

Requirements for Honorable Service Lapel Button. Service between 8 September 1939 and 31 December 1946, both dates inclusive.

Air Force Lapel Button. A small metal replica of the Air Force star and wings. It is authorized for wear by all active duty and reserve (including AFROTC cadets) personnel.

Air Force Retired Lapel Button. Presented to each retiree.

Wearing Lapel Buttons. Lapel buttons may be worn on civilian clothes only.

Supply of Appurtenances. Only the following appurtenances will be supplied by the Department of the Air Force:

- Valor V's.
- Service stars.
- Arrowheads.
- Clasps.
- Service ribbons.
- Lapel buttons for United States military decorations (except Medal of Honor).
- Lapel button for World War I Victory Medal.
- Honorable service lapel button (in lieu of a lapel button for World War II Victory Medal).
- Air Force lapel button.
- Air Force Retired lapel button.

An initial issue of the above appurtenances will be made with the corresponding service medals. Replacements for military personnel on active duty will be supplied to commanders on requisition in the usual manner. Replacements for others will be made at cost price upon request to USAFMPC/DPMSAA, Randolph AFB, Texas 78148.

The following appurtenances for service medals will not be sold by the Department of the Air Force:

- Miniature service medals and appurtenances.
- Miniature service ribbons.
- Lapel buttons, except the lapel button for the World War I Victory Medal, the Honorable Service Lapel Button, and the Air Force Lapel Button.

Bronze Letter "V" Device indicates an award for valor in combat. The *Bronze Arrowhead* (Army-Air Force) is awarded for participation in an initial assault landing. The *Bronze Oak-Leaf Cluster* (Army-Air Force) is issued for each succeeding award of the same decoration. The *Silver Oak-Leaf Cluster* (Army-Air Force) equals five bronze oak-leaf clusters.

The *Berlin Airlift Device* was awarded for 90 consecutive days' service in direct support of the Berlin Airlift, 26 June, 1948—30 September, 1949. The *Hour Glass Device* is awarded for each succeeding award of the Armed Forces Reserve Medal. The *Foreign Service Clasp* (Army-Air Force) is worn on suspension ribbon of the American Defense Service Medal.

APPURTENANCES.

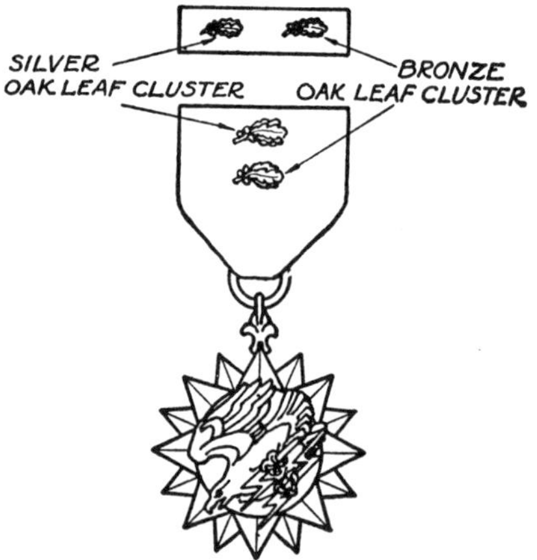

OAK LEAF CLUSTERS ON DECORATIONS AND ON SERVICE RIBBONS OF DECORATIONS
(Silver oak leaf clusters are placed to the wearer's right of bronze oak leaf clusters.)

Manufacture, Sale and Illegal Possession. AFR 900-3 prescribes:

Restrictions on manufacture and sale of service medals and appurtenances by civilians.

Penalties for illegal possession and wearing of service medals and appurtenances.

When Decorations, Service Medals, and Badges Are Worn. (See also United States Military Service Medals, above.)

Wearing Individual Awards. Decorations, service medals, and badges will be worn on the service coat when not equipped for combat or simulated combat. Only aviation, parachutists, glider, and Combat and Expert Infantryman badges may be worn when equipped for combat or simulated combat. Marksmanship badges may be worn on the USAF service uniform (Air Force Distinguished and Excellence-in-Competition badges).

Decorations, service medals, and badges will not be worn on the overcoat.

Wearing Unit Awards. An individual assigned, or permanently attached, to and present for duty with a unit in the action for which a unit award was awarded may wear the award as a permanent part of the uniform.

Wearing Badges. Badges of military societies may be worn on the uniform only when attending meetings, ceremonies, and functions of such societies, and they may then be worn with decorations, service medals, or substitutes therefor.

On Civilian Clothing. The wearing of decorations, service medals, or miniatures on civilian clothes should be limited to ceremonial occasions, and then only when strictly appropriate to the occasion.

When permitted by the employing agency, employees of Federal, State and city governments who served honorably on active service, and whose duties require that they wear special uniforms to denote their authority, may wear U.S. service ribbons.

How Decorations and Service Medals Are Worn. Decorations and service medals are worn only on the service coat as authorized or prescribed in the first part of this chapter.

The Medal of Honor is worn pendant from the ribbon placed around the neck outside the shirt collar and inside the coat collar, the medal proper hanging over the necktie near the collar. Other military and naval decorations and service medals are worn in order of precedence from right to left of the wearer, immediately above the pocket on the left breast in one or more lines which are overlapped. The top line consists of those decorations and service medals highest in the order of precedence. At least one United States decoration or service medal must be worn at the same time that a foreign service medal is worn. Order of precedence is as follows:

Decorations, in order shown in table earlier in this chapter; US unit citations; achievement awards; service medals in order earned; Philippine ribbons, foreign decorations; United Nations service awards; foreign service awards.

Emblems denoting unit decorations are worn as prescribed in AFM 35-10.

A bronze oak-leaf cluster is authorized for wear for each additional Distinguished Unit Citation received by a unit. A silver oak-leaf cluster is authorized for wear in lieu of five bronze oak-leaf clusters.

Wearing Badges. Only one United States aviation badge may be worn at a time. It will be worn centered ½ inch above the left breast pocket of the coat, jacket, or shirt when worn as an outer garment. When ribbons or medals are worn, the aviation badge is worn centered ½ inch above them. The Parachutist, Combat Infantryman, and Medical badges, also are worn centered above the ribbons and medals. When any of these badges are worn, the aviation badge will be centered above it. Foreign aviation badges may be worn only while in foreign countries or while engaged in special protocol activities. Foreign badges are worn centered ½ inch above the right breast pocket. The wearing of miniature aviation badges is optional.

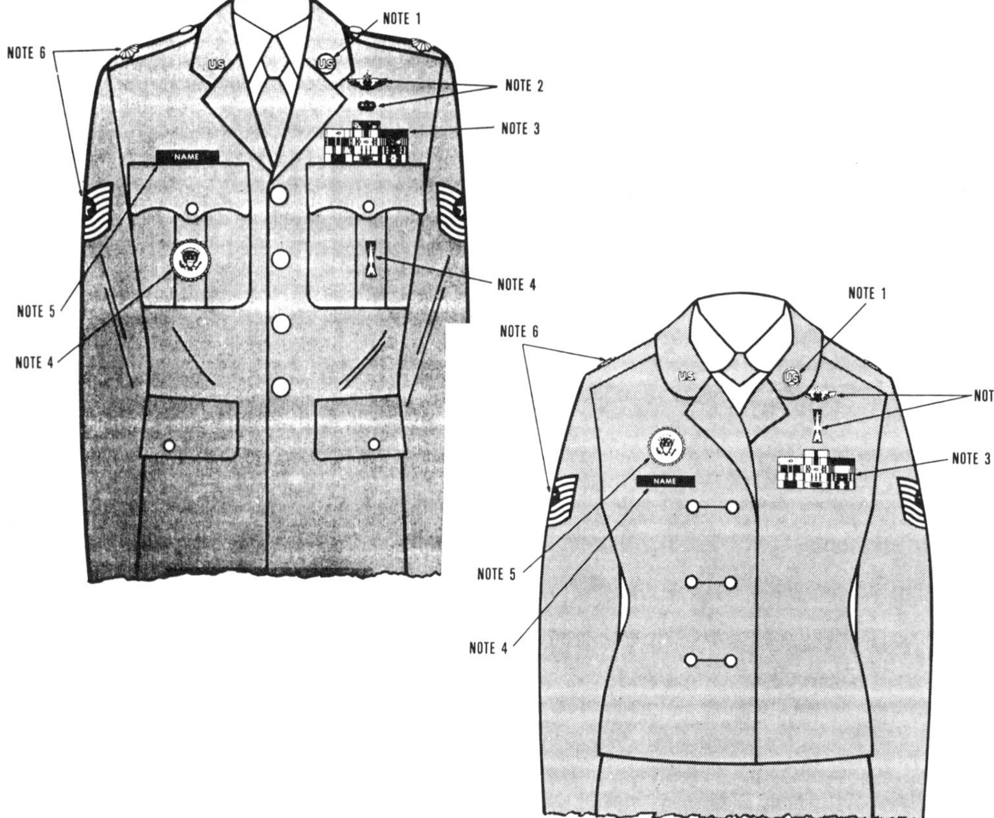

WEAR OF ACCOUTERMENTS ON MEN'S COMBINATION 1.

NOTES:

1. Place approximately halfway up the seam, resting on but not over it. Bottom of U.S. is horizontal with ground.
2. Center badge ½ inch above the top row of ribbons. When no ribbons are authorized, center ½ inch above top of pocket. Center additional badge or insignia ½ inch above the other one worn.
3. Ribbons must be resting on, but not over, top edge of the pocket and centered between the left and right edges. If worn four-in-a-row, left edges of the ribbons are aligned with the left edge of the pocket.
4. Center badge on the lower portion of the pocket between left and right edges and bottoms of flap and pocket.
5. Name tag rests on, but not over, top edge of the pocket, centered between the left and right edges.
6. Center regular size grade insignia ⅝ inch from end of epaulet. Enlisted personnel center 4-inch sleeve chevron halfway between shoulder seam and elbow when elbow is bent.

WEAR OF ACCOUTERMENTS ON WOMEN'S COMBINATION 1.

NOTES:

1. Centered with bottom of U.S. horizontal with the ground.
2. Center badge ½ inch above the ribbons. When no ribbons are authorized, center ½ inch above imaginary row of ribbons. Center additional badge or insignia ½ inch above other one worn.
3. Centered on left side between lapel and arm seam, 2 to 2½ inches higher than top buttons, horizontal with ground. Bottom of the ribbons will be even with the bottom of the name tag.
4. Centered on right side between lapel and arm seam, 2 to 2½ inches above the buttons, horizontal with ground.
5. Center badge ½ inch above name tag.
6. Officers: Center regular size grade insignia ⅝ inch from end of epaulet (imaginary seam line). Airmen: Center 3-inch sleeve chevron halfway between imaginary shoulder seam and elbow when elbow is bent at 90 degree angle.

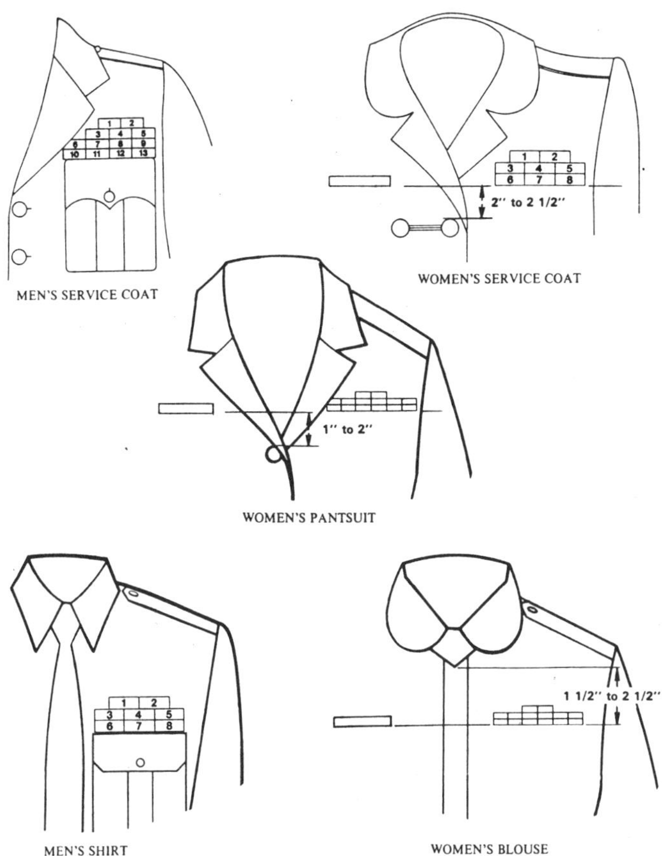

WEAR OF RIBBONS ON SERVICE UNIFORM COMBINATIONS (men's and women's).

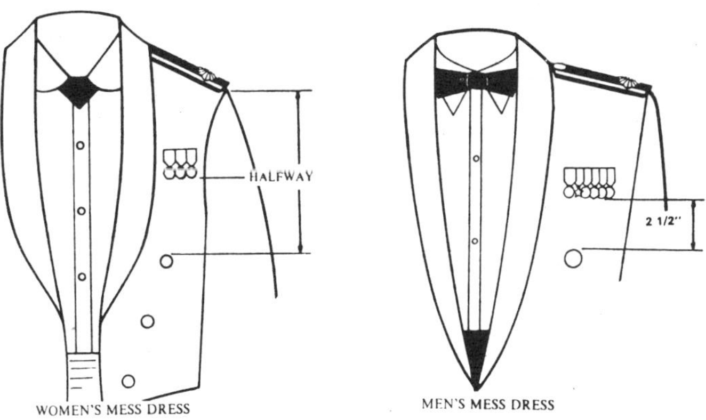

WEAR OF MINIATURE MEDALS ON THE MESS DRESS UNIFORM (men's and women's).

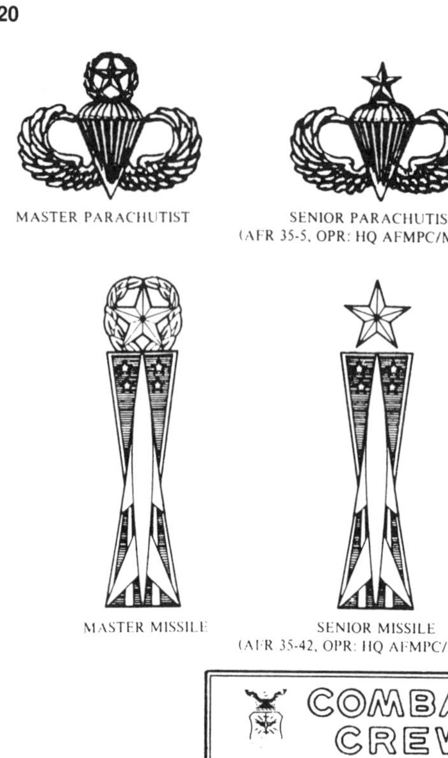

MASTER PARACHUTIST SENIOR PARACHUTIST PARACHUTIST
(AFR 35-5, OPR: HQ AFMPC/MPCAJD)

MASTER MISSILE SENIOR MISSILE MISSILE
(AFR 35-42, OPR: HQ AFMPC/MPCAJD)

COMBAT CREW MEMBER
(AFR 900-48, OPR: HQ AFMPC/MPCASA)

RECRUITING SERVICE (ATCR 35-1, OPR: ATC/RS) ATC INSTRUCTOR INSIGNIA (ATCRs 51-25 & 52-8, OPR: ATC/TT) SECURITY POLICE BADGE (AFR 125-3, OPR: HQ AFOSP/SPP)

PARACHUTIST, MISSILE, COMBAT CREW, RECRUITING SERVICE, ATC INSTRUCTOR, AND SECURITY POLICE BADGES.

Identification Tags. All military personnel of the Air Force will wear identification tags when engaged in field training or when outside the continental limits of the United States.

Name Plates. It is mandatory to wear name plates just above the right pocket of coat or shirt. The plates are blue plastic, 3³⁄₁₆″ long by ⅝″ wide, and the last name only is shown in block white letters.

PRESIDENTIAL SERVICE BADGE.

VICE-PRESIDENTIAL SERVICE BADGE.

OFFICE OF THE SECRETARY OF DEFENSE BADGE.

JOINT CHIEFS OF STAFF BADGE.

PERMANENT PROFESSOR USAF ACADEMY BADGE.

Presidential Service Badge. This badge is issued to members of the Air Force, other than Presidential Aides, who are detailed to duty in the White House. It is worn during the period of detail and is centered on the right breast pocket.

The Office of the Secretary of Defense Identification Badge, Lapel Button, and Pin. The Badge, Lapel Button, and Pin are authorized for wear by military and civilian personnel while on duty with and after departure from The Office of the Secretary of Defense.

Personnel are authorized to wear the badge after completion of one year's service. The badge is worn in the center of the left upper pocket of the outer garment, except overcoats and raincoats. The badge is worn by military personnel only. Both military and civilian personnel may wear the lapel button and the pin (miniatures of the badge) on civilian clothes only.

The Joint Chiefs of Staff Identification Badge. This badge is worn by military personnel after they have served one year in the Organization of the Joint Chiefs of Staff (JCS), and may continue to be worn after reassignment. It is worn centered on the upper left pocket of the coat, jacket, or outer garment shirt.

FOREIGN DECORATIONS, SERVICE MEDALS, ETC., RECEIVED DURING WORLD WAR I AND WORLD WAR II

Decorations, medals, badges, and wound chevrons or stripes awarded by any country associated with the United States in World War I or World War II for service as a member of the military forces of such country in World War I or World War II, may be worn at such times as the corresponding United States decorations, medals, badges, and wound chevrons or stripes are worn.

As far as may be consistent with the foregoing, foreign decorations and medals will be worn as nearly as practicable in accordance with the regulations of the country concerned.

The authority and procedure for acceptance of foreign decorations and service medals are prescribed in AFM 900-48.

Members of the Air Force on duty within the continental limits of the United States are not authorized to wear on the uniform badges of foreign nations for service as a member of the armed forces of the United States, except when all of the following conditions have been met:

(1) The individual was authorized by competent orders to take and has successfully completed the prescribed course of training for such badge as provided by the nation concerned.

(2) The badge was awarded by the national government of a co-belligerent nation or an American Republic during World War II or within one year thereafter.

(3) Acceptance of the badge has been approved by the Department of the Air Force or by a commander designated by the Department of the Air Force.

Fourragere. Normally when a unit is cited, only the organizational color, standard, and/or guidon is decorated. Unless specifically authorized by orders of the foreign government, no emblem is authorized for wear on the uniform, either permanently or temporarily, by individuals of the unit decorated. The only emblems so far authorized for wear on the uniform to indicate a foreign decoration received by a unit are the French and Belgian Fourrageres and the Netherlands Orange Lanyard.

BRASSARDS

Special duty brassards are worn centered on the left sleeve of the outer garment—overcoat, raincoat, topcoat, coat, jacket, or shirt when worn as an outer garment—halfway between elbow and sleeve shoulder seam.

WEAR YOUR RIBBONS PROUDLY—AND PROPERLY
(Worn in order listed)

	Medal of Honor	
Air Force Cross	Distinguished Service Cross	Distinguished Service Medal (Air Force)
Distinguished Service Medal (USA)	Silver Star	Legion of Merit
Distinguished Flying Cross	Airman's Medal	Soldier's Medal
Bronze Star	Meritorious Service Medal	Air Medal
Joint Service Commendation Medal	Air Force Commendation Medal	Army Commendation Medal
Purple Heart	Presidential Unit Citation	Air Force Outstanding Unit Award
Air Force Good Conduct Medal	Good Conduct Medal	American Defense Service Medal
American Campaign Medal	Asiatic-Pacific Campaign Medal	European-African Middle Eastern CPN Medal
World War II Victory Medal	Army of Occupation Medal	Medal for Humane Action
National Defense Service Medal	Korean Service Medal	Antarctica Service Medal
Armed Forces Expeditionary Medal	Vietnam Service Medal	AF Longevity Service Award
Armed Forces Reserve Medal	Air Reserve Meritorious Service Award	USAF NCO Academy Graduate Ribbon
Small Arms Expert Marksmanship Ribbon	Philippine Defense Service Ribbon	Philippine Liberation Ribbon
Philippine Independence Ribbon	Philippine Presidential Unit Citation	Republic of Korea Presidential Unit Citation
United Nations Service Medal	United Nations Medal	Republic of Vietnam Gallantry Cross
		Republic of Vietnam Campaign Medal

THE MILITARY COURT—RIBBONS ARE WORN AS PART OF THE SERVICE UNIFORM.

PROVIDING UP-TO-THE MINUTE WEATHER INFORMATION. THE SERVICE UNIFORM, SHOWING WEARING OF RIBBONS.

8

Your Rights, Privileges and Restrictions

I believe that every right implies a responsibility; every opportunity, an obligation; every possession, a duty. —John D. Rockefeller, Jr.

When a citizen enters military service he or she undergoes a change in legal status. Some civilian rights are restricted or modified. He or she takes on additional hazards and obligations. They are balanced by additional benefits which are not enjoyed by the civilian. In a lesser degree officers in inactive status and retired officers enjoy benefits and are also subject to restrictions. Some former members of the armed forces whose separation was honorable enjoy very important benefits which are administered by the Veterans Administration.

JUSTIFICATION FOR BENEFITS

There are strong reasons justifying the granting of military rights and privileges, which herein are called benefits. Those citizens who are members of the Army, Navy, Marines, or Air Force have the primary mission of protecting and preserving the Constitution, including our free institutions and way of life; the prosecution of wars with the incident hazard; and the service of the Federal Government wherever duty is directed. They give up many freedoms of choice which the civilian takes for granted.

The hard core of the armed forces consists of the officers and members of the Regular components who are volunteers. They are backed up by the several categories of Reserves, also volunteers. Volunteers will not be obtained in the number required or the quality necessary unless the conditions of their life and lot are acceptable. Beset as we are by international strains and recurrent wars, our country needs as its first essential the armed forces necessary to protect itself; and this armed force must be strong enough, brave enough, and proud enough to do its job. Let no thoughtful person attack the principle of rights and privileges for the military.

A phase of this subject invariably overlooked by the critic of things military is that the bulk of officers of the Regular and Reserve components are beyond the age of being subject to the draft. Even in wartime the officer is a volunteer. As long as the nation has need of the best military leadership of all grades and ages, it will be wise to recognize this condition by granting appropriate benefits, first to attract good people to service, and then to hold them.

There is an inescapable difference between the individual in civilian employment and the member of the service. The civilian may quit or refuse a task with no greater penalty than loss of employment, being thereafter free to choose another job. But the wearer of the uniform can do so only at the peril of punishment by action of courts-martial which, if refusal to obey or cowardice before any enemy is involved, may result in a death penalty.

Justification for Restrictions. There is a sufficient case also for imposing restrictions upon military people, especially commissioned officers, which are not borne by civilians.

The Government must have a clearly defined power to deploy its forces and require individuals to perform specific missions, however unpleasant or hazardous such locations or duties may become.

The Government must insist upon full service of its officers and thus is justified by defining and prohibiting improper outside activities of individuals.

Since procurement officers and others in the business end of Government have many prerogatives incident to the letting of contracts, the Government must require high standards of ethics as well as clearly codified methods of conducting these affairs.

In order to assure fair treatment for all, and prevention of abuses in the exercise of Federal power, limitations must be placed on authority especially in the field of punishments, sentences of courts-martial, and the like.

RIGHTS

Let us be sure of our meaning in this matter of rights. To do so, if followed by consideration of some examples, will identify many benefits in their true perspective. It will be seen that while there are many obligations of service there are not many actual rights.

Definition. A right in the sense of this discussion is a benefit established for military people by Federal law. Unless a benefit is established by law, in contrast to a departmental regulation which is subject to administrative change or withdrawal, it is something less than a right.

Acquisition of Military Rights. A citizen who has subscribed to his oath of of-

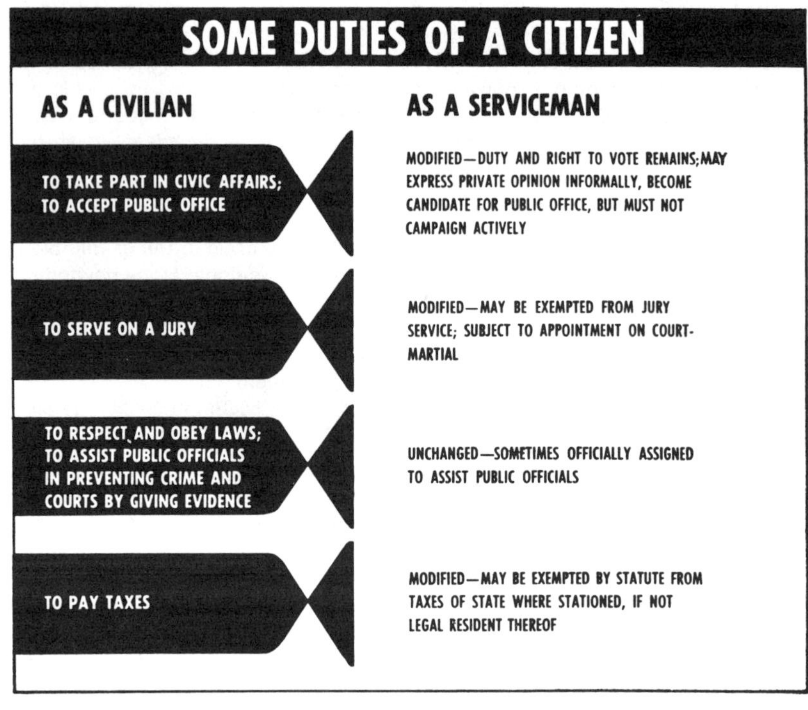

COMPARISON OF DUTIES

fice as an officer, or oath of enlistment if an airman, becomes entitled at once to certain rights of military service. For example, the right to wear the uniform. Other rights accrue only by completing specified requirements; for example, the right to retire after completing a stipulated period of service.

The Right to Wear the Uniform. Members of the military service have the right to wear the uniform of their service. That the department may require the wearing of the uniform off duty as well as on duty is beside the point. First of all it is a right.

Members of the Reserve components on inactive status, retired personnel, and former members of the service who have been honorably separated have the right to wear the uniform only at stipulated times or circumstances and unless these conditions exist the right is denied. Chapter 6 describes these conditions fully.

The Right of Officers to Command. In the commission granted an officer by the President will be found these words: ". . . And I do strictly charge and require all Officers and Soldiers under his command to be obedient to his orders. . . ." The commission itself may be regarded as the basic document which gives military officers the right to exercise command and to exact obedience to proper orders.

Air Force Regulation 35-54 establishes this right in further detail, along with definite restrictions on this right.

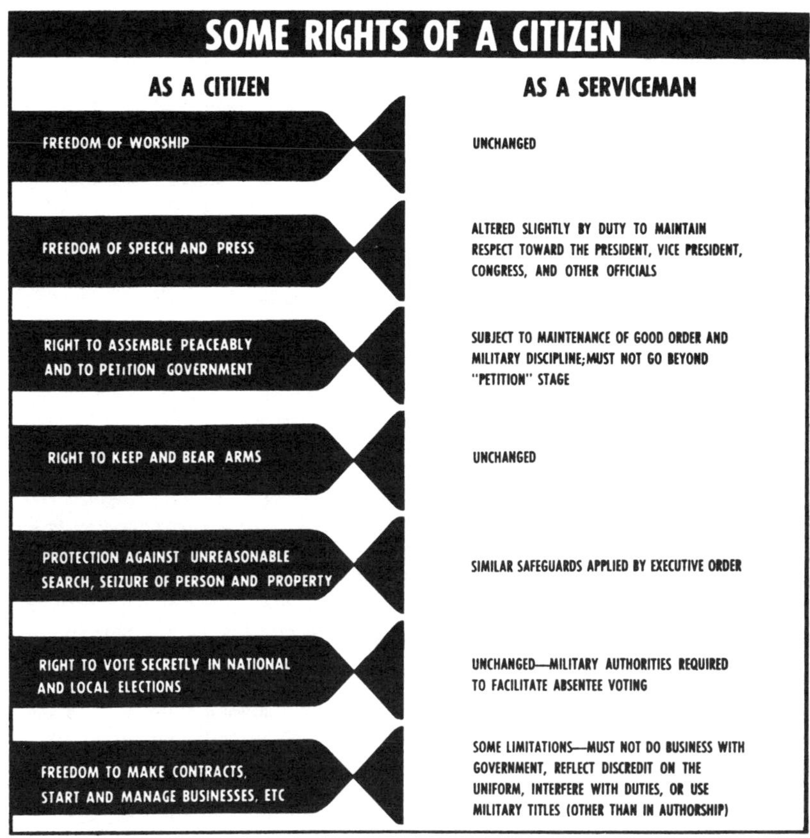

COMPARISON OF RIGHTS.

The Right to Draw Pay and Allowances. Pay scales for grade and length of service are established by law. See chapter 5, *Pay and Personal Allowances.*

The rights as to pay and allowances may be suspended, in part, by action of a court-martial, or forfeited in part by absence without leave.

Tho Right to Receive Medical Attention. Members of the military service and in general their dependents are entitled to receive appropriate medical or dental care for the treatment of their wounds, injuries, or disease. In fact, refusal to accept treatment ruled to be necessary may be punishable by court-martial. For rights of dependents to receive medical care see chapter 21.

The Right to Individual Protection Under the Uniform Code of Military Justice. All members of the military service are under the jurisdiction established by the Articles of the Uniform Code of Military Justice. Many persons regard the Manual for Courts-Martial, which contains this code, merely as the authorization of courts-martial and the implementation of their procedures as a means of maintaining discipline or awarding punishment for crime. This is a shallow view.

Except for the punitive articles, the Uniform Code of Military Justice pertains in considerable measure to the protection of individual rights. Here are examples:

(1) No person may be compelled to incriminate himself before a military court.

(2) No person shall without his consent be tried a second time for the same offense.

(3) Cruel and unusual punishments of every kind are prohibited.

(4) While the punishment for a crime or offense is left to the discretion of the court, it shall not exceed such limits as the President may from time to time prescribe.

The Soldiers' and Sailors' Relief Act. The Soldiers' and Sailors' Relief Act, passed in 1940 and still in effect, has for its purpose the relief of draftees, enlistees, and reservists on active duty of some of the pressure of heavy financial obligations they may have assumed in civil life. See chapter 21, *Personal Affairs and Aid for Your Dependents.*

Redress of Wrong. Each of the armed services provides a procedure by which any member of the military service may seek redress of wrong. The officer should acquaint himself fully about this matter. He or she may have occasion to register an official objection, or complaint, with respect to his or her own treatment, although such occasions should be rare since most officers complete their entire service without finding it necessary to use this privilege. But the officer should certainly know that his juniors also enjoy this right and if he or she takes action which is grossly injurious to an individual, or action that is so considered, he or she may be obliged to endure the process as the injuring party rather than the injured.

There may come a time in the service of any person when he feels that he or she has been wronged by a superior. A few examples among many may be listed as follows: An unfavorable ruling regarding pay or allowances; unsatisfactory living conditions; undeserved stigma; treatment by a superior which is directly contrary to military laws or regulations.

Any person contemplating registering an official complaint should first get the facts straight and be quite certain of sustaining them. It is contrary to regulations to punish anyone for filing a complaint. But they can be punished for knowingly making statements that are unfounded, untruthful, or unjustly harmful to another person's good name. No person need fear the after-effects of filing an honest complaint or stating a genuine grievance. But the basis of the grievances must be factual and subject to proof.

After getting the facts straight, the individual should normally state the entire situation to his or her commander and seek advice. In most instances this will end the matter since, if the grievances are real, the commander may take action which will correct the situation.

In the event the conference with the commander is insufficient, or for some sufficient reason is not held, a person may bring the problem to the attention of the inspector general who serves the organization. This may be done by going to this officer, or writing to him or her. It is likely that he will require a written statement in any event. This officer will advise the complainant of the final action taken on the complaint or grievance. Nor need this be all. The officer or airman may write directly to the inspector general of the major command, or he may write directly to The Inspector General of the Air Force.

Right to Request Correction of Military Records. Military records serve as the basis for recording the fact, nature, or character, and duration of military service. Accordingly, the military records that pertain to each individual are of inestimable value. When, therefore, any individual military record contains an error of commission or omission means have been made available to petition for correction.

Each Service Secretary, acting through boards of civilian officers or employees of their respective Services, is authorized to correct any military record where in his judgment such action is necessary to correct an error or remove an injustice. In the Air Force this board is known as the Air Force Board for Correction of Military Records.

Application for correction should be submitted by the person requesting corrective action on DD Form 149 "Application for Correction of Military or Naval Record." (Application may be made by other persons as authorized by the Board when the person himself is unable to submit application.) In general a statute of limitations of three years applies to applications. There must be a showing of exhaustion of normal administrative remedies. Hearing on an application may be denied if a sufficient basis for review has not been established or that effective relief cannot be granted. When an application has been found to be within the jurisdiction of the Board and sufficient evidence has been presented indicating probable error or injustice, the applicant will be entitled to a hearing before the Board either in person or by counsel of his own selection or in person with counsel. Detailed instructions concerning the entitlement to hearing, notice, counsel, witnesses, and access to records are set forth in AFR 31-3 and AFR 31-11.

The Right to Vote. Legislation enacted by the Congress establishes the right of voting by members of the armed forces, and commanders are required to establish facilities for absentee voting for members of their commands. See AFR 211-19. Exercise of the right to vote, is of course, subject to State laws with which the serviceman must comply.

The Right to Retire. After satisfying specific requirements of honorable service, or having endured physical disability beyond a fixed degree, commissioned officers of the armed forces have the right to retire. See chapter 12, *Retirement.*

The "GI Bill of Rights." Former members of the armed forces and retired personnel may obtain benefits under the "GI Bill of Rights." Veterans Administration benefits for eligible veterans, their dependents, and beneficiaries include:

Servicemen's Group Life Insurance.

Educational Aid.

Guarantee of loans for the purchase or construction of homes, farms, or business property.

Readjustment allowances for veterans who are unemployed.

Disability compensation.

Vocational rehabilitation.

Physical examinations, hospital care, and outpatient medical and dental treatment.

Domiciliary care and guardianship service.

Pensions.

Death benefits to survivors.

These are discussed in chapter 21, *Personal Affairs and Aid for Your Dependents.*

Social Security Benefits of Retired Personnel. Retired officers including those who are employed or "self-employed" may now qualify for Social Security benefits in the same manner as other citizens. See chapter 12, *Retirement.*

The Right to be Buried in a National Cemetery. The rights of a deceased serviceman to be buried in a National Military Cemetery are discussed in chapter 21, *Personal Affairs and Aid for Your Dependents.*

PRIVILEGES

The Privileges of Rank and Position. That "rank has its privileges" is a saying as old as armed forces. It is the deference in all walks of life to one's elders or seniors. It is no more nor less pronounced, although it may be more codified, than among faculty members, or in a business establishment, a legislative body, or among doctors, lawyers, ministers. Throughout chapter 3, *Military Courtesy,* and chapter 4, *Customs of the Service,* will be found numerous examples.

Leave. Under current laws and regulations military people become entitled to accumulate leave and to take it when their duties permit. See chapter 9, *Leave.* But the person in uniform must apply for permission to take leave from station and duties, regardless of its accumulation to his or her credit. His or her application may be denied. The training or tactical situation will govern the decision. If he or she absents himself without permission he is subject to forfeiture of pay and to disciplinary action. Hence it is a privilege, rather than a right.

Election to Public Office. Members of the Air Force other than the Regular Air Force, while on active duty, may become candidates for election to public office, without the tender of resignation, and may file such evidence of their candidacy as required by local laws. The candidacy must not interfere with duty. If elected the officer must not, while in active service status, act in his official capacity as the holder of the office, or perform any of its duties. Members of the Regular Air Force on the active list do not enjoy this privilege.

As this particular privilege has been altered several times in recent years any individual who considers becoming a candidate for office should consult local military authority for otherwise he may vacate his commission.

Membership in Officers' Clubs and Messes. All officers assigned at a station have the privilege of club membership, if clubs are provided. They must follow the rules of the club or mess as to payment of dues, bills, and other matters and unless they do so this privilege may be curtailed or denied.

The Privilege of Writing for Publication. One way an active duty officer or a retired officer may with propriety attempt to add to his income is to write for publication. Official regulations permit it. There is ample precedent for it. Articles of military interest are sought by service journals. Stackpole Books, publishers of *The Air Force Officer's Guide,* has produced a large number of books by military authors.

When will the military author do his writing? He or she will regard writing as a hobby and devote such time to it as others devote to gardening, woodworking, photography, golf, bridge, the movies, just sitting around, and so on. If time does not permit, the service member does not write. Certainly it must not interfere with the performance of duty.

Publishing Articles on Military Subjects Is Authorized. Subject to restrictions which are stated below, any member of the military service may publish articles on military subjects which contain nothing prejudicial to military discipline.

If publication is not objectionable such permission will be granted, but no reference to approval by the Department of the Air Force will appear in the publication.

If the author offers the work to the Department of the Air Force, and it is accepted and published by the Department of the Air Force in original form, proper recognition will be given to the individual.

The inclusion of classified military information (secret, confidential, or restricted data) in any article published by a member of the Air Force is prohibited.

Manner of Obtaining Clearance on Proposed Addresses and Military Publications. The Security Review Branch, Office of Public Affairs, Department of Defense, is responsible for reviewing for security and coordination with existing policies of the Department of Defense, and its components, information disseminated to the public on a national scale by individuals or agencies of the Departments of the Army, Navy, and Air Force.

This agency is charged with reviewing manuscripts, speeches, advertising material, radio scripts, still and motion pictures, submitted for clearance by individuals and agencies of the Department of Defense and its components.

Active Duty Personnel. All material containing information about the Air Force prepared by officers or airmen on active federal service, will be submitted to appropriate public affairs authority for review prior to publication. Material not relating directly to the Air Force is governed only by the dictates of propriety and of good taste. Review of such material is not required.

Retired Personnel. Retired military personnel are responsible for the security and propriety of material prepared by them.

Inactive Personnel. Inactive Air Force Reserve and National Guard personnel assume civilian status upon completion of terminal leave and are not required to submit material prepared by them for review.

Attention is called to a long-standing policy prohibiting military and civilian personnel assigned to public affairs divisions from accepting direct remuneration from civilian agencies for any type of public affairs activity without topside approval. Exceptions may be granted in cases where the subject matter is in no way related to the official duties of the personnel.

Inventions and Rights to Profit Therefrom. Another way in which active duty personnel may seek to add to their incomes is through an invention. The armed forces have many people with inventive minds and good ideas. It is likely that many good ideas are unexploited because of lack of information on how to proceed.

An officer who makes an invention has the same right as any citizen to profit therefrom provided it does not refer to and is not evolved in the line of duty.

A free pamphlet on the subject is issued by the Commissioner of Patents, Washington, D.C. The embryo inventor should obtain it before disclosing anything about his project.

Army and Air Force Exchange Privileges. The following persons, identified as prescribed by the regulations (AFR 147-7) when purchases are made, and organizations, are entitled to all Exchange privileges.

All uniformed personnel on active duty in excess of 72 hours and their dependents.

Officers and airmen of foreign nations when on duty with United States armed services under competent orders issued by one of the armed services.

Widows of servicemen of the categories listed below, who have not remarried: Members of the uniformed services who at the time of death were on active military duty for a period exceeding 72 hours; members of the Reserve components of the armed forces who died in line of duty while on active duty; and retired personnel.

Retired personnel entitled to the privilege include:

All personnel carried on the official retired lists of the armed services.

Contract surgeons during the period of their contract.

Uniformed personnel of the Red Cross assigned to duty within an activity of the armed services.

Patrons Entitled to Limited Privileges. Those entitled to make limited purchases at Exchanges are stated below. The items which may be purchased by individuals who enjoy restricted privileges are stated in the departmental Exchange regulations.

(1) Honorably discharged veterans who are receiving medical treatment at a facility where exchanges are operated.

(2) Exchange employees.

(3) Fountain, snack bar, and restaurant privileges may be extended when civilian facilities are not conveniently available to the following: Civilian employees; Red Cross non-uniformed personnel working in offices within an activity of the armed forces; visitors; Reserve components members on duty for periods less than 72 hours.

(4) Members of Reserve components not on active duty are entitled at all times to purchase such necessary articles of uniform clothing, accouterments, and equipment as would be required immediately when called to active duty.

(5) Unlimited Exchange privileges are authorized for Reserve component personnel who participate in regularly scheduled inactive duty training on the basis of one day of Exchange use privileges for each day of inactive duty training performed. This training must be appropriately documented by the individual's unit/activity commander.

Identification. It is noteworthy that all patrons are required to identify themselves as a person entitled to full or restricted privileges at the time purchase is made. Dependents and nonuniformed people are required to obtain an official DD Form 1173, Identification and Privilege Card. Widows and retired personnel will be obliged to establish their right to purchase by the same card. Uniformed people should be prepared to display their official Identification Card or official orders as may be required by the Exchange officer before a sale is made.

Private Practice by Medical and Dental Officers. Private practice by medical and dental officers is strictly regulated by the Air Force. Established application procedures must be followed. Approval is not ordinarily granted unless there is demonstrated community need, including a letter in support of such need from the local medical or dental society. In no case may such practice be in conflict with military requirements, and the establishment of an office for the purpose of engaging in civilian practice is prohibited. Medical officers engaging in private practice must be licensed in that state and must provide their own professional liability insurance.

The same provisions govern private practice for dental officers.

Concessions and Scholarships at Civilian Educational Institutions. Several civilian educational institutions offer concessions to service children and grant scholarships to discharged airmen whose records and educational qualifications warrant the action. Any person in the military service who desires information relating to scholarships should apply directly to the Director of Administrative Services, Hq, USAF.

Privileges Extended by Civilian Organizations. In many localities it is quite common for civilian professional, civic, golf, or social clubs to extend the privilege of membership to officers stationed at a nearby military establishment. This custom (which is by no means universal) is of great importance to officers. It enables some among them to obtain a desired membership without the costly entrance fees charged in some instances.

Abuse of Privilege. The evil which has been practiced by the few and discredited many of the officer corps is *abuse of privilege.* It consists of taking advantage of position or rank to secure pleasures or facilities to which he is not entitled by law, regulation, or custom. It is the "getting away with something." This chapter is an attempt to clarify what is meant by proper benefits so that the proper officer may observe and evaluate alleged rights and privileges before practicing them. The selfish and grasping may not be deterred. But the officer who might offend innocently may be helped on a correct course.

Here is a simple way to determine whether an alleged benefit or privilege is genuine or spurious. Find the answer to these two questions:

(1) Can you establish authorization in any current departmental or major command document?

(2) Observe the five or ten best officers of experience known to you whom you observe frequently. They must have high standing as good officers among their fellows. Is the questioned privilege utilized or practiced by half or more of them?

THE TREATMENT OF PRISONERS OF WAR

In the foregoing discussion, with examples, definition has been made of the terms rights and privileges. The application of this analysis to the status and situation of a prisoner of war is especially important to all who wear their country's uniform.

Prisoner of War Code of Conduct. Quoted below is the Executive Order establishing rights and restrictions applicable to servicemen under the code for American fighting men.

EXECUTIVE ORDER 10631

"Code of Conduct for Members of the Armed Forces of the United States. By virtue of the authority vested in me as President of the United States, and as Commander-in-Chief of the armed forces of the United States, I hereby prescribe the Code of Conduct for Members of the Armed Forces of the United States which is attached to this order and hereby made a part thereof.

"Every member of the armed forces of the United States is expected to measure up to the standards embodied in this Code of Conduct while he is in combat or in captivity. To ensure achievement of these standards, each member of the armed forces liable to capture shall be provided with specific training and instruction designed to better equip him to counter and withstand all enemy efforts against him, and shall be fully instructed as to the behavior and obligations expected of him during combat or captivity.

"The Secretary of Defense (and the Secretary of the Treasury with respect to the Coast Guard except when it is serving as part of the Navy) shall take such action as is deemed necessary to implement this order and to disseminate and make the said Code known to all members of the armed forces of the United States.

"Dwight D. Eisenhower

"The White House
"August 17, 1955.

CODE OF CONDUCT FOR MEMBERS OF THE UNITED STATES ARMED FORCES

I

"I am an American fighting man. I serve in the forces which guard my country and our way of life. I am prepared to give my life in their defense.

II

"I will never surrender of my own free will. If in command I will never surrender my men while they still have the means to resist.

III

"If I am captured I will continue to resist by all means available. I will make every effort to escape and aid others to escape. I will accept neither parole nor special favors from the enemy.

IV

"If I become a prisoner of war, I will keep faith with my fellow prisoners. I will give no information or take part in any action which might be harmful to my comrades. If I am senior, I will take command. If not, I will obey the lawful orders of those appointed over me and will back them up in every way.

V

"When questioned, should I become a prisoner of war, I am required to give my name, rank, service number, and date of birth. I will evade answering further questions to the utmost of my ability. I will make no oral or written statements disloyal to my country and its allies or harmful to their cause.

VI

"I will never forget that I am an American fighting man, responsible for my actions, and dedicated to the principles which made my country free. I will trust in my God and in the United States of America."

The Secretary of Defense has directed training under the code in three main phases:

1. A general citizenship type of orientation.

2. A formal course for combat crews likely to fall into enemy hands. This could include training in avoiding capture, survival while evading or during captivity, resistance to giving information and making false "confessions," and escape attempts.

3. A specialized course for selected students. This could include prisoner interrogation and treatment courses of definite duration designed to toughen the individual student and to let him test his resistance ability and determine his physical and mental weaknesses.

RESTRICTIONS

There are many *Thou Shalt Nots* in the military life. They consist for the most part of restrictions or standards of conduct inapplicable to the civilian. Some are contained in Federal laws. Others are in departmental regulations. A few are included in observed customs. They need not be regarded as onerous. They have come about through experience and necessity. In any event they are well balanced by the military benefits which have been discussed. Since their violation would be regarded as a serious matter, at the worst resulting in trial by a court-martial, officers should know of them.

Effect of Conduct Unbecoming an Officer. The 133d Article of the Uniform Code of Military Justice reads as follows: *Any officer, cadet, or midshipman who is convicted of conduct unbecoming an officer and a gentleman shall be punished* as a *court-martial may direct.* (Manual for Courts-Martial, United States, 1951.)

There are certain moral attributes common to the ideal officer, a lack of which is indicated by acts of dishonesty or unfair dealing, of indecency or indecorum, or of lawlessness, injustice, or cruelty. Not every one is or can be expected to meet ideal standards or to possess the attributes in the exact degree demanded by the standards of his or her own time; but there is a limit of tolerance below which the individual standards of an officer or cadet cannot fall without him or her being morally unfit to be an officer or cadet. This article contemplates such conduct by an officer or cadet which, taking all the circumstances into consideration, satisfactorily shows much moral unfitness.

This article includes acts made punishable by any other article of the Code, provided such acts amount to conduct unbecoming an officer; thus, an officer who embezzles military property violates both this and the article dealing with embezzlement.

Instances of violation of this article are: Knowingly making a false official statement; dishonorable neglect to pay debts; opening and reading another's letters without authority; giving a check on a bank where he knows or reasonably should know there are no funds to meet it, and without intending that there should be; using insulting or defamatory language to another officer in his presence, or about him to other military persons; being grossly drunk and conspicuously disorderly in a public place; public association with notorious prostitutes; cruel treatment of airmen; committing or attempting to commit a crime involving moral turpitude; failing without good cause to support his family.

Removal From the Active List for Cause. AFR 36-2 provides procedures by which the Secretary of the Air Force may remove from the active list any officer who is derelict in respect to meeting standards of moral conduct or of professional performance, or otherwise in the interests of the national security.

Liability Regarding Classified Documents. By the very nature of their duties, officers are required to have possession of and to utilize classified documents. Officers must be mindful of the restrictions placed upon such documents and the punitive action which may be taken against them for improper handling or use.

AFR 205-1, *Information Security Program,* is the principal source of instructions. The inclusion of classified military information in any article, speech, or discussion by a member of the Air Force of the United States is prohibited unless specifically authorized.

"Whoever, being entrusted with or having lawful possession or control of any document, writing, code book, signal book, sketch, photograph, photographic negative, blueprint, plan, map, model, note, or information, relating to the national defense, through gross negligence permits the same to be removed from its proper place of custody or delivered to anyone in violation of his trust, or to be lost, stolen, abstracted, or destroyed, shall be punished by imprisonment for not more than ten years and may, in the discretion of the court, be fined not more than $10,000. (June 5, 1917, c. 30, Title I, Sec. 1; 40 Stat. 217. Act of March 28, 1940; Public Law No. 443, 76th Congress, 3d Session.)"

Officer in Arrest. An officer in arrest cannot exercise command of any kind. (Par. 20a, Manual for Courts-Martial, U.S., 1951.)

Officers Suspended from Rank or Command. Officers suspended from either rank or command, and airmen serving sentence of confinement, are prohibited from wearing decorations, medals, or substitutes therefor.

Effect of Disrespectful Language Concerning Certain Government Officials. The 88th Article of the Uniform Code of Military Justice reads as follows:

Any officer who uses contemptuous words against the President, Vice President, Congress, Secretary of Defense, or a Secretary of a Department, a Governor or a legislature of any State, Territory, or other possession of the United States in which he is on duty or present shall be punished as a court-martial may direct.

Restrictions on Outside Activities. Officers will not engage in or permit their names to be connected with any activity, participation in which is incompatible with the status of an officer.

There are limitations upon the activities of officers and other personnel subject to military law outside of their military duties in which they may properly engage and upon the outside interests they may have without impropriety. Some outside activities and interests are specifically prohibited by statute or regulation or both; but there are many others from which certain military personnel are barred by the high standards of conduct required of persons in the military service. The general principle underlying the limitations mentioned above is that every member of the military establishment, when subject to military law, is bound to refrain from all business and professional activities and interests not directly connected with military duties, participation in which activities or interests would tend to interfere with or hamper in any degree his or her full and proper discharge of such duties or would normally give rise to a suspicion that such participation would have that effect. Any substantial departure from this underlying principle would constitute conduct punishable under the articles of the Uniform Code of Military Justice.

It is impossible to enumerate all the various outside activities to which the above restriction applies. The following examples may be regarded as typical: (1) Acceptance by an officer or, with the approval of the officer, by a member of his or her immediate family of a substantial loan or gift or any emolument from a person or firm with whom it is the officer's duty as an agent of the government to carry on negotiations. (2) Acquisition or possession by an officer of a financial interest in any concern whose business includes the manufacture and sale of articles of a kind of which it is the duty of the officer to make purchases for the Government.

An officer who is engaged or who contemplates engaging in outside professional or business activities should inform himself of pertinent laws, regulations, and standards of the service, in order to determine conscientiously and impartially whether or not such activities or interest might be considered as being in any way incompatible with the proper performance of his official duties or in any sense adverse to the interests of the Government. If after such investigation there is any doubt, the individual concerned should report all pertinent facts to the Department of the Air Force and request instructions. An officer who has certain outside interests which have no bearing upon the performance of his military duties at the time of acquiring such interests, and is later assigned to duties in the performance of which the possession of such interests might normally be suspected of having an influence upon him adverse to the interests of the Government, will immediately dispose of such outside interests and report the facts to superior military authority, or without disposing of such interests, he or she will report all pertinent facts and circumstances to superior authority with a view to his relief from his assignment or such other action as may be deemed appropriate.

Acting as Attorney or Agent. No member of the military establishment on the active list or on active duty, or a civilian employee, whose official duties are concerned with patent activities shall act as agent or attorney in connection with the inventions or patent rights of others, except when such action is a part of the official duties of the person so acting.

Acting as Consultant for Private Enterprise Prohibited. No member of the military establishment on the active list or on active duty, or a civilian employee of the Air Force or of the Department of the Air Force shall act as a consultant for a private enterprise with regard to any matter in which the Government is interested. (AFR 30-30.)

Assistance to Persons Preparing for Civil Service Examinations Prohibited. No officer or employee of the Government will, directly or indirectly, instruct or be concerned in any manner in the instruction of any person or classes of persons, with a view to their special preparation for the examination of the United States Civil Service Commission or of the boards of examiners for the diplomatic and consular services. *The fact that any officer or employee is found so engaged will be considered sufficient cause for his removal from the service.*

Solicitation for Contributions for Gifts Prohibited. No officer, clerk, or employee in the United States Government employ shall at any time solicit contributions from other officers, clerks, or employees in the Government service for a gift or present to those in a superior official position. *Every person who violates this section shall be summarily discharged from the Government employ.* (R.S. 1784.)

Acceptance of Gifts from Subordinates Prohibited. No official or clerical superior shall receive any gift or present offered or presented as a contribution from persons in Government employ receiving a lesser salary than himself. *Every person who violates this section shall be summarily discharged from Government employ.* (R.S. 1784.)

Receiving Gifts from Civilian Sources in Recognition of Services Rendered Not Approved. The practice of receiving presents from persons not in the military establishment or in the employ of the Government in recognition of services rendered, or from firms, or their representatives, with whom the officer has negotiated as an agent of the Government is prohibited. Officers will exercise their influence over members of their immediate families to insure that contributions or gifts are not received or accepted from such persons or firms. (AFR 30-30.)

Effect of Refusal of Medical Treatment. An officer or airman may be brought to trial by court-martial for refusing to submit to a surgical or dental operation or to medical or dental treatment, at the hands of the military authorities, if it is designed to restore or increase his fitness for service and is without risk of life.

Communication With Members of Congress. Communication with Members of Congress by personnel of the Air Force on matters of personal interest is restricted only by the provisions of AFR 11-7, as amended, and the provisions of 10 USC 1034, which states "No person may restrict any member of an armed force in communicating with a member of Congress, unless the communication is unlawful or violates a regulation necessary to the security of the United States." However, all members of the Air Force are advised that all legislative matters affecting the Department of the Air Force program shall be conducted through the Office of the Secretary of the Air Force or as authorized by him.

Action on Attempts to Secure Personal Favor. Except when properly made by the officer himself, requests for personal favor or consideration for any officer will be referred to the officer in question for statement whether he directly or indirectly procured the request to be made and whether he avows or disavows the request as one on his behalf.

Writing Checks with Insufficient Funds and Payment of Debts. When considered necessary, commanders concerned will take action under Article 133 or 134, UCMJ, where an individual under their command issues a check against an account with insufficient funds therein or fails to clear his personal accounts prior to departure from his station. When information of indebtedness is received subsequent to the departure of an individual, action will be taken to effect prompt settlement of such accounts by direct correspondence between the commander of the station at which personal accounts remain unsettled and the individual's new commander.

Restrictions on Using Penalty Envelopes and Letterheads. All Department of the Air Force letterheads, envelopes, and other stationery are for OFFICIAL use only and will not be used as personal stationery.

9

Leave

The privilege of taking leave is a valuable attribute of military service. The taking of leave is encouraged. Duties should be so arranged as to enable officers to avail themselves of leave due them. The wisdom of taking leave should be particularly impressed upon any officer who is exhibiting signs of an impending physical or mental breakdown incident to close application to duty.

Commanders should bear in mind that the persistence of conditions within their commands in time of peace which preclude the granting of leave to their officers is a most direct indication of either poor organization or poor administration. No military organization should be built around indispensable people. Moreover, the commander who grumbles that he or she is so short of officers that he "can't afford" to permit his or her officers to take leave is admitting he or she has no workable plan to meet the occurrence of sickness, accident, or death among his officers. Ultimately, such a commander will receive a cold letter from the Inspector General pointing out his or her failure to comply with Air Force policy.

AFR 35-9 is designed to require using leave as it is earned and therefore to discourage hoarding of leave so that large cash payments for unused leave are not due upon discharge or retirement.

AIR FORCE POLICIES

Each commander will insure that members of his or her command are afforded an opportunity and encouraged to take leave. The practice of accumulating leave to the maximum permitted is undesirable and is to be discouraged. Periods of cessation from routine work for the purpose of travel, healthful recreation, and diversion are essential to the efficiency of individuals in the military service. Commanders in all echelons will encourage individuals of their command to avail themselves frequently of accrued leave and, subject only to military necessity, all commanders will approve such request of leave.

Holidays. The following days in each year are public holidays established by law, and will be observed in the Air Force except when military reasons prevent:

New Year's Day—January 1
Washington's Birthday—Third Monday in February
Memorial Day—Last Monday in May
Independence Day—4 July
Labor Day—First Monday in September
Columbus Day—Second Monday in October
Veterans Day—November 11
Thanksgiving Day—Fourth Thursday in November
Christmas Day—25 December

When a holiday falls on a Sunday, the following day is a holiday. When a holiday falls on Saturday, the preceding day is a holiday.

TYPES OF LEAVE

Ordinary Leave. Leave granted upon request of the individual at any time during a fiscal year to the extent of the leave which may be earned during that fiscal year, plus leave credit from prior years.

Sick or Convalescent Leave. Leave granted for absence because of illness or convalescence upon recommendation of the surgeon. It is not chargeable as leave.

Advance Leave. Advance leave may be granted in anticipation of the future accrual of leave. In the case of officers, such leave would apply in case of emergency and pre-embarkation leave.

Emergency Leave. Leave granted upon assurance that an emergency exists and that granting of such leave will contribute to the alleviation of the emergency. Emergency leave may be granted in such an amount that the total leave advanced will not exceed 45 days. It will not be prejudicial to granting of future leave but will be charged against present or future accrued leave.

Excess Leave. Leave granted in excess of that amount accrued by the individual and except for such advance or ordinary leave as is specifically authorized, is without pay and allowances, and will be taken only under exceptional circumstances upon authority of commanders up to 30 days and of the Department of the Air Force for more than 30 days.

Graduation Leave. Graduates of the service academies are granted 30 days leave from date of graduation if they are commissioned in the Air Force. This leave is not charged against accrued leave credits.

Prenatal and Postpartum Leave. An individual who becomes pregnant while on active duty normally is placed in "sick in quarters" status approximately 4 weeks before delivery, as determined by the attending physician. Time spent in the hospital for delivery is duty time. Following completion of inpatient care, the member will be granted convalescent leave until her medical condition permits her to return to duty, normally not to exceed 6 weeks after her release from the hospital.

Delays En Route in Executing Travel. Authorized delays stated in travel orders will be counted as leave and so charged.

COMPUTATION OF LEAVE CREDITS

Leave credits will be accounted for on a fiscal year basis. In any case where only a part of the fiscal year is considered, earned leave will be prorated at the rate of 2½ days for each month of active (honorable) service.

In all computations, leave shall be credited for portions of a month as follows:

1 to 6 days duty inclusive—½ day leave.
7 to 12 days inclusive—1 day leave.
13 to 18 days inclusive—1½ days leave.
19 to 24 days inclusive—2 days leave.
25 to 31 days inclusive—2½ days leave.

Leave may not accumulate in excess of 60 days (except that personnel in a combat zone may accumulate 90 days). Members of components of the Air Force of the United States ordered to active duty for periods of 30 days or more will be granted leave in accordance with the above computations. Those ordered to active duty for periods of less than 30 days will not be granted leave.

COMPENSATION WHILE ON LEAVE

Personnel shall receive the same pay and allowances while on leave which they receive if on a duty status: when absent on sick or convalescent leave; absent with leave not exceeding the aggregate number of days leave standing to his or her credit or authorized to be advanced to his credit; when by direction of the Secretary of the Air Force the individual is absent awaiting orders on disability retirement proceedings in excess of the number of days leave accrued or authorized to be granted; absent awaiting orders assigning them to their initial duty station for newly appointed officers of the Regular Air Force (AFR 35-9).

Personnel granted excess leave shall receive no pay or allowances while so absent from duty. However, advance leave will be granted with pay and allowances, but if the individual is separated before he or she has accrued sufficient leave to cover the advance leave, the unaccrued portion will be considered excess leave without pay and allowances and will be so recorded.

APPLICATION FOR LEAVE

Applications for leave will be made by use of AF Form 988. They will state the amount of leave desired and amount of accumulated leave due. Generally it is also required that the address while on leave be stated in the application. For form and channels officers are advised to consult their personnel officers.

Day of Departure; Day of Return on Duty Days. Both the day of departure and the day of return will be charged as leave unless the member was present for duty all or nearly all of the normal working day on either the day of commencement or termination of the leave period.

Day of Departure; Day of Return on Nonduty Days. When a member signs out on a nonduty day, the day will be charged as leave. When a member signs in on a nonduty day, the day will not be charged as leave.

Leave to Visit Outside the United States. Air Force officers may visit foreign countries on leave, either from the United States or from their oversea station, in accordance with procedures set forth in AFR 35-9. Such leaves are chargeable as ordinary leave.

At this time, however, officers may not visit the following countries on leave except by special authority of Hq, USAF: Albania, Bulgaria, China, Czechoslovakia, Estonia, Hungary, Latvia, Lithuania, Manchuria, Mongolia, Poland, Rumania, Russia, Tibet, Yemen, North Korea, Vietnam, and Cuba.

Warnings. Officers are responsible to report to their *duty* station, not just any Air Force station, at the expiration of leave. Military air transport may be used on a space available basis during leave, but a delay in securing a flight is no excuse for failing to return from leave on time.

Above all, use your leave as a time for mental and physical rejuvenation. Toward this end, avoid killing yourself on the highway, drinking all the liquor available, or spending next year's income. Have fun, but take it easy.

THE A-10 CLOSE-SUPPORT AIRCRAFT

10

Leadership

Leadership is the most important consideration, if any one thing is more important than another.
—General George C. Marshall

Leadership is *essential* to a successful officer. If you glanced at the other chapter titles of this book, you would note topics of *importance* to officers, such as the art of training, the preparation of effectiveness reports, the proper way to wear the uniform. Increased proficiency in these matters will make a good officer a better officer. But such proficiency is not *essential* to an officer. *Leadership is essential.* When we consider other qualities normally the attributes of successful officers, we can recall specific exceptions to all save leadership. U.S. Grant almost never wore his uniform properly, and used liquor to excess, yet he knew how to win a war. General Forrest, he of the "git thar fustest with the mostest" philosophy, was nearly illiterate. Yet he was a genius in battle. Robert E. Lee, the model, combined every desirable attribute of an officer within himself. But of all his galaxy of military attributes, that which shone brightest in Lee was *leadership.* Renowned Air Force officers such as Arnold, Vandenberg, and LeMay were men of many talents, but their outstanding talent was leadership.

The young officer should look forward with pleasure and keen anticipation to the opportunity to lead. It is the most exciting and interesting experience the service can offer you. It is not expected that you will equal an Arnold, Vandenberg, or LeMay, but who can tell? The leaders wanted so badly and in such large numbers have been described as "the good, common, garden variety of leaders; officers who can inculcate those under them with knowledge of their tasks and imbue them with the spirit to do them well." That is not an impossible goal. With application, normal individuals who meet the standards for commissioned officers can be that sort of good leader.

Definition. The word "leader" is defined as follows: "One fitted by force of ideas, character, or genius, or by strength of will or administrative ability to arouse, incite, and direct individuals in conduct and achievement."

"Leadership" has this definition: "The art of imposing one's will upon others in such a manner as to command their respect, their confidence, and their wholehearted cooperation." The meat of the definition is that *you must impose your will on others;* they must do what you direct them to do.

The martinet accomplishes this by simple reliance upon his position of authority and his power of punishment. The *leader* accomplishes it through that quality of leadership we seek. In the case of the martinet, the people commanded do what they are directed to do, but often in a sullen and grudging manner. In the case of the leader, the people carry out directions with "zip" and enthusiasm, contributing their own native American ingenuity and drive to the task at hand.

WHAT MAKES LEADERSHIP?

The young officer is constantly told: "You must learn to be a leader." Too often no one tells him *how.* There are many facets of a person's outward bearing which influence his or her ability to lead. A firm, resonant voice is helpful. An erect, athletic carriage also contributes. A warm, friendly smile goes far. But these are details. At the core of leadership are three fundamentals, and on these three the young officer must work with all his or her might. They are:

1. Character.
2. Knowledge.
3. Power of Decision.

Character. By *character* is meant integrity, courage, morality, humility, and unswerving determination. Character is a spiritual force. It is a reflection of a person's *grip upon himself or herself,* the degree to which he or she is able to dominate the baser instincts that beset us all. Because people know that the conquest of one's own weaknesses is a far more difficult task than any other which could be set before us, they tend to believe that he or she who can conquer himself or herself can also conquer whatever problem is at hand. Therefore, the masses look to a man of character to lead them. For the young officer, the job is: Know your own weaknesses and conquer them.

Knowledge. Knowledge is power. People will seek a leader who *knows what to do.* Why is it so easy for a doctor to take charge at the scene of an accident? Why are his instructions obeyed, and people so quick to act upon his orders? Because the doctor has knowledge. He or she knows what to do and how to do it. For the young officer, the job is: Learn what to do and how to do it with respect to each position you hold. Learn it thoroughly. Remember it. That is *knowledge.*

Power of Decision. The power of decision is actually an outgrowth of character. It is, however, of such importance as to warrant being cited separately. What good can result, though a person possess the honesty and wisdom to see the right course of action, and the knowledge of the means to accomplish the action, if he or she *does* nothing? It would be far better to make an occasional honest mistake, and learn thereby, than to fritter away the benefits of wisdom and know-how by vacillating indecision. If you would be a leader, learn to make up your mind, though it be between the devil and the deep blue sea.

Once decided, drive your decision home to the hilt. For the young officer, the job is: Drill yourself in small, routine decisions to make an *unqualified* decision. Watch yourself about hedging your decisions. Then follow your decisions through and make them stick.

MECHANICS OF LEADERSHIP

Keep Everyone Informed. It is all-important in seeking to lead Air Force personnel that they be told what it is all about. Considerations of security may restrict the leader's freedom of action somewhat, but he must overcome this obstacle. The officers and airmen must know as much as possible about what they are expected to do, when, where, how, and above all, why. Americans resist being driven and resent being treated as mental incompetents. They perform best when they know the purpose of their effort and believe the purpose worth their effort. To merely tell the men of an air wing, "You must keep more aircraft in commission," will probably accomplish little good. On the other hand, if the same men are told, "A maximum sustained effort is going to be made during the period of the next two weeks. The object is to destroy a major force of enemy who have engaged our ground forces. The fate of the Army units engaged depends on the magnitude and success of the Air Force operations coming up. Everyone must work as fast as he can without being careless. Many lives will depend on us. Let's all have at it." On the basis of such a proposition, few if any officers or airmen will fail to give their absolute best.

An example of a lapse in this facet of leadership and its instant correction by the commander arose in connection with a bomber group operating during the Korean War. The high command had directed the dropping of leaflets calling on the enemy troops to surrender. The group in question was assigned the task. Air crews and maintenance crews grumbled at the seemingly fruitless labor of preparing huge bombers, flying long hours over enemy territory, not to a satisfying result of "target destroyed," but to the end that merely a lot of paper floated down on the Korean terrain. The group commander acted. He requested higher headquarters to give him facts on the number of enemy personnel who surrendered because of the leaflets. He told these facts to his men, pointing out the valuable intelligence garnered by questioning the surrendered personnel. In short, he took action to "sell" the leaflet dropping operation to his group. That, too, is leadership.

The Leader Must See and Be Seen. The officer who would lead people must be close to them. The door of his or her office must be literally and figuratively open to any person who may wish to talk. Of more importance, however, is the habit of getting out of the office and observing your people at their work. During World War II, General LeMay was especially noteworthy for being seen by his men at the point of action. In the Vietnam War, Air Force generals were present at every place of crisis. What American will fail to follow a leader who himself leads hazardous missions? The leader cannot be personally present at all times, but he or she can and should be present *when the chips are down.* Do not wait, however, for some critical issue before observing your people at their work. Make it a matter of practice to find out by first-hand observation the one who does his work well, who does it badly, and why.

The Leader Must Display Capacity to Get Things Done. About the finest reputation an officer can build is to have it said of him or her that he "gets things done." Such individuals are always sought, they are so few and so highly appreciated. *You cannot build a reputation on things you are going to do.*

Ability to organize men and means so as to accomplish a mission in the most efficient manner, with genuine economy in manpower and equipment, can be developed by study, planning, application, and continued practice. While there is danger of over-simplifying the steps involved, it can be approached in this manner: Understand exactly and fully the mission to be accomplished. Break it down into steps or phases in order to identify its elements. Determine the number of subordinate leaders, or supervisors, who are required to assume personal direction of the phases and who must be given responsibility and authority. Determine the tools or equipment or materials which will be required and if the desired or preferred equipment is unavailable rack your brain to improvise an acceptable substitute from that which is at hand. Issue instructions to each subordinate leader. Allocate to each, the personnel, equipment, and materials each will require. Arrange for the necessary coordination between the subordinates. Set a time for the start and a goal for its completion, allowing time prior to the start for subordinate leaders to lay their plans, and issue their own instructions. Thereafter permit your organization to go ahead under your instructions with a minimum of interference, using your own time for supervision, coordination, and future planning. Most will agree that such a procedure will give results on small tasks. It is interesting that with only slightly increased complexity it works as well with large ones. Big responsibilities break down into smaller tasks, and they in turn into elements. Success in organization requires the *vision* to see that which is to be done; the *wisdom* to plan and order the steps in execution; and the *courage* to act to gain the end sought. *Ability to organize* is a very important part of leadership.

It is strange, but the novice or the unskilled seem always to choose the hard way rather than the easy way to get things done. The easy, simple path becomes obvious only after the development of real skill. The officer must be adept at finding the easy ways to achieve results. This will include encouraging people to study their tasks to discover better methods. When a person finds a better method, by all means reward him according to its worth. A suggestion and reward system is a smart step in leadership.

The Leader Sets the Example. The military leader must set the standard in all things. Do you want your people to look well? Then you must be meticulous in your dress, throw aside the non-standard articles of uniform, refrain from wearing soiled and rumpled clothing, keep your clothing clean and neat. Do you want them to be courteous? Then you must be courteous, to your juniors as well as your seniors. When your seniors are present you must be careful to do exactly the things which are expected of you. Do you expect your people to be loyal? Then you must be loyal to those junior to you as well as to those senior to you.

Do you want your people to be drunks? Then drink in their presence. To shirk their duties? Then shirk yours. Your organization will mirror your strength and weakness.

The Leader Must Know the Men to be Led. A careful understanding of the characteristics of the people to be led is essential. They are volunteers, with the traits, expectancies, and viewpoints of those who choose military service as a permanent or temporary livelihood. The leader must prepare to cope with them. Most will respond to inspiration, others must be prodded; a few must be driven. Methods of leadership must be adapted to gain the results desired by applying to each person the sort of approach which will produce the highest attainments.

The leader's knowledge of his or her people must include their traits, education, intelligence, physical capacity, skills, and all manner of things about them. The leader must watch for those who display even the faintest signs of leadership capacity, nurture it, and help it grow. Having acquired this knowledge he or she must use it continually. Such an officer will get the most from his last and most unpromising man.

Development of Pride. In defining leadership it was emphasized that the airman must obey. But the definition goes on to other things which are also important. Obedience must be obtained in a manner which will evoke confidence and wholehearted cooperation.

The airman must be helped to develop pride in himself, the work he does, and in his organization. Of all influences over mankind, none are so far reaching as pride. It is pride in his good name, in his fair reputation, and the standing of his unit that holds him to the mark.

The officer who can turn this trait to account has won half the battle. Do and say those things which foster the development of pride. Start at once, when your new people arrive, by telling them the service they are to perform is of the greatest importance to the nation. It is, isn't it?

When they do something well, observe it, and make some brief comment about it. The greatest satisfaction we have is the knowledge of a task well done. Don't be lavish with praise, nor slop over; just let the person know he or she has done a good job at some assigned task. Then, later, when you have occasion to tell the individual his or her work is not up to standard in another task it will come easier to you and much better for him or her.

THE HUMAN SIDE OF LEADERSHIP

There is a human side of leadership we shall try to define. While by virtue of the rank and authority vested in you by your government, you will at all times be in a position to enforce your will upon those under you, it will be well to rely only a little upon this law. Depend first of all upon yourself, your ability, your personality. Keep rank and authority in the background. It is a "reserve." Be human, with all of the interest, the sympathy, the pride, and satisfaction in your men you take in your best friends.

Good Will. You should wish to obtain the good will of your people. It is an important factor in human contacts.

Good will has been defined as "the sum of an infinite number of favorable impressions." Many years ago a large industrial bank purchased the Dodge Motor Car Company. In the price paid was the sum of fifty million dollars as the "good will" of the company; it was the value of the belief in the minds of millions of people throughout the world that Dodge cars were good cars.

Good will is not a substance that can be locked in a vault, or written into a ledger, nor a tangible something you can see or hear. Paradoxically, it is a "something" you can feel. You will know, after you have started, whether you have the good will of the people you command, the officer-associates beside whom you will labor, just as you will know also when you do not have it.

The good will of the people who work under or with you is dependent on how they size you up after a period of observation. Are you competent? Do you know your job and do it, despite difficulties? Are you fair and consistent in your attitude and actions relative to people under your supervision? Are you reasonable and human in respect to the faults and mistakes of others? Are you compassionate and sensitive to the troubles of people with whom you associate in your work? Are you open-minded regarding suggestions of better ways that you or others can accomplish their tasks? Are you decent and honorable in your performance of duty and in your way of life? If you can answer all or most of these questions "yes," you will gain and hold the good will of those who observe you.

Morale. The morale of a squadron is the state of mind of the average officer and airman with respect to the mission of the squadron. If this average state of mind is one of confidence, courage, determination, and enthusiasm, morale is high. If the average state of mind is one of pessimism, dissatisfaction, despondency and anxiety, morale is low. Practically every facet of an airman's life and work affects, and is affected by, his morale. It is often a decisive factor with respect to the mission of a squadron.

The evidence of high morale may be observed in the smooth, seemingly effortless operations of an air depot, in the efficient manner in which an air base is maintained, in the continuous, adequate, timely flow of supplies. High morale is the priceless dividend of good leadership.

Though there are many factors adversely affecting morale—such as poor housing conditions, lack of promotion or other recognition when earned, poor food, inequitable treatment—perhaps the most aggravating of these to the American airman is stupidity. When the airman sees himself called upon to do an arduous task, to sacrifice his own interests, or to refrain from actions he desires to take for no apparent good reason, as the airman sees it, this is stupidity. To have high morale in a squadron it is not necessary to coddle the people, but it is necessary to do the best you can for them and to make them believe it *is* the best you can do, or anyone could do. As their leader, you expect the best efforts of your people; as their leader, they have a right to expect from you your best. If you convince the people of your squadron that their conditions of life, their chance of accomplishing the squadron mission, their means at hand, and their prospects for the future are the best obtainable under the circumstances, your squadron will have a high morale.

Reward and Punishment. A goal of leadership is to obtain the very highest standard of performance of duty by each individual. The attainment of this necessary end is most difficult. Some battle leaders of proven competence have concluded that in combat the really important work is performed by not more than twenty-five percent of the men. The others press but lightly upon the wheel. In order to increase to the maximum the efficiency of an organization to the end that all men contribute their maximum to the success of the whole, the leader must utilize fully each of the tools placed in his hands. Some perform their duty in superb degree merely because they see their duty and are determined to do it. Others may be inspired by fine leadership and perform beyond their natural bent or inclination. Many can be encouraged to maximum effort through rewards, including small rewards such as favorable comments and recognition, or the granting of special individual or group privileges, or unusual provisions for comfort after arduous duty, or a well-earned leave of absence, as well as the award of decorations and promotions. The judicious use of rewards is a powerful tool of leadership.

But others do not respond strongly to inspiration or rewards. They must be coerced. For in the Air Force we must not permit the wastage of manpower entailed in allowing the backward to withhold their talents from the work of the command. For this reason the commissioned officer is provided with power to secure obedience by award of punishment in minor transgressions, or to resort to trial by court-martial in more serious offenses. However, too frequent use of his punishing power is an indication of defective leadership. Such punishment is a last resort.

The civilian leader has at least some degree of choice of hiring and firing the men he leads. The military leader must take the individuals assigned to him and work out his destiny with them. Therefore, he must get the maximum of accomplishment from each person.

The exercise of the art of leadership requires the leader to make wise and judicious use of the several tools with which he is equipped. He or she should inspire those who will respond to the higher appeals of duty, honor, and achievement. He or she must encourage people to do their best and reward them in an appropriate manner. The leader must prod when individuals hold back or fail to give their best. The leader must drive when such demands are necessary. He or she must not shrink from awarding proper punishment, even trials before military courts, when no better way remains to enforce his will. Reward and punishment in their varying degrees are tools in the hands of the military leader with which he must be adept.

OVERCOMING DIFFICULTIES

One of the earliest judgments to be made by airmen of their leader is his or her attitude toward difficulties. This is so because difficulties are so much with us, so constantly with us. Difficulties come in two general categories: those which are nuisances, and those which are bars to the accomplishment of your mission. In the case of the former, make light of them publicly—but try your best to remove them. In the case of the latter, *do not* make light of them to your airmen (don't try to kid an American airman), and attack these difficulties with every means at your command. Give them over-riding priority in your thinking. Seek help from all sources, *including* your airmen. Never, never let it be thought you have any idea of succumbing to difficulties which prevent the accomplishment of your mission. Exude the attitude that the question is not one of whether critical difficulties will be overcome, but when and how.

OFFICER-AIRMAN RELATIONSHIPS

If we are to have the fine Air Force which is sought (and a lesser kind is more handicap than benefit) the officers and airmen who comprise its strength must be proud of their membership and their status. The airmen must have the highest confidence in the integrity, professional capacity, and administrative ability of officers as a group and their own officers in particular. The officer must have an abounding faith in the ability, courage, and loyalty of his airmen. The officer and airman must constitute a smooth-running, efficient, harmonious team.

This imposes a very heavy responsibility upon the officer. He or she must convince his airmen by day-in and day-out demonstration that he or she knows the job and can be expected to continue to do it well. The attainment of missions and objectives must become habitual. Juniors, whether officers or airmen, must be treated in a manner which will encourage pride and self-respect. The officer must be objective, impartial, and just in his contacts and decisions. Favoritism or the suspicion of favoritism will wreck an organization. The best qualified individuals must be selected for promotions or the favored assignments. All people must be studied, their aptitudes discovered, and placed in line to increase their usefulness to themselves and their organization. The environment under which the organization lives must be lifted as high as possible. The organization and people must be first in the mind and heart of the officer.

HOW NOT TO BE A LEADER

During remarks on leadership by Lt. General Seth J. McKee, then Assistant Vice Chief of Staff, USAF, to the graduating class of the Squadron Officers School, Maxwell AFB, Ala., the general said about leadership techniques:

"Here are some styles of leadership I've observed that may serve as a handy guide on how *not* to do it.

"First, *The Anemometer:* He makes sure of wind direction and velocity before setting a course. Always flies downwind.

"*The Gyroscope:* This man maintains a rapid, stable rate of rotation around a fixed axis, with imperceptible forward movement.

"*Attila the Hun:* Leads through fear. Apparently suffers chronic indigestion or badly decayed molars. Guaranteed to discourage new ideas and dampen initiative.

"*Linus, the Less-Than-Lion-Hearted:* He's the officer who uses his staff as a security blanket. This may be okay when you begin a new and strange assignment, but you'd better not go that route for very long.

"*The Super-Ego:* Believes he was favored above all men when brains were issued. Creates bottlenecks with curves in them by redoing everything his staff does.

"*The Green-Eyed Monster:* The guy who doesn't trust anyone and won't give credit to his subordinates. He has forgotten that without their support, his ladder to success has no rungs.

"*The Hermit Crab:* Lives in splendid isolation, gathering more and more ignorance of what goes on in his organization.

"*The Madison Avenue Minion:* Thinks leadership is a popularity contest to be won on transparent superficialities. Avoids the hard decisions that won't move the applause meter off the high scale.

Finally, *The Mushroom Grower:* Keeps everyone in the dark and fertilizes regularly.

"Whatever your own style of leadership may be," the General went on, "There are two attitudes you should never tolerate in your organization. The first is 'Don't rock the boat'; the other, 'Don't stick your neck out.' There wouldn't be an Air Force as we know it today—or maybe even a country—if it weren't for some of your predecessors who were willing to pay a price for constructive change.

"Every one of us has to be prepared at some time to lay his career—or at least his ER—on the line for what he believes to be right. This is not something to be done frivolously or impulsively—certainly not to the prejudice of discipline. Knowing when the end justifies a risk of great personal sacrifice—either in combat or in management—is the mark of a true professional."

LEADERSHIP IN STAFF POSITIONS

Many officers erroneously assume that leadership is a quality needed by only those who are in command of Air Force units. This is not the case. There is a place and a need for leadership in almost every type of assignment an Air Force officer may receive. When working in a staff, particularly in positions where one lacks the authority of a commander with respect to other officers of the staff, a very high degree of leadership is necessary in order to achieve the tasks assigned. This form of leadership has as its principal characteristics the arts of persuasion and logical presentation of points of view. Even the chief of a staff will always find it more desirable to persuade than to order. When officers of a staff accept the point of view of another officer, they are much more likely to enthusiastically assist in the staff work flowing from such a point of view than they would be were they simply ordered to support the idea in question. As a junior officer working within the staff organization of a squadron or group, you will frequently find it necessary to obtain the coordination and assistance of other officers in positions parallel to yours. If, for example, it is your conviction that the pilots of your unit should receive more training in dropping napalm, in order to cause this conviction to be implemented you will find it necessary to obtain the support of the officer whose responsibility it is to schedule the training, and any officers in positions of authority who may believe that more training in rocket firing rather than with napalm would provide a greater return. Thus, within the confines of your own organization, you must exercise leadership in order to contribute the maximum of which you are capable to the betterment of your unit.

LEADERSHIP IN JOINT ASSIGNMENTS

Frequently Air Force officers find themselves assigned to duties involving other services. A typical case is the situation of Forward Air Controllers or Air Liaison Officers with U.S. Army divisions in combat. Here the Air Force officer must not only provide leadership for the Air Force officers and airmen working with him or her in performing tactical duties assisting the ground forces, but he must also exert influence on the Army officers with whom he cooperates, so as to arrange agreeably for the best performance of air power in coordination with our ground forces. The Air Force officer must utilize tact and persuasiveness, as well as exhibiting flexibility of mind and a sure, professional knowledge of tactical aviation. Such duties demand courage, resourcefulness, professionalism, and a very high order of leadership.

CONCLUSION

If we are to have a strong Air Force we must have strong military leadership. Wars, whether in Europe, Korea, or Vietnam, are won more by people than by weapons. It is the function of leadership to bring out the best capabilities of the people led and to direct those capabilities in support of the assigned mission. If Air Force officers perform this task of leadership well, our readiness for war and our accomplishments in war will be sufficient to the security of the United States.

11

Responsibilities of Command

If officers desire to have control over their commands, they must remain habitually with them, industriously attend to their comforts, and in battle lead them well.

—Stonewall Jackson

To the more junior officers of the Air Force nothing seems more remote from current or prospective duties than command of an Air Force unit. Command is indeed a rare assignment for Air Force officers of any grade. The officer strength of the Air Force is more than 100 times the number of Air Force units available to be commanded. Hence the rarity of command jobs. Yet, another aspect of command should be considered. Command is not a function exercised solely by one officer, as for example the squadron commander. The squadron commander cannot exercise command other than through the officer and noncommissioned officer members of his squadron. Each officer of the squadron is not only subject to command, he also helps exercise command. Thus each and every officer participates in the function of command. For this reason, all officers, whatever their grade or assignment, should understand the basic elements and objectives of the command function. He or she is a part of the command operation and it is one of the most important responsibilities. The command function has to do with the *mission* of the squadron or unit, with *training,* with *discipline,* with *supply,* with *morale.* How could an officer not be involved in these factors? It is also important that each officer understand command responsibilities for quite another reason than his or her own personal involvement. That is: each officer should appreciate the responsibilities of his or her squadron or unit commander, so as to better comprehend the reasons for the decisions of the commander. This is one of the hallmarks of a professional: to appreciate the responsibilities of other members of the team, and to assist in the discharge of those responsibilities.

IMPORTANCE OF THE MILITARY MISSION

In the exercise of command, the very first necessity is to accomplish the mission. The mission may be assigned in orders. If so, just go ahead and do it. Often the assignment is general, with an overall objective. The leader may then be obliged to select and adopt successive missions, the sum of which will accomplish the requirement. Successful commanders are the ones who accomplish their missions, on time, with minimum expenditures of personnel and means, and in a workmanlike, professional manner.

THE WELFARE OF PERSONNEL

Next to the mission, the welfare of personnel is a leader's most important responsibility because of its effect on their morale and their consequent ability to perform the mission. All commanders and leaders must interest themselves in all matters affecting the welfare of their personnel, including food, career guidance, adequate clothing and equipment, health and sanitation, recreation and entertainment, and personal problems. The measures to be used include planning, procurement and distribution of necessary supplies, training, frequent inspections, and corrective measures, which may include disciplinary action.

The Air Force has undertaken the task of shaping careers of officers, warrant officers and airmen, to the end that the greatest value is obtained from their services, and further to provide an upward course for the best qualified individuals. It is a program of great importance. Its furtherance is a function of command. (See Chapter 17, *Career Development.*)

Food. Food must be well prepared, special efforts being made to insure the best possible meals even under the most adverse conditions. The storing of food, the health and cleanliness of kitchen workers, the control of insects, the disposal of wastes, and the cleaning of utensils and other equipment, are matters of strict daily supervision.

Clothing. Clothing must fit properly, must be properly worn and cared for, and promptly salvaged and replaced when no longer usable. The uniform must be worn with pride. Care of clothing must be stressed as a necessary means of conserving national resources and Government property, and as a habit which will insure dependable functioning under battle conditions when failure may prove costly.

Health. The sick and venereal rate are indications of a unit's efficiency. Sanitary conditions in kitchens, mess halls, sleeping quarters, wash rooms, latrines, and other sources of infection must be watched carefully. Instructions in health habits, provisions of the means for protecting health, and insistence on development of proper habits are equally important, especially under oversea conditions. Constant vigilance and frequent inspections are necessary.

Recreation. Wholesome recreation and entertainment must be provided, especially that which encourages active participation, physical development, and the team spirit.

Leaves and passes must be granted impartially and as freely as the situation permits.

Consideration and, when possible, help must be given in dealing with personal problems. Young people away from home for the first time, may need unsolicited advice.

The spirit behind the leader's interest must be one of genuine helpfulness and thoughtfulness. In matters involving the proper care of Government property, protection of health, and fitness of the airman to perform assigned duties, stern measures are taken, if necessary.

MILITARY SECURITY

Always there is the problem of security. Personnel must be ready to detect attack from the air or ground. The repeated sudden attacks on Air Force bases in Vietnam are sufficient examples of the necessity for vigilance. Teach your people thoroughly the difficult art of security.

IMPORTANCE OF UNIT ADMINISTRATION

Administration begins in the squadron. That is where the airmen live and work, where changes in their status take place. That is where they are fed, housed, disciplined—and trained to become efficient, high-spirited fighting teams. The entire process of administering their affairs depends in large measure upon the initial action of the commander. If that first action by the commander and lieutenants is speedy and accurate, the administrative process may then become a smooth-running operation of which the tactical commander is scarcely conscious.

Command of a unit includes a diversified responsibility over several different but interdependent phases of operation. The breakdown of one will have an injurious effect upon the others. Balance is required. These phases are: *organization; morale; discipline; training to develop battle efficiency; administration; food management; supply.*

Classification and Assignment. The Air Force has an excellent classification system which identifies special skills possessed by our airmen. After classification, people are assigned to units or other assignments in accordance with these skills. This knowledge is of inestimable value, since the utilization of skills reduces the training burden. Record is made of the classification, and it follows the airman wherever he goes.

There is no system of classification and assignment which cannot be nullified or destroyed by thoughtless commanders. Unfortunately, this happens too often. If a unit needs ten truck drivers, and gets them, and ten cooks, and then assigns the truck drivers to the kitchen and the cooks to the trucks, the entire system is rendered useless. This may sound silly, or improbable, but search your own unit carefully, and see the extent to which you may be guilty!

Look at a table of organization. Note the numbers along the edge; the Air Force specialty classification numbers, which refer to a particular classified skill or Air Force Specialty (AFS). Take the service records of your people; ascertain the classification number of their skill. Determine the exact job which each person performs. Decide whether the skill of each individual is being fully used. As you proceed, study the training of each person to see whether he or she is assigned to render the maximum value to the Government. Check to see whether you have people of special skills which you do not require in your organization. You may be able to uncover some of the "rare birds" so badly needed by other units. If you find such individuals, report them, and perhaps by a little judicious trading you can obtain people of skills you need, in exchange for people who have skills you do not need.

The Air Force is vitally interested in placing each person where he can use his best talents. In so doing we utilize the training of our industry, our schools, and our colleges. Don't be guilty of hiding a radio operator, or airplane mechanic, for example, when the Air Force is running great schools to develop such individuals without prior experience.

THE SQUADRON COMMANDER

The commander is responsible for the execution of all activities pertaining to the unit. This includes the successful accomplishment of all missions assigned either in training or war. He is responsible in every way for all that his organization does or fails to do.

To a greater degree than other commanders, he must be in intimate touch with his people, and know their individual characteristics and capacities, their degree of training, their morale, and discipline. He must provide for the welfare of his people in all ways and under all conditions. In training, he must prepare his own people to undergo the most rigorous hardships. In war or even on maneuvers, he must save them from all unnecessary trials to preserve their stamina. The conditions which the airman experiences in his or her own squadron determine his or her opinion of the Air Force as a whole.

There are several duties which the commander cannot decentralize. These duties include: Responsibility for organization property; responsibility for the unit fund; the award of "non-judicial" punishment under the 15th Article of the Uniform Code of Military Justice; the authentication of many official records.

Use of Subordinate Leaders. The commander who attains high success will be adept at obtaining maximum efficiency from subordinate leaders. Their responsibilities should be clearly defined so that each may proceed with confidence.

Events which depart from routine must be disclosed to subordinate leaders in time to permit them to make their plans and prepare to do their several parts. A meeting may be held at which the requirement is discussed. Thought should be given to the manner of presenting the project before the group assembles. At the assembly the commander should state the mission with its time and place of execution. Opportunity to ask questions should be provided. At the end of the meeting the commander will have announced his decision and plan for the joint execution of a mission for which he alone is responsible.

Meetings of the key officers of a squadron should be held frequently. This is an opportunity for the commander to announce new work to be undertaken, specific small tasks to be completed in anticipation of a later action, and to announce inspections and other matters which affect the squadron as a whole. Opportunity to make suggestions for improvement in conditions should be extended. A healthy reaction is obtained by asking opinions as to matters of general concern and interest. These meetings should be informal and brief. They provide an opportunity for the commander to obtain the cooperation of all subordinate leaders because they will know exactly what is wanted.

Understudies and Replacements. The services of key individuals are often lost to an organization with disconcerting suddenness. The Air Force must maintain itself so that any officer, noncommissioned officer, or specialist may be immediately replaced without affecting the efficiency of any unit or activity. The training of understudies must be continuous.

The best people must convince themselves that they will receive consideration for appointment to higher responsibility or position if they prepare themselves for the task. Some regular procedure must be followed for the training of ambitious airmen who wish to demonstrate their capacity for advancement. As opportunity offers, airmen should be given a chance to exercise command or control of others to provide a means of further development. When this is done the selection of the best people for advancement or for replacements is made more certain.

Preparations to be Relieved of Command. The conditions of military service often result in a limited tenure of command by the individual officer. While it is thought to be the wisest policy to approach each current duty as if it would continue for a lifetime, the officer must realize that the day will come when he must turn his organization over to a successor.

It has been said that the efficiency of a commander can be better judged after relief from an organization than while in the exercise of command. This observation is subject to many objections, but has some merit for consideration. When command is relinquished it should be "clean" with no hidden skeletons, no loose ends to confound a successor unfamiliar with the problems. Property should be in good condition with no shortages not adjusted by the responsible officers, all required entries should be made in official records, the unit fund should be built up and its exact status made entirely clear. The principle of *Noblesse Oblige* should be observed to the utmost.

It would seem to be a questionable practice for an officer about to be relieved to embark upon a program of expenditure of the unit fund. The successor may have his own ideas on these matters and be prevented from carrying them into execution.

The new commander should not be embarrassed by requests which are just within permissible practices. It is an imposition to ask a new commander to receipt for "overages," "shortages," and if such request is made it should be refused. The successor should not be asked to complete transactions which should be finished by the officer being relieved.

It is reprehensible in the extreme to criticize the predecessor in command. Even in those few instances when it may be well deserved it should be strictly avoided. But the officer being relieved can eliminate much of the chance of such criticism being leveled at him or her by being very certain that he or she has completed every required transaction, informed the successor of every helpful bit of information (not forgetting the special desires and requirements of his immediate commander), and finally has transmitted to the new commander an organization of which he or she is proud.

FOOD SERVICE MANAGEMENT

Food service management is the supervision and control exercised over every phase of the operation of a mess. The term *mess* is applied to those military groups who for convenience, sociability, or economy, eat together. As used herein the term *mess* applies to squadron messes. Mess supervision is a function of command and is exercised, in some degree, by all commanders over the messes within their respective organizations. Direct control is exercised by the food service officer, who is assisted by the steward, cooks, and other food service personnel. Within the squadron the food service officer may be either the commander or a designated officer. The phases of operation involved are the preparation of menus; the procurement and storage of food; the preparation, cooking, and serving of food; the proper use of the mess equipment; the economical and efficient use of rations; sanitation; and mess accounting.

Object of Good Management. The object of good food service management is to build and maintain an efficient, economical, and attractive mess. Nothing contributes more to the morale of an organization than the fulfillment of this mission. *A good mess is the sum of good food ingredients, good cooks, good tools with which to work, and painstaking supervision.* It is not necessary for a food service officer to be an expert in cooking and nutrition to accomplish this task; only to apply fundamental principles and check the operation daily to see that a variety of good food, balanced and properly cooked, is served under sanitary conditions, without waste, in an attractive manner.

SUPPLY

The United States has been lavish with money and effort to make our armed forces the best equipped of any. Most individuals understand this point clearly when they consider outstanding single items. The total number of separate items is infinite and the cost prodigious. The job of the commander is to obtain the items allowed for his or her unit, see that they are issued to his or her people, and thereafter see that they are maintained in top-notch condition.

Tables of Organization and Tables of Allowances show the articles authorized. The unit supply officer can provide all the necessary information. The unit commander must supply the energy, care, and attention to detail which is necessary to attain the results which are required. This is what the unit commander must do: Find out the equipment authorized for his unit; obtain it; distribute it; maintain the required records with accuracy; inspect the equipment and maintain it always ready for use.

Supply Personnel. Responsibility for property is a function of command. The squadron administrative unit is the place of final distribution of supplies to airmen. Supplies are of small use unless they reach this ultimate point of distribution at the right time, in the right amounts, and in good condition. Tables of Organization provide personnel for this function. Zealous, efficient, capable supply personnel are worth their weight in gold and will be much appreciated. Select the best individuals available for this assignment.

Supply in the United States. Unit commanders constantly face supply problems. The supply procedures require considerable bookkeeping and responsibility. It cannot be otherwise. The whole problem resolves itself into accomplishing two primary objectives: first, to equip the unit with the items needed; and second, to keep every last item of equipment in thoroughly serviceable condition.

The procedures for supply in the United States are adequately covered in instructions which are distributed to all organizations. Unit commanders should be certain they have in their possession the necessary current regulations and consult their unit supply officer to make certain they are fully informed as to the regulations pertaining to their own unit. Thereafter it is only a matter of execution.

Supply in the Field. As soon as a unit enters field conditions, as in Vietnam, it faces a different sort of supply problem. But if any officer concludes that his "responsibility" has been left behind he or she is in for a rude awakening. An officer is responsible that the enlisted personnel are provided with the items of equipment they require, in combat especially, and the penalty for carelessness or inefficiency may be to do without, which in combat may be the difference between victory and defeat.

Let us assume that a unit reaches an oversea area fully equipped. At once things will happen which must be handled promptly. People are careless and they lose things. Property is lost by pilferage. It has been stated authoritatively that during the early years of World War II, losses from pilferage exceeded losses from enemy submarine action. There is also the temptation for the sympathetic and liberal American airman to give articles of clothing, of food, and other necessities of life to poverty-stricken natives. The airman may feel that he can obtain replenishment, or has more equipment than he or she will need, anyway. There is even the temptation to sell articles to natives for disposal on thriving black markets. In any event, the commander of a small unit is responsible for having personnel of the unit fully equipped and these are some of the matters that must be solved lest precious equipment vanish into thin air. The saying that lost articles will be found "on the payroll" may be exploited to curb carelessness. It will not solve the problem. Training to impress personnel with the importance to themselves of guarding their equipment jealously will help. Frequent and thorough property checks will also reduce losses and inform the commander of the unit's situation.

Wear and tear is a most difficult problem, especially in combat zones. Clothing especially wears out quickly. Lack of normal laundry facilities is an important factor. Higher commanders will provide for reserve stockage in bases and depots, but the unit commander must reduce wastage and present the needs in time for the supply or repair machinery to operate.

The commander of a unit must gain a complete understanding of the system of supply. The supply of rations will be automatic and quantity will depend upon the number of men he is required to feed. However, he or she must know the time and place of deliveries and take the steps necessary to obtain them. The supply of fuel and ammunition will be automatic in the sense that it will be available at distribution points. But he or she must know where these points are and the things that must be done to replenish stockage.

The problems of supply of food, ammunition, fuel, and clothing are with the unit commander day and night. He or she must provide for them and plan for them.

In combat supply lines are life lines. Every item is in short supply. Squadrons which are unaccustomed to frugality with supplies pay for this failing in inconvenience, inefficiency, and sometimes in lives.

Importance of Maintenance. The American citizen has become so accustomed to the garage around the corner that he has never taken the trouble to learn how to repair and maintain mechanical devices. In combat any failure to maintain equipment in perfect condition may be disastrous. A gun which fails to fire may be the difference between life and death. Rockets or bombs which fail to explode can waste the effort and increase the risk involved in a mission. Continual training and inspection are necessary. A landing gear which is improperly lubricated will be subject to unnecessary breakage and wear. It may fail in a crucial moment. Adequate maintenance will reduce the necessity for returning vast quantities of supplies to the U.S. for extensive overhaul. The unit commander is in direct charge of the equipment and the people who use it; must understand the whole problem; must see the need for maintenance; must acquaint himself with the condition of his equipment by unremitting inspections and corrective action. Maintenance may be the difference between success and failure.

MORALE PROBLEMS

Personal Problems of Airmen. All airmen have left interests, or roots, or problems behind them which may require their attention or action while in the military service. At home he or she would turn for advice to a parent, a friend, a lawyer, a minister or priest, or other person in whom he or she has trust and confidence. In the service he or she will usually turn to his commander.

These occasions provide a fine opportunity for the commander to show deep interest in the welfare of his men and women. He should adopt an impersonal and kindly attitude in hearing these problems. If the matter is confidential it must never be divulged improperly. When it is proper to do so, he or she should give the counsel he or she feels is correct, or obtain the necessary information, or direct the airman to an authority who can supply the information.

A good leader must have a genuine understanding of human relations. His or her tools are people, and therefore he or she must be able to deal with people. The necessary warmth of military leadership may be demonstrated when individuals carry their baffling personal problems to their commander for advice or solution.

The Complaint Problem. The commander must be accessible to members of the unit who wish to state a complaint. Some of the complaints will be petty. Others may be deliberate attempts to injure the reputation of another with a charge which is without foundation. There are other occasions, however, even in the best organizations, where genuine cause for dissatisfaction may occur. This is information which the commander must obtain lest the morale of his unit be seriously impaired. Personnel must know that they may state a cause for complaint to their commander with the knowledge that he or she will give them a hearing and correct the grievance if convinced of its truth. They must believe that he or she will wish to remove causes for dissatisfaction.

Poor airmen find many causes for grievance. Their petty or unfounded accusations must be heard, but they may receive the treatment or disposition which they merit. Good airmen will seek to avoid making a complaint. When ideas, or suggestions, or reports are stated to the commander which affect the welfare, the efficiency, or the morale of the unit, or any individual therein, they must be heard sympathetically. If the condition reported can be corrected, or deserves correction, the action should be taken at once.

Absence Without Leave. Commanders must face and solve the problem of unauthorized absences from duty. Particularly, commanders must analyze the causes and find solutions to the AWOL situation in their own unit. It is not a new problem and it exists during times of peace as well as in time of war. It is defined as, "the status of a person subject to military law who has failed to repair at the fixed time to the properly appointed place of duty, or has gone from the same without proper leave, or has absented himself from his command, guard, quarters, station, or camp without proper leave." In time of war the punishments authorized in the 86th Article of the Uniform Code of Military Justice depend upon the circumstances of the absence; but in a crucial situation in contact with the enemy there is no military offense more serious.

But the punitive approach is inadequate and will not provide the cure. The commander who places his reliance on punishment will fail to eliminate the cause of unauthorized absence and while guardhouse population may increase he or she will not maintain the on-duty strength of the organization, which is the vital thing. The officer must get acquainted with unit personnel. It is important for reasons of administration and leadership, but is especially necessary in connection with understanding the AWOL problem. He or she may facilitate this process by a "get-acquainted" interview preceded by an analysis of the airman's service record and related forms so that he or she will know in advance of the interview the individual's age, education, aptitudes, length of service, training status, dependency situation, and conduct as shown by the disciplinary record. With that information in mind the commander should proceed with an informal interview to get acquainted with the airman and let the airman get acquainted with him or her. Seek to find the views of the airman as to family responsibilities and any worries he or she may have with respect to his or her family. By so doing the commander may be able to correct oversights in dependents' benefits provided by the government and thus remove at once the cause for people absenting themselves without leave. Inquire of the airman as to his or her progress and determine whether the individual is satisfied that he or she is contributing a maximum of talents. Find out about civilian training and background to determine whether the individual should be assigned to duty for which he or she is better fitted. Encourage the individual to talk freely and take the time to explain to the airman those misunderstood things which may confuse or disappoint the young person. Knowledge of pass and leave regulations should be tested for it is true that some personnel absent themselves without leave when by asking they might have received an authorized absence. In this interview the airman should be invited to bring problems to the commander when in need of help or advice so that the individual will feel he or she has someone to lean upon when confused by personal problems. Concentrate on personnel of the lower intelligence ratings who are more likely to go AWOL than their more intelligent or better educated comrades.

The Department of the Air Force authorizes commanders to grant leaves so that people receive an average of 30 days leave per year. Like leaves for officers, they are a privilege and not a right. The requirements of training, or work, or operations come first. These requirements must take precedence. The commander should be informed thoroughly on the regulations governing leaves prescribed by the Department and local senior commanders. The commander should study the program of the organization and determine periods most suitable for granting leaves, and those periods when they should not be granted except for important emergency reasons. He or she must *plan* his leave procedures. Personnel who are designated for shipment overseas are required to be granted leaves before departure unless they have had a recent absence or military necessity prohibits a delay in departure. To the extent practicable, persons in units likewise should receive leaves before departure overseas. Military necessity must govern, and it will not always be possible to grant these absences. The personnel must be informed. It is most unwise for a commander to announce a leave policy in a manner which can be interpreted as a promise. Broken promises are themselves a cause for some absence without leave. The commander can inform unit personnel of the regulations; can tell them of his plans and his intentions; can inform them that circumstances at the time the leave is desired must determine his action; can seek opportunities to send personnel on leave; and can see to it that opportunities are equitably distributed to all personnel. These measures will replace ignorance with understanding and serve to reduce the causes for unauthorized absence.

The regulations for passes over week ends, and in evenings, are likewise important. Local policies will govern in this matter, and the people must be thoroughly instructed. Just as in the case of leaves, the granting of passes must not interfere with requirements or training of work. Liberality in granting short passes within the scope of prescribed policy is a wise practice. Where this is combined with sincere efforts to provide recreational activities on the base for men who do not wish to visit neighboring communities, many of the small but human dissatisfactions can be prevented.

The whole subject of eliminating unauthorized absences is of primary importance in building an effective Air Force. It is likewise of importance in developing pride in an organization. Success will identify the commander who studies causes of AWOL and removes them. Few people will absent themselves without leave from a well commanded organization.

Delivery of Mail. In priority, next to food, the airman might place the prompt delivery of mail. Indeed, for the airman with a spouse, a son or a daugher, a sweetheart or a beloved father or mother, mail may be of far greater lasting importance than food. Commanders of units must pay particular attention to mail delivery. They must see to it that the airmen's mail reaches them quickly and safely. They must make very certain that none is lost and that chance of malicious opening and violation of privacy is reduced to the vanishing point. A letter from home is of the highest importance to the airman as it serves to unite him for a moment with those he loves. For the most part, letters will quiet his apprehensions and dispel the worries which beset those who are far removed from their homes. Knowledge that all is well at home secures a peace of mind so strong that it justifies every effort a good mail system requires.

The commander is urged to check the method of receiving, guarding, and distributing the mail as it reaches his organization. Are the individuals charged with handling it completely trustworthy? Are there unsolved complaints of rifled or stolen letters? Is the mail guarded scrupulously from the time of receipt until it is handed to the airman? Is it allowed to lie about, subject to scrutiny and mishandling? Is it tossed promiscuously into a milling throng with small regard to its actual delivery? Is it handed to the individual to whom addressed? Is it held for long hours after receipt to be distributed at the whim of some individual? Is the mail delivered as promptly as circumstances permit after receipt? Are registered and special delivery pieces handled with due regard to postal regulations? Are your receipts for these classes of mail maintained exactly as prescribed? The airman treasures a package from home. Are packages zealously guarded so that petty pilfering is surely prevented? These are all matters which are very important in the daily life of the airman.

Importance of Letters to Parents. In a number of organizations, commanders make it a practice to write letters to parents of their people when important personal events have occurred in which they may take pride. The people of the United States are tremendously and vitally concerned about the progress of the Air Force to which they have given the services of their sons and daughters. Their impressions are formed by the reports they receive from their children and their neighbors and friends. No amount of big-name announcements as to morale and conditions will offset the local effect of an unfavorable report from a personal acquaintance. The interests and satisfaction which can be developed by contact, even by letters, with the folks at home are worth the effort required.

Many leaders write these letters. "Dear Mrs. Brown," they may write, "I am pleased to tell you that I have recommended your son for promotion to the grade of Airman 1st Class. He has worked hard here and his record is splendid. He is quite a good airman." Or this: "As you know, your son has been confined in the base hospital. I have visited him several times and have had frequent reports about his condition from his medical officer, the chaplain, and others. I can tell you now that he is well on the road to recovery, and his return to his organization will occur very soon." The American airman is a young person. In many instances he or she is away from home for the first time. Parents are anxious that he or she perform creditably.

Have you been present when a letter is received from a young airman? Do you know how it is discussed and passed around or read aloud to others? Can you not imagine the effect of a letter from an airman's commander? Form the habit of writing personal letters to the parents of your airmen on proper occasions. It places a human touch on a relationship which is otherwise impersonal and detached.

Personal Appearance and Morale. The uniform is supposed to enhance the appearance and standing of the wearer. It will do so only to the extent that it is well designed, clean, neatly pressed, and well fitted. There is nothing glamorous about a soiled, rumpled, ill-fitting uniform. No airman can feel the pride in his service we strive to develop unless he or she looks the part of a good airman.

Health as an Aid to Morale. It is an accepted principle that the commander is responsible for the health and physical well-being of the members of the unit. In this responsibility he or she is assisted by the medical officers. The medical officer serves as a staff officer and adviser of the commander in all matters which pertain to the care of the sick and evacuation of wounded, as well as in matters of sanitation or hygiene, all in addition to his professional duties within the fixed or mobile hospitals which serve the needs of the armed forces. There is nothing so destructive of the morale of service personnel as the knowledge or even the suspicion that the sick or wounded are being inadequately attended. In the execution of normal command functions the officer of the Air Force has, in the medical officer, a material source of assistance upon which he or she may draw.

Religion and Morale. Freedom of religion and of religious belief is one of the most important doctrines in our governmental philosophy. This principle must be extended to the airman and proper respect extended to religious convictions.

The religious and spiritual welfare of the members of a command is an important factor in the development of individual pride, morale, and self-respect, all essentials in a military organization. While Kipling may have over-emphasized the sordid side of the life led by military personnel in saying that men in barracks do not grow into plaster saints, it is a fact the environment in some instances leaves a little to be desired. We are a religious people and our airmen are subject to a wholesome, religious influence. There is a relation between morals and morale just as there is a relation between fair and just treatment and morale, or cleanliness and morale. The organization commander who is mindful of the religious and spiritual environment of his or her people will be able to make a strong appeal to the better instincts of all.

The Chaplain. The chaplain of a command is charged with the religious responsibility and a host of others. While he conducts the routine religious services, and attends as a matter of course to baptisms, marriages, and funerals, his most important work may lie in the close and personal human contacts which he enjoys throughout a command. Unless he receives the cooperation of the commander his work cannot possibly achieve the utmost. The task of the chaplain is always a difficult one. A word of suggestion to the chaplain as to a needed service for an individual will often result in the solution of a baffling problem. A word of encouragement to the chaplain to let him know his service is recognized and valued will stimulate further effort. The commander who cooperates with the chaplain and seeks thereby to improve the spiritual, religious, and moral standard of his unit will receive important dividends for his efforts.

Special Services. The Special Services Officer leads the recreational services including voluntary athletics. The closest coordination and support should be developed between commanders and the Special Services Officer.

The overall mission of Special Services is to provide an attractive off-duty environment for personnel. Such a program should have a direct effect upon reenlistment rates which are a matter of direct continual concern to commaders.

Athletic programs which encourage maximum participation by individuals constitute the heart of all Special Service activities. The Airmen's Club should be the hub around which many activities of high interest to airman will revolve. The library service is taken for granted at most stations, and its popularity accepted, but in prewar days it was not generally of high order. Organization commanders should inform their people of its location and facilities and encourage its use. Hobby shops with manual arts programs should be encouraged. Music programs and shows in which airmen take part constitute a part of these activities.

The goal is to increase the degree of satisfaction experienced by all military personnel. The effect should be that the airmen find interesting things to do at their stations near their units, rather than seek every opportunity to visit nearby civilian communities.

BASE CHAPEL, LANGLEY AIR FORCE BASE, VIRGINIA.

The commander of a small unit is his own Special Services officer. He should be thoroughly familiar with facilities available for his use, and make it possible for his men to make use of them in the fullest degree.

The Public Affairs Officer. The commander who learns the objectives, opportunities, and value of the Public Affairs program has added an important tool to his or her own professional capacity. The Public Affairs Officer is trained to control the program. The time has come when the Air Force must recruit an even higher type of individual than it has in the past. It must provide social and mental environment with full opportunity for intellectual development and professional advancement.

The Armed Forces Information and Education Division operates three major activities: The Troop Information Program; the Armed Forces Radio and Television Service; and service newspapers. The Armed Forces Institute is the core of the educational program. It offers correspondence courses, self-teaching courses, group study materials, and testing services. The young officer is well advised to investigate these matters in his or her own behalf, as well as in the interests of the airmen.

The American Red Cross and Morale. The American Red Cross acts as the medium of communication between the military and the civil community. This organization has chapters or representatives in all parts of the United States, and its foreign possessions. At many stations there is a Red Cross Field Director to serve the needs of the base. While this nation-wide organization stands ready always to perform a wide variety of worthy missions, its services are often useful in cases where airmen are troubled about conditions at home. In such instances it is common practice to call upon the representative of the Red Cross, acquaint him fully with the problem, and ask him to make an investigation of the actual conditions. The situation can often be adjusted. In any event the officer will obtain reliable information which will enable many baffling human problems concerning the airmen of his organization to be handled in a judicious manner based upon facts as they exist.

Air Force Aid Society. The work of the Air Force Aid Society is another valuable adjunct to the efforts of the commander to provide for the welfare and morale of his or her airmen. For a fuller discussion of the Society, see chapter 21.

Use of Allocation of the Organization Fund. Through non-appropriated fund channels, the base commander receives an allotment of funds applicable to units on his base for the purpose of enhancing morale. The squadron commander can apply for an allocation from this fund for specific morale-building activities of the squadron. A typical utilization of such a fund allocation would be to hold a squadron picnic or beer party. Any worthwhile, enjoyable entertainment for squadron personnel can be financed in this manner. It should be noted that the squadron commander must make application for the allocation of funds, indicating the intended use.

Members of a unit are entitled to know the status of the fund at all times because, after all, it is *their* fund. They should be informed of the intent to make future purchases. However, the commander who announces that a purchase will be made and fails to make it soon nullifies any beneficial effect such announcements can provide.

Athletics and Morale. Provisions for "athletics for all" is an important factor in the physical development of the individual airman and in the problems of maintaining a high morale. Officers must sponsor and develop athletic facilities. For the most part, airmen are young, strong, and energetic people. They need an outlet for these energies. A variety of athletic facilities should be available to meet the varying preferences of members of the unit. The use of the facilities should be encouraged by all proper means. Ask your Special Services Officer for help.

Housing. One of the most frequently overlooked factors affecting morale is housing. When an officer or airman reports to your base, the first consideration in his mind is where he is to live. If a bachelor, his concern is less acute than if married. Nevertheless, married or unmarried, there must be housing for him or her. Unit commanders should consider housing their first and most important duty relative to a newly-arrived person. Do not wait until the incoming airman or officer has arrived. Contact him or her by mail as soon as you know about the orders to your unit. (Check with the Base Family Services Center; perhaps they have already done this.) Determine the person's housing needs through your communication. Establish whether there will be quarters provided on the base. If so, check them for adequacy and readiness. If he or she must live off-base, seek out several suitable rentals for him or her to consider. While awaiting his arrival, arrange temporary quarters for him, on base if possible. Meet him or her at the base Family Services Center. Give your assistance in getting settled temporarily, including arranging for items needed until his or her furniture arrives. (Play pen? Bottle warmer?) Later, see to any help needed in occupying the permanent housing. You cannot expect to have effective officers or airmen until they and their families are adequately housed.

DISCIPLINE

Discipline Defined. Military discipline is intelligent, willing, and cheerful obedience to the will of the leader. Its basis rests upon the voluntary subordination of the individual to the welfare of the group. It is the cohesive force which binds the members of a unit, and its strict enforcement is a benefit for all. Its constraint must be felt not so much in the fear of punishment which it evokes as in the moral obligation it imposes on the individual to heed the common interest of the group. Discipline establishes a state of mind which produces proper action and prompt cooperation under all circumstances, regardless of obstacles. It creates in the individual a desire and determination to undertake and accomplish any mission assigned by the leader.

Discipline is a controlling factor in the combat value of airmen. The combat value of a unit is determined by the military qualities of the leader and members and its "will to fight." A point too often overlooked is the fact that poorly trained and poorly disciplined men reap the defeats and greatest losses. The greater the combat value of the organization, the more powerful will be the blow struck by the commander.

People are and always will be the vital element in war. In spite of the advances in technique, the worth of the individual is still decisive. They must be disciplined and must have morale.

In war, discipline is a matter of the gravest importance, affecting both the questions of life and death of individuals, and also the question of localized or global war. Those who disobey sanitation regulations risk poisoning. Pilots may be briefed to avoid violating the Chinese border. An incident leading to open world conflict may result from infringements of this order, whether intentional or unintentional. Discipline is the cement in the structure of the Air Force: without it, an Air Force is merely a mob. We *must* have a disciplined Air Force; else, ultimately, we shall have none.

Relationship of Superiors Toward Subordinates. Superiors are forbidden to injure those under their authority by tyrannical or capricious conduct or by abusive language. While maintaining discipline and the thorough and prompt performance of military duty, all officers, in dealing with airmen, will bear in mind the absolute necessity of preserving their self-respect. A grave duty rests on all officers and particularly upon organization commanders in this respect. Officers will impress upon the young airmen lessons of patriotism and loyalty, and above all will impress upon them the necessity of obedience in the service. These lessons will be repeated again and again. The difference in the status of an airman compared with that of a civilian will be carefully explained to make him or her understand that in becoming an airman he or she is subject to a new control and has assumed obligations that did not rest upon him or her as a civilian. Officers will keep in as close touch as possible with the personnel under their command, will take an interest in their organization life, will hear their complaints, will endeavor on all occasions to remove the existence of those causes which make for dissatisfaction, and will strive to build such confidence and sympathy that the people will come to them for counsel and assistance, not only in military and organizational matters but in personal or family distress or perplexity. This relationship may be gained and maintained without relaxing the bonds of discipline.

Punishment. The maintenance of control requires the promulgation of regulations for the guidance of members of the organization. These regulations must include positive matters, things to do, and the negative matters, things to avoid doing. Good management requires that these regulations be few, entirely justified by conditions, easily understood, necessary for the accomplishment of unit objectives, and at least reasonably in harmony with those in use by neighboring organizations. There is small occasion for the adoption of extra-official restrictive orders which serve only to gratify the aversions of the commander. The application of restrictive regulations on an entire organization because of the transgressions of one, or of a small number, is a lazy and ineffective form of control. The process of "Beating down the bush to kill the snake" is poor leadership. An error of leadership method is the use of threats or naming a penalty in issuing an order. Such a practice causes antagonism. The *Manual for Courts-Martial*, regulations of the Air Force, and the orders of higher authority prescribe the methods of control, and they should be well and carefully studied. An understanding of the requirements which must be enforced is an important part of the maintenance of control.

In any organization there will be violations of standing orders, and offenses committed which are punishable under the Uniform Code of Military Justice. The proper exercise of command requires that personnel be fully informed of the regulations which they are to observe. The commander must be alert to prevent offenses, for very often a kindly word of caution may prevent an offense. Offenses may be committed in ignorance, through carelessness, or with deliberate intent. The skilled commander can eliminate many of those which might be done through ignorance or carelessness.

An immediate recognition and reward of the deserving does much to prevent the commission of offenses. The officer need not be voluble. A softly spoken

"Good work," may be more stimulating to pride than a flowery and lengthy oration. The award of extra privileges, promotion, or mere recognition of merit has much to do with the development of both a sound discipline and a high morale.

The commander of a unit should prefer to take disciplinary action under Article 15, Uniform Code of Military Justice, rather than submit charges which will result in a trial by court-martial. For first offenders or for petty offenses this will usually suffice. Careful investigation must be made before any decision as to punishment is made. This investigation must be impersonal, fair and just, and it must include an opportunity for the accused airman to refute the charge.

The method of maintaining discipline must be such that it will stand the test of battle. Reliance upon trial by court-martial as the sole means of securing obedience to law and regulations will certainly fail. A trip to the "guardhouse" may be considered, by some, far preferable to participation in a combat mission.

Military Courts-Martial. There are three kinds of courts-martial, the summary court, the special court, and the general court. The summary court consists of one commissioned officer, the special court of three or more members, and the general court of five or more members. Officers may serve as members of any court-martial. Warrant officers may serve as members of a general or special court-martial for the trial of warrant officers and airmen. Airmen may serve on a general or a special court-martial for the trial of airmen when, prior to the convening of the court, the accused requests in writing the airmen be so detailed. When so requested, the membership of the court must include at least one-third airmen. Punishing power of the two inferior courts, summary and special, is limited by statute, the summary not being empowered to adjudge confinement in excess of one month and forfeiture of more than two-thirds of one month's pay. The special court may not adjudge confinement in excess of six months and forfeiture of more than two-thirds pay per month for six months. The punishing power of the general court is usually, by the wording of the Uniform Code of Military Justice denouncing a particular offense, left to the discretion of the court. That apparently unlimited power does not, however, exist. In a few instances the Article itself prescibes the punishment for a particular offense. In all other cases the President under authority given him by an Article of the Uniform Code of Military Justice has prescribed a table of maximum punishments. The punishment of nearly all offenses which are denounced by the common law and by non-military codes has been similarly limited by the President. In the rare event that an offense is committed which is not covered by the President's limit of punishment order, the punishment may not exceed that fixed as a maximum for the offense by the United States Penal Code or the Criminal Code for the District of Columbia. The sentence of a court-martial has no validity until it has been approved by the officer appointing the court. The sentences of all general courts-martial are subject to a series of reviews and approvals or disapprovals in which the record of trial is examined not only to determine its legal sufficiency but also to insure that no sentence of unnecessary harshness is finally executed.

An officer may be tried by a special court-martial as well as by a general court.

DISCIPLINARY PROCEDURES

Officers repeatedly find it necessary to determine what procedure to adopt in disciplinary and quasi-disciplinary matters. This recurrent problem most often takes one of the forms that will be hereafter discussed:

Courts-Martial
Non-Judicial Punishment
Administrative Discharges

Commander's Disciplinary Measures. The Uniform Code of Military Justice provides two procedures by which commanders may enforce discipline: the courts-martial procedure, and the application of Article 15 of the Uniform Code, which authorizes punishment without resort to courts-martial.

Courts-Martial. *Restraint.* Any person in the military service may sign and swear to charges that are an offense under military law. Usually, however, the person having knowledge of the alleged offense makes his information known to the immediate commander of the accused airman or officer. Ordinarily this commander would be the squadron commander. The commander must then make an immediate decision as to whether the accused person should be put under any degree of restraint, as for example, confined to barracks. If restraint seems indicated, the commander should issue the necessary orders at once.

Informal Inquiry. The commander (or a representative) must promptly make a preliminary inquiry of the accusations. This inquiry may be no more than a consideration of the charges and such evidence as is offered by the accuser. It may include a search of barracks, quarters, or other pertinent areas. Collection of documentary evidence may be desirable, depending on the nature of the charge. Individuals may be questioned informally. It is most important to remember, however, that before questioning an accused person, that person must be informed of the accusations made, and warned that any statement he or she utters may be used against him or her if a trial results. (Article 31, UCMJ.)

The commander must reach decisions as matters proceed. If it appears that courts-martial action is probable, the commander should obtain written, signed testimony from all persons involved. If the preliminary inquiry leads to a conclusion that courts-martial action is not indicated, the commander may wish to adopt lesser punitive procedures, or none at all. In many cases, accusations will be found to lack substance, and may simply be dismissed.

Charges for Courts-Martial. If the commander believes that courts-martial procedure is proper, he or she should have formal charges (DD Form 458) executed at once and so inform the accused. In carrying out this action the commander is well advised to consult with the Office of the Base Judge Advocate. See AFM 111-1.

Punishment Under Article 15. Article 15 of the UCMJ provides authority to the squadron or unit commander to impose punishment of a degree less than that for which courts-martial action is required. This authority, however, is qualified by the fact that any airman may demand trial by courts-martial rather than to accept punishment under Article 15. Nonetheless, when a commander has conducted a preliminary inquiry or is otherwise satisfied that courts-martial action is not desirable, he or she should inform the offender in writing of the intent to impose punishment under Article 15. The offender must state, in writing, his willingness to accept Article 15 punishment, or his demand for courts-martial. If the offender demands courts-martial, the matter is then handled by preferring formal charges. If the offender certifies his willingness to accept Article 15 punishment, the commander proceeds to impose the punishment. In this correspondence, the commander does not indicate what punishment is intended until after the offender has accepted punishment under Article 15.

Under Article 15, the commander may impose a wide variety of punishments. A commander in the grade of major or above may adjudge punishment to an airman of his command as follows: not more than 30 days correctional custody, forfeiture or half of his pay for two months, extra duty of 45 days, restriction for two months, detention of half his pay for three months, or reduction in grade (an E-4 or higher may not be reduced more than two grades.)

A commander below the grade of major may under Article 15 punish an airman of his or her command by imposing not more than: one week of correctional custody, forfeiture of one week's pay, extra duty for two weeks, restriction to limits for two weeks, detention of two weeks pay, or reduction of one grade (provided the commander has the authority to promote to the grade from which the airman is to be demoted).

A commander has authority to admonish or reprimand an officer under his or her command.

A commander in general officer grade may under Article 15 punish an officer under his command by imposing not more than: suspension from duty for two months, arrest in quarters for one month, forfeiture of half his pay for two months, or detention of half his pay for three months.

The commander should consult with the Staff Judge Advocate if the punishment intended approaches the levels for which courts-martial action is required. In general, the use of Article 15 is preferred by commanders, as against courts-martial, when it is an appropriate procedure. The courts-martial procedure is costly to the government and consumes the time of valuable men. Article 15 should be employed when the offender accepts it, when the punishments available under Article 15 are considered adequate, and when courts-martial is not mandatory.

Administrative Discharges. Under AFM 39-12 commanders are given authority to eliminate the unfit and the undesirable from the Air Force. AFM 39-12 authorizes the commander to get rid of the airman who is mentally or psychologically unable to perform acceptably as an airman. Such a person may well want to perform properly, but cannot. He or she should be eliminated.

AFM 39-12 provides for the separation of those who can be useful airmen but will not do so. These are the alcoholics, the rebels, the repeating offenders, the undisciplined. They should be removed from the Air Force as quickly as possible.

AFM 39-12 authorizes the separation of airmen who are sentenced by civil authorities to confinement for more than one year. AFM 39-12 expedites the separation of those who show homosexual tendencies.

In preparing action to eliminate an airman, the commander must formulate a record of offenses, instances, or other evidence of unsuitability. Usually it will be necessary to show that efforts have been made to correct the situation, through counseling or other methods. This record is attached to a letter to the Group Commander, asking that a Board of Officers be formed to review the case and determine disposition.

Officers dealing with cases of unsuitability should be guided entirely by the needs of the Air Force. An unfit airman stands in a space which might be filled by a productive airman. There is no room for the unfit or the undesirable. Get rid of them. Each person in the total strength of the Air Force must be first class.

Crime and Punishment. In dealing with offenders against military law, officers should be guided wholly by a desire to serve the best interests of the Air Force.

Vengeance has no place in such a consideration. The object is not to punish for the sake of "an eye for an eye." The object is to deter airmen from offenses, to exact recompense when property of the government has been lost, to rehabilitate offenders when possible. It is usually wise to be lenient with first offenders, particularly when the subsequent attitude of the offender is commendable. It is equally wise to be rough on second offenders; usually it is best to evict them from the Air Force. Seek no reputation for mercy, nor one for ruthlessness. Seek, instead, to advance the good reputation of the Air Force for level-headed action in the interests of the national security.

TRAINING

Purpose of Military Training. The ultimate purpose of all military training is the assurance of victory in war. We want a fighting Air Force composed of officers and enlisted who can do their complex jobs better than the enemy. If we train well, the task of winning the victory will require far less time than that which will be necessary if we train poorly.

One of the outstanding lessons derived from experiences in World War II, Korea, and Vietnam is that the state of training of units is the governing factor determining the combat capability of an Air Force unit. Air Force equipment in general is excellent in quality and adequate in quantity. Air Force morale is high. Supply and maintenance standards are superior. Discipline, while not perfect, is very commendable. Leadership, from the top down to the lowest unit, is superb. In the field of training, however, lapses may be noted. Every commander should heed this lesson: *training is never good enough; improvement is always possible.* Training is a round-the-clock function for the squadron. A great deal of it may seem dull and tiresome. Nonetheless, it must be accomplished thoroughly, for it pays off in lives saved, in battles won.

War also points up the tremendous importance of the less dramatic aspects of Air Force training. The officers in control of service activities, for example the supply of munitions and aviation fuel, actually hold the throttle of the Air Force. It is of critical importance that those officers in command of service squadrons realize the controlling nature of their units' performance, and bring those units to the highest possible state of training. Aircrews cannot soar off into the wild blue without fuel, ammunition, spare parts, operating weapons, well maintained air bases, in-commission aircraft. The Air Force is a chain involving many links, only one of which is the operating combat force. The air depots, the engineers, the researchers, the administrators, the hospitals, the maintenance crews, the tactical control parties, the intelligence agencies—all of these and many more are absolutely indispensable links in the chain of air power. Every one has a crucial job to do; every one must be highly trained to do it.

Joint Training. Every squadron commander will sooner or later find his unit engaged in some form of joint training with the Army or Navy or both. In the past there has been too much of a "to hell with that" attitude about joint training. Some of this attitude stemmed from a belief that since the Air Force could win wars alone, joint training was superfluous. Some stemmed from the feeling that there was already too much to be accomplished in the training program without adding joint training. If any single fact stood out of the Korean war and has been again proven in Vietnam, it is the *need for more and better joint training.* In the Vietnam war, as it was in Korea, the Air Force, Army, Marines, and Navy fought in close coordination. Each depended upon the others.

For the unit commander, the point is: Wars are won by teamwork between the services; your unit *must* not only want to take part in this teamwork, but it must be well trained in joint operations. Mere willingness to cooperate in an emergency will never take the place of a practiced know-how. It would be a sickening thing to think that Americans were dying because of a low state of joint training of *your* unit; nor would you find it any solace to reflect that such Americans were in the Army, Navy, or Marines rather than in the Air Force.

Analysis of the Training Process. Training doctrines, principles, methods, and management are covered in various training publications. A point which stands out is that the training of individuals is essentially identical. The qualities to be developed in the individual or unit are the same for all. Differences in training are confined to subjects relating to *tactical* and *technical proficiency.* All must be good airmen. And all must have the same high standard of *discipline; of health, strength,* and *endurance; of morale, initiative,* and *adaptability;* and all must have the same fine *leadership* in order to produce the final goal of *teamwork.*

Estimate of the Training Situation. Higher headquarters prescribe the training objectives, allot the use of facilities to units in turn, and designate periods for maneuvers and tactical exercises. The commander of a subordinate unit must make a careful analysis of the objectives to be met, the steps required to attain each objective, survey his or her own problem with reference to the existing state of training of his or her unit, determine the facilities which will be required, and make detailed plans for the best use of the time which is available to him or her to execute all of his missions. Foresight combined with careful planning is required from unit commanders who succeed in executing with smoothness and efficiency the training missions which are prescribed.

The commander must make an estimate of the training situation just as he or she would make an estimate of the tactical situation. It is a logical process of thought which assists the commander in arriving at sound conclusions in view of all of the facts. This estimate includes consideration of the following subjects:

Mission (training objective).
Essential subjects (relative importance of each).
    Basic.
    Technical.
    Tactical and logistical.

Time available.
Equipment and facilities available.
Personnel available as instructors.
Local conditions.
Existing state of training.
Organization for training.
Obstacles: administrative, physical, human.

After the commander has made the estimate and arrived at a plan he or she is ready to promulgate the plan in orders, or in a schedule, which may well include delegation to subordinate leaders the task of conducting phases of the training process. The next and final step is that of carrying out the plan. The commander must supervise the instruction given by others to make very certain that it is given correctly and that it will be completed satisfactorily within the available time.

THE CONDUCT OF INSTRUCTION

Every officer has responsibilities for the conduct of instruction. Some officers are engaged in formal instructional activities, such as in the schools of the Air Training Command. Others are confronted with the regular influx of new airmen into their units or offices. These latter airmen will have received instruction in their specialties, through the formal training processes of the Air Force. But instruction never ceases. Each airman should be brought to a higher state of competence as rapidly as possible, particularly in respect to the specific tasks he or she must perform in the unit of assignment. Officers should engage themselves in this process, either through direct participation or through supervision of the noncommissioned instructors who perform the training tasks.

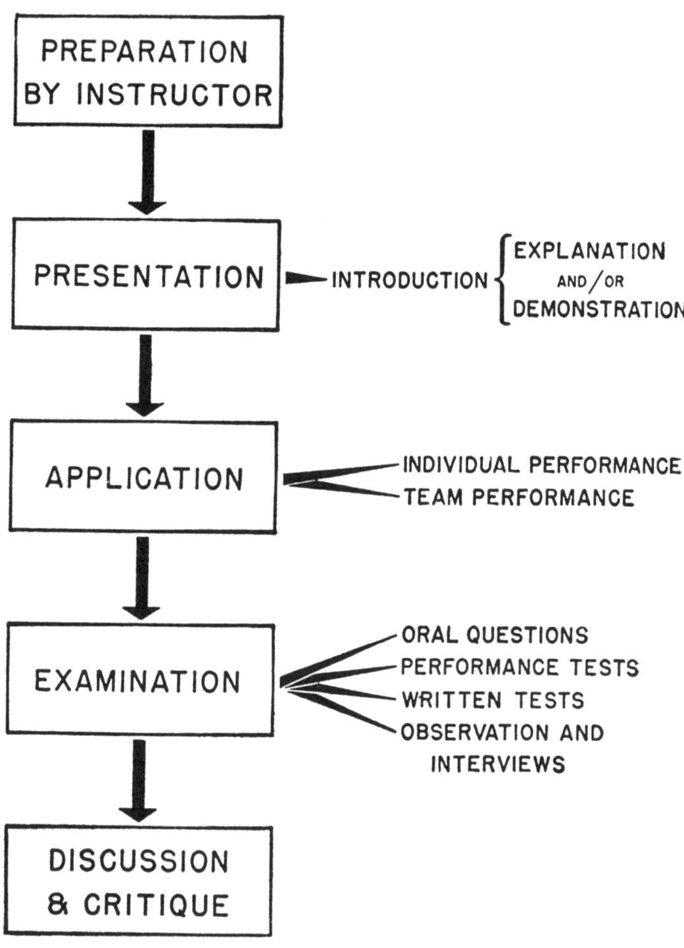

STAGES OF INSTRUCTION.

Instruction Should Be Planned. Officers should examine the plans of instruction to ensure that they (1) exist, and (2) are properly geared to the objective. Further considerations are:

The timing and duration of instruction periods;
The availability of competent instructors;
The availability and utilization of adequate instructional tools;
The procedures applied in carrying out instruction.

There is no single method of instruction which works best in all circumstances. But all instruction reduces to four basic moves:

Explanation.
Demonstration.
Practice.
Correction.

Whatever and wherever may be the instructional objective sought—in the United States, in the Pacific, in Europe—instruction must accomplish the four phases listed above. It is the responsibility of each officer to ensure that within his area of responsibility instruction is carried on constantly and well. He cannot safely assume that this will be automatically accomplished by noncommissioned officers or that the formal prior training of airmen is sufficient. In the Air Force change is constant and almost every change requires new instruction of airmen. This is a prime responsibility of command.

THE SOUND OF COMBAT

Only a very few words will be devoted to the subject of combat duty. Officers of the Air Force who have seen the sights and heard the sounds of battle express no undue concern over the reactions of unblooded Air Force officers. Most experienced officers state that inexperienced officers show more aplomb, more bravery, more unconcern for enemy fire than do those who have sweated through more deadly circumstances on a number of earlier occasions.

Combat is too often described as being only the event of deadly encounters between two adversaries, each possessing and directing an attacking vehicle, such as an airplane. Not so. Combat is much more often the tiresome struggle to supply aviation material, such as fuel, at the proper time and place in the theater of war. Combat is frequently the effort to produce sanitary water and food in an area where it is impossible to do so; to build revetments around fighter aircraft; to operate a decent mess in indecent surroundings; to predict weather from totally inadequate soundings; to furnish needed parts for inoperational aircraft when the parts are 10,000 miles away. Combat is doing the best you can under the circumstances; but the "best" means *best.* It is a sharp realization that *your* life or death, or someone else's, someone else who you value, hangs in the balance between your performance and that of your enemy counterpart.

No officer of the United States Air Force need be unduly afraid of combat, provided he has done his best to prepare himself for the test.

Statement by Lt. General Joseph Moore, then Commander Seventh Air Force, Vietnam.

"The story of the battle of Plei Me would have had a different ending without the men and aircraft of the Second Air Division (now Seventh Air Force) reacting instantaneously to the tactical situation. Fighter bombers, flying low level

missions around-the-clock under intense ground fire, were able to disperse and rout heavy concentrations of the enemy whenever called upon. Our airlift forces, under the most adverse conditions day and night, flew weapons, food, and equipment without which the Special Forces outpost could not have survived. Airpower, working in close coordination with the U.S. and Vietnamese Army ground forces, saved the day at Plei Me."

EO GUIDED BOMB

12
Retirement

I have done the state some service, and they know't. —Shakespeare, *Othello*

Retirement from the Air Force is an event which, in most cases, terminates a long and successful military career, and opens the way to a new career or to years of modest leisure. The majority of mandatory retirements of Air Force officers are after 28 years service, in the grade of lieutenant colonel, on retirement pay equal to 70 percent of the officer's basic pay as a lieutenant colonel. However, there are many other mandatory or voluntary versions of retirement which will be discussed later in this chapter. Some such retirements are the result of dire physical disability: dangers inherent in Air Force life, sharpened by wars such as Vietnam, make disability retirements not uncommon. Other early retirements, as after 20 years service, are voluntarily sought by Air Force officers who wish to shift to a civilian career field.

VOLUNTARY RETIREMENT

After 20 Years' Service. Any commissioned officer on the active list of the United States Air Force who has completed not less than twenty years or more than thirty years active Federal service in the armed forces of the United States, at least ten years of which have been active commissioned service, may in the discretion of the Secretary of the Air Force be retired upon his or her own application with annual pay equal to 2½ percent of the annual basic pay of the rank with which retired, multiplied by the number of years of service credited for pay purposes and not to exceed a total of 75 percent of such annual basic pay. For members who entered the service after 7 September 1980, retired pay is determined using a monthly retired pay base. See 10 U.S.C. 1407. In computing the number of years of such service for the purpose of determining the percentage of active-duty annual pay, service credit shall be computed to the nearest whole month actually completed for any portion of a year that is six months or more.

After 30 Years' Service. When an officer has been thirty years in the service, he or she may, upon his or her own application, in the discretion of the President, be so retired, and placed on the retired list. 10 U.S.C. 8918. AFR 35-7.

Application for Retirement. Application for retirement will be made by AF Form 1160 over the signature of the officer concerned, as prescribed in AFR 35-7.

Early Retirement Policy. The Air Force follows a liberal retirement policy for Regular and Reserve officers who have a minimum of 20 years active Federal service. At least 10 years of the 20 years active Federal service must have been in a commissioned status. For time in grade requirements, see the section on retired grade later in this chapter. 10 U.S.C. 1370.

RETIREMENT FOR PHYSICAL DISABILITY

The Career Compensation Act established a very important departure in remuneration following separation from active service as a result of physical disability. See 10 U.S.C., chapter 61. No individual is retired for physical disability if the disability is less than 30 percent, unless he has at least 20 years service.

The provisions of the act which affect disability retirement become less difficult to understand if each is considered separately. The intention is to treat commissioned and enlisted persons, Regular and non-Regular, on the same basis; to relate the pay given to the degree of disability or the length of service.

All personnel subject to disability retirement fall into one of two groups. These are (1) Regulars and non-Regulars called or ordered to active duty *for more than 30 days*; (2) Regulars and non-Regulars ordered to active duty, including training duty, *for 30 days or less.* (See 10 U.S.C. 1201, 1202, 1204, and 1205.)

The first stage in any proceeding for separation for physical reasons is a finding by the Service that the person, by reason of a disability, is not qualified to perform his duties. If a person is kept on duty, there are, of course, no separation proceedings.

But if a finding is made that the person *cannot* be retained in Service, the proceedings enter a second stage. If the disability was due to his "intentional misconduct" or "willful neglect" or incurred during unauthorized absence, the Government gives him nothing, merely separates him or her.

Third, if the disability was *not* due to misconduct or neglect, the next question is: Is the disability 30 percent or more under the Veterans Administration standard rating? (Loss of an eye or loss of use of a limb, and chronic, severe, high blood pressure are disabilities of 30 percent or more; loss of one or two fingers or one or two toes, loss of hearing in one ear, or defects of scars which do not seriously interfere with functions, are not.)

If the disability is less than 30 percent, and the member has less than 20 years service, no retirement is given. Instead, the Service person is given *severance* pay, which is 2 months' basic active duty pay for each year of active service, to a maximum of 2 full years' active pay. Half or more of a year counts as a full year.

From this point the disposition of disabled personnel in the two groups varies in some respects, depending upon the group to which each person belongs. The three stages in separation proceedings, just mentioned, are common to all personnel, *except* for one group—the non-Regular ordered to active duty, including training duty, for 30 days or less. With this group, the disability normally must result from an injury.

Types of Retirement. Two types of retirement are possible. If the disability is obviously permanent, retirement is final. But if the PEB has any question about the permanency of the disability, the Service person goes on a "temporary disability retired list."

If *permanently retired,* the Service person is paid by one of the following pay plans, whichever is more beneficial:

(1) Two and one-half percent of active-duty basic pay of the rank held at time placed on retired list, multiplied by the number of years of service creditable for pay purposes. (This benefits those of long service whose disabilities may be less than total. A half year or more of service counts for a full 2½ percent of active pay.)

(2) The same percentage of active-duty pay as the percentage of disability. (This benefits individuals of lesser service who have grave disabilities. Since one cannot retire with less than the 30 percent disability, nor receive more than 75 percent of active pay, the percentage of active pay under this option will run from 30 to 75 percent, depending on degree of disablement.)

The same pay provisions apply while on the *temporary retired list,* except that, in recognition of the adjustments to civil life and the employment handicaps faced by a person subject to recall to duty, the minimum pay *while on the temporary list* will be 50 percent of active pay.

For those who entered service after 7 September 1980 and become either permanently or temporarily retired, retired pay will be determined using a monthly retired pay base. See 10 U.S.C. 1407.

Special provisions of the act, which will not be discussed here because of their limited application, but which can be read from 10 USC by those interested, provide for (1) basing of retired pay on a temporary rank previously held satisfactorily, (2) recomputing retired pay of disabled persons who incur further disability while on a post-retirement active-duty tour, and (3) extending to non-Regulars retiring for disability the same commissary, base exchange, military hospital, and other benefits as are enjoyed by retired Regulars.

Maximum period on the temporary retired list will be five years. During this period, examinations will be given at least every 18 months, the actual frequency probably depending on the nature of the disability. At the end of the five years, or earlier if one of the examinations definitely settles the permanency of the disability, one of the following things can happen:

(1) The retirement may be made permanent. The percentage of disability will be recomputed as of time of permanent retirement. Thus a change in retirement pay may result. Of course, change in pay *will* result if the final pay under the most favorable option is less than the half-pay the person was assured on the temporary list.

(2) The disability may be found to be less than 30 percent. Retirement pay will be stopped and the person given severance pay if he has less than 20 years service.

(3) The person may be qualified for military duty. If so, his retirement pay will be stopped and he or she will be, if he desires, reappointed a Regular or Reserve officer or reenlisted, with a status as much as possible like that he would have attained had he or she never left the active list. If the individual does not consent to return to duty, he or she is dropped from both active and retired rolls for good.

Persons who have completed 20 years' active service are entitled to retirement even if their disability is less than 30 percent.

Other retirement provisions contained in AFR 35-4 should be examined by

Armed Forces personnel. If you are facing retirement for disability, you should study all methods of compensation—including Veterans Administration compensation—weighing the income-tax factor, if any, before making a decision.

STATUTORY AGE RETIREMENT

At 62 or 64 Years of Age. Unless retired or separated earlier, each regular commissioned officer (other than an officer who is a permanent professor or registrar of the United States Air Force Academy) shall be retired on the first day of the month following the month in which he or she becomes 62 years of age. An officer who is a permanent professor or registrar at the United States Air Force Academy shall be retired on the first day of the month following the month in which he or she becomes 64 years of age.

The President may defer the retirement of an officer serving in a position that carries a grade above major general, but such a deferment may not extend beyond the first day of the month following the month in which the officer becomes 64 years of age. Not more than ten such deferments of retirement may be in effect at any one time. 10 U.S.C. 1251.

MANDATORY RETIREMENT

Major Generals. Unless provided otherwise by some provision of law, each officer in the regular grade of major general in the Regular Air Force shall be retired on the first day of the first month beginning after the date of the fifth anniversary of his or her appointment in that grade in the Regular Air Force, or on the first day of the month after the month in which he or she completes 35 years of active commissioned service, whichever is later.

Brigadier Generals. Unless provided otherwise by some provision of law, each officer in the regular grade of brigadier general in the Regular Air Force who is not on a list of officers recommended for promotion to the regular grade of major general shall be retired on the first day of the first month beginning after the date of the fifth anniversary of his or her appointment in that permanent grade in the Regular Air Force, or on the first day of the month after the month in which he or she completes 30 years of active commissioned service, whichever is later. 10 U.S.C. 635.

Colonels. Unless provided otherwise by some provision of law, each officer in the regular grade of colonel in the Regular Air Force who is not on a list of officers recommended for promotion to the regular grade of brigadier general, shall be retired on the first day of the month after the month in which he or she completes 30 years of active commissioned service. 10 U.S.C. 634.

Lieutenant Colonels. Unless provided otherwise by some provision of law, each officer in the regular grade of lieutenant colonel in the Regular Air Force who is not on a list of officers recommended for promotion to the regular grade of colonel, shall be retired on the first day of the month after the month in which he or she completes 28 years of active commissioned service. 10 U.S.C. 633.

Majors, Captains, and First Lieutenants. Any promotion list major, captain, or first lieutenant who has been considered and not recommended by a selection board for permanent promotion will be designated a "deferred" officer. If he or she is not recommended by the next consecutive selection board convened for the selection of officers of his or her grade, he or she shall be discharged from the Regular Air Force on the date requested by him or her and approved by the Secretary of the Air Force, but not later than the first day of the seventh calendar month beginning after the month in which the President approves the report of the last board that did not recommend him or her for promotion. If he or she is eligible for retirement under any provision of law in effect on that date, he or she shall be retired. If on the date on which the officer is to be discharged he or she is not eligible for retirement under any provision of law, and is not within 2 years of becoming entitled to retirement under some provision of law, he or she shall be honorably discharged. 10 U.S.C. 631-632.

Deferment of Discharge. If on the date on which an officer is to be discharged as prescribed in the above paragraph for the various grades, an officer has not completed 20 "years' service" and is not eligible for retirement under any provision of law in effect on that date, but is within 2 years of becoming entitled to retirement under some provision of law, his or her date of discharge shall be the date on which he or she becomes entitled to retirement, rather than that prescribed above, and he or she shall be retained on the active list in the permanent grade then held until qualified for retirement and then be retired, unless sooner retired or separated under some other provision of law. 10 U.S.C. 632.

Removal from Active List. If an officer is removed from the active list of the Regular Air Force pursuant to the provisions of AFR 36-2 for failure to achieve such standards of performance as the Secretary of the Air Force may by regulations prescribe, and, if on the date of removal is eligible for voluntary retirement under any provision of law then in effect, he or she shall be retired in the grade and with the retired pay to which he or she would be entitled if retired upon his or her own application. If on the date of removal the officer is not eligible for voluntary retirement, he or she shall be honorably discharged in the grade then held. 10 U.S.C. 1186.

RETIRED GRADE, RANK, AND STATUS

A commissioned officer shall be retired in the highest grade in which he or she served on active duty satisfactorily, as determined by the Secretary of the Air Force, for not less than six months. In order to be eligible for voluntary retirement in a grade above major or below lieutenant general, a commissioned officer must have served on active duty in that grade for not less than three years.

An officer whose length of service in the highest grade he held while on active duty does not meet the service in grade requirements specified shall be retired in the next lower grade in which he or she served on active duty satisfactorily for not less than six months. Upon retirement, an officer who is serving in or has served in a position of importance and responsibility designated by the President to carry the grade of general or lieutenant general may, in the discretion of the President, be retired, by and with the consent of the Senate, in the highest grade held by him or her while serving on active duty. 10 U.S.C. 1370.

Physical Disability. Any officer of the Regular Air Force who may be retired for physical disability determined or incurred while serving under a temporary appointment in a higher grade shall have the rank and receive retired pay computed as otherwise provided by law for officers of such higher grade.

Any officer of the Regular Air Force on the retired list who shall have been placed thereon for reasons other than physical disability shall, if he or she incurs physical disability while serving on active duty under a temporary appointment in a higher grade, be promoted on the retired list to such higher grade and receive retired pay computed as otherwise provided by law for an officer of such higher grade retired on account of physical disability incident to service.

Any officer of the Regular Air Force on the retired list who shall have been placed thereon by reason of physical disability shall, if he or she incurs additional physical disability while serving on active duty under a temporary appointment in a higher grade, be promoted on the retired list to such higher grade and receive retired pay computed as provided by law for officers of such higher grade, provided that the Secretary of the Air Force, or such person or persons as he or she may designate, shall find that the additional physical disability is incident to service while on active duty in the higher grade and not less than 30 percent permanent.

Any officer of the Regular Air Force on the retired list who shall have been placed thereon for reasons other than physical disability, shall, if he or she incurs physical disability while serving on active duty in the same grade as that held by the officer on the retired list, receive retired pay computed as otherwise provided by law for officers of such grade retired on account of physical disability incident to the service.

Increases of Retired Pay. Retired pay is increased in accord with rises in the Consumer's Price Index, the pay increase being implemented annually. Thus between June 1977 and March 1982 the pay of retirees increased 55.4 percent. This increase will be the computed percent change adjusted to the nearest $1/10$ of one percent. 10 U.S.C. 1401(a).

MISCELLANEOUS

Date Retirement Becomes Effective. Except in cases of officers retired for disability, retirement must be "effective" on the first day of a month. The last day of a month is their last day of active service and the following day is their first day of retired status. Retirement for disability is effective on the date specified in the retirement order. 5 U.S.C. 8301.

Government Employment of Officer After Retirement. In 1964 the Congress passed legislation authorizing the Federal Government to employ retired Regular Officers on a basis of compensation making it feasible for the retired Regular officer to accept such employment. It is now possible for the retired Regular officer to receive the full pay of a civilian government job, plus a stipulated portion of retirement pay. Retired military officers may not be hired by the Defense Department for at least six months after their retirement.

Residence and Travel Abroad. Permission to travel and reside in a foreign country is no longer required of retired Air Force personnel, except for personnel who occupied "sensitive" positions or acquired "sensitive" information prior to retirement.

An officer's request for retirement at a foreign service station, if otherwise appropriate, normally will be approved by the Department of the Air Force. Such approval will not be given should it become necessary to return him or her to the United States for hospitalization or other purposes of the Government, and he or she will not be returned to an oversea station solely for the purpose of retirement thereat. However, the officer may obtain authority for foreign residence or travel, after retirement, as indicated above.

Travel of Retired Personnel by Transport. An important privilege for specified retired personnel is "space available" transportation on government owned aircraft. The term "space available" means space unassigned after all space requirement travel assignments have been made, and which would otherwise be unused if not authorized and assigned to the use indicated. (AFR 75-48.)

There is a nominal charge to cover subsistence and service.

Applications for space available travel are to be submitted to the appropriate terminal authority who acts upon them on a first come, first serve basis. Consult any Transportation Officer or Personnel Officer. Return transportation cannot be assured and return by commercial transportation may be necessary at personal expense.

Travel to Home. An officer of the Regular Air Force is presumed to have no established home. He or she may select and proceed to a home at Government expense, so far as authorized, at any place in the world at which he or she desires and intends to establish a bona fide home at any time within one year after retirement, provided he or she—

(1) Actually proceeds thereto and establishes a home;

(2) Submits, thereafter, a mileage voucher for land travel, certifying thereon that such place is home, or previously executes such certificate;

(3) Previously obtains authority of the theater commander in writing in event residence is not within a territory of the United States, and meets host government residency rules;

(4) Previously obtains a passport or statement that passport will be issued from the Department of State if required for residence at the place where retirement is requested.

Retired Officers, Status. An Air Force officer placed on the retired list is still an officer of the United States (31 Ct. Cl. 35).

Change in Status After Retirement. In the absence of any showing of fraud, the retirement of an officer under a particular statute exhausts the power of the President and the Secretary of the Air Force, and the record of executive action cannot be revoked or modified so as to make retirement relate to another statute, even though the case were one to which more than one statute properly applied at the time retirement was accomplished; and, further, the statutes relating to retirement apply only to officers on the active list, and there is no authority for the restoration of a retired officer to the active list for the purpose of being again retired. Sec. 326 (1) Dig. Op. JAG 1912-40 (See also AFR 31-3).

United States Air Force Retired Lists. The Secretary of the Air Force maintains officers' retired lists, upon which are placed the names of all commissioned officers of the Regular Air Force and the Air Force Reserve retired from active service.

RETIREMENT OF AIR NATIONAL GUARD AND AIR FORCE RESERVE OFFICERS

Retirement for Physical Disability. The laws governing retirement for physical disability apply equally to all officers on active military duty whether of the Regular Air Force or civilian components.

Retirement for Age and Length of Service. Chapter 67, Title 10, United States Code establishes retirement opportunities for AF Reserve and Air National Guard officers who attain age 60 and who satisfy stated requirements of service or service credits described in the law as "points." (See AFR 35-7.)

A minimum of 50 points must be earned in any year in order for that year to count for retirement purposes. However, all points are credited in computing retirement pay, if eligible for retirement. Points accrue and are accredited on the following basis:

(1) One point for each day of active Federal service;

(2) One point for each drill or period of equivalent instruction, such drills and periods of equivalent instruction to be restricted to those prescribed and authorized by the Secretary of the respective service for the year concerned, and to conform to the requirements prescribed by other provisions of law;

(3) Fifteen points for membership in a reserve component for each year of Federal service other than active Federal service.

All periods of active Federal service are counted for retirement purposes. One day of retirement credit is credited for each point earned while on inactive service with a limitation of 60 days for retirement purposes in any year.

The amount of retirement pay is computed at a rate equal to 2½ percent of the active duty annual basic pay (excludes allowances for quarters and rations, hazardous duty and other special pay) which he would receive if serving, at the time granted such pay, on active duty in the highest grade, temporary or permanent, satisfactorily held by him during his entire period of service, multiplied by a number equal to the number of years and any fraction thereof for which retirement credits have been granted. For this purpose the year is established at 360 days.

Retired pay for members who entered the Air Force after 7 September 1980 is determined using a monthly retired pay base. In computing this pay base, the rates of basic pay to be used are those most favorable to the member. 10 U.S.C. 1407 (a).

Officers of the Air National Guard and AF Reserve should have their own carefully checked and verified record of service establishing their retirement credits up to a proper date. Advice of Regular instructors or their assistants should be solicited in developing this record. Thereafter they should enter currently all credits earned so that they will be continually informed of their retirement status. This retirement pay does not reduce other retirement benefits such as may accrue from Social Security legislation, civil service retirement pay, nor should it serve to reduce retirement pay earned by participating in retirement programs of corporations or other employers.

THE RETIRED OFFICERS' ASSOCIATION

The Retired Officers' Association has as its purpose the aid of retired personnel of the various services and components in every proper and legitimate manner. It is located at 201 N. Washington Street, Alexandria, Virginia 22314. Dues are $15 per year.

AIR FORCE ONE OVER WASHINGTON

13

Voluntary and Involuntary Separations

The officer should wear his uniform as the judge his ermine, without a stain.
—Rear Admiral John A. Dahlgren

Separation of officers other than by retirement is covered in AFR 36-2, AFR 36-3, and AFR 36-12. Their provisions apply to all officers of any of the components who are serving on active duty.

INVOLUNTARY SEPARATION

Under the provisions of pertinent sections of Title 10, United States Code, as implemented by AFR 36-2 and AFR 36-3, Regular Air Force officers may be involuntarily discharged for moral or professional dereliction, in interests of national security, or substandard performance of duty. Air Force policy extends the same safeguards to nonregular officers. The following are reasons for initiating action to determine if a Regular or nonregular officer should be retained in the Air Force:

(1) Financially irresponsible.

(2) Mismanaging personal or Government affairs.

(3) Recurrent misconduct.

(4) Drug abuse (AFR 30-2).

(5) Failure at any school when attendance at the school is at Government expense if the failure can be reasonably traced to factors over which the officer has control.

(6) Failure to conform to prescribed standards of dress, personal appearance, or military deportment.

(7) Defective attitude.

(8) Retention is not clearly consistent with the interests of national security (AFR 36-2).

(9) Misrepresenting or omitting material fact(s) in official written or oral statements or documents.

(10) Failure to demonstrate acceptable qualities of leadership required of an officer of his grade.

(11) Failure to demonstrate acceptable standards of professional (including technical) proficiency required of an officer of his grade.

(12) A progressive falling off of duty performance resulting in an unacceptable standard of efficiency.

(13) Sexual perversion (AFR 36-2).

(14) A record of marginal service over an extended period of time as indicated by effectiveness reports.

(15) Failure to properly discharge assignments commensurate with grade and experience.

(16) Fear of flying.

Any commander may recommend initiation of action when he or she determines that action is appropriate. The recommendation is forwarded through channels to the wing or base commander (or commanders of groups directly subordinate to a major command) for initiation of action if determined that action is appropriate. This does not preclude a commander above wing or base level or HQ USAF from initiating action. Initiation of action is accomplished by preparing a letter of notification and forwarding it to the officer concerned.

Officers against whom action is initiated may submit written statements or other documentary evidence they feel should be considered in evaluating their case or they may apply for retirement, if eligible, or tender their resignations. Upon receipt of officer's reply to the letter of notification of initiation of action, the major commander determines if further processing of the case is warranted. If the officer applies for voluntary retirement or tenders his resignation, the major commander forwards the application to HQ USAF.

If an officer does not apply for voluntary retirement or tender his or her resignation, and the major commander determines that further action is warranted, the case is referred to a Selection Board. If the Selection Board determines that the officer should be required to show cause for retention in the Air Force, the major commander convenes a board of inquiry. The board of inquiry examines witnesses and documentary evidence and then recommends that the officer be retained or removed from active duty with the character of discharge. When an officer is recommended for retention, the major commander may recommend that a Reserve officer be released from active duty. The board of inquiry report is sent to HQ USAF where the case is reviewed by a Board of Review and final action taken by the Secretary of the Air Force. The cases of certain probationary officers and officers who waive consideration by a board of inquiry are referred to HQ USAF for consideration by the Air Force Personnel Board and final action by the Secretary of the Air Force.

A regular or nonregular officer who has completed five or more, but less than twenty, years of active service immediately before his or her discharge or release from active duty is entitled to separation pay equal to:

(1) 10 percent of the product of his or her years of active service, and 12 times the monthly basic pay to which he or she was entitled at the time of discharge or release from active duty, or $30,000, whichever is less; or,

(2) one-half of the above amount, but in no event more than $15,000,

as determined by the Secretary of the Air Force, unless the Secretary determines that the conditions under which that member is discharged or separated do not warrant such pay. In determining the number of years of active service for computing separation pay, a part of a year that is six months or more is counted as a whole year, and a part of a year that is less than six months is disregarded.

Rights of Officers Required to Show Cause for Retention. The recorder invites witnesses, both for the officer and the government, to appear if, in the legal advisor's opinion, they are both reasonably available and their testimony can contribute materially to the case.

Any officer may apply for voluntary retirement, if eligible, or tender his or her resignation before a final decision is made by the Secretary of the Air Force.

An officer may appear in person before the Board of Inquiry, may present evidence in his or her own behalf, be represented by counsel, and be allowed access to records the legal advisor to the board of inquiry considers relevant to the case and which are unclassified.

Any officer may be granted such time as is reasonable and necessary to prepare and present his case before the board of inquiry.

Involuntary Release of Nonregular Officers. Air Force policy requires the involuntary release from active duty of nonregular officers not promoted to the next higher temporary grade, due to insufficient retainability for permanent change of station, and because of failure to complete flying or technical training (AFR 36-12).

Except for ROTC Scholarship or AECP graduates whose education was funded at Government expense, AFROTC or OTS graduates with less than five years active duty immediately before date of separation are released from active duty as soon as possible after elimination from undergraduate flying training or technical training, unless a determination is made that their continued service on active duty is required. Officers are ineligible for readjustment pay.

Officers with an established date of separation (DOS) who have less than five years active duty immediately prior to DOS may be released when they are surplus, upon movement or inactivation of units or bases, or upon return to CONUS from an oversea assignment (AFR 36-20 and AFR 36-12). Officers are ineligible for readjustment pay.

Officers serving in the grades of second lieutenant, first lieutenant or captain are released from active duty when they are not promoted to the next higher temporary grade. Second lieutenants are released when they are found not qualified for promotion. First lieutenants and captains are released when they are twice passed over.

Involuntary Discharge of Regular and Nonregular Officers—Failure of Promotion to Their Next Higher Permanent Grade. Title 10, United States Code, requires the involuntary discharge of Regular and nonregular first lieutenants, captains and majors who are twice passed over for promotion to their next higher grade, provided they are not eligible for retirement or for retention to qualify for retirement (AFM 35-7).

Regular and Reserve second lieutenants are honorably discharged when they are found not qualified for promotion. A Regular second lieutenant is discharged as soon as possible. Reserve second lieutenants are discharged as soon as possible after completing three years of promotion service in the grade of second lieutenant and provided they have no unfulfilled military service obligation.

Lieutenant colonels not recommended for promotion are retired upon completing 28 years' service. See also Chapter 12.

Dates of Separation for New Officers. Beginning in the fall of 1970, dates of separation were assigned to graduates of OTS—four years following their date of commissioning for nonrated officers, and five years for those engaged in flying training. Graduates of the AFROTC since 1971 receive dates of separation in the same manner. The Air Force convenes annual selection-in boards which choose a number of officers each year for career Reserve status.

SELECTIVE CONTINUATION PROGRAM

Captains and Majors. An officer who holds the regular grade of captain or major and who is subject to discharge or retirement because of failure of selection for promotion may, subject to the needs of the service, be continued on active duty if he or she is chosen by a selection board with the approval of the Secretary of the Air Force. A captain may not be continued on active duty under this program for a period which extends beyond the last day of the month in which he or she completes twenty years of active commissioned service unless he or she is promoted to the regular grade of major. A major may not be continued on active duty under this program for a period which extends beyond the last day of the month in which he or she completes twenty-four years of active commissioned service unless he or she is promoted to the regular grade of lieutenant colonel.

Each officer who is selected for continuation on active duty, is not subsequently promoted or continued on active duty, and is not on a list of officers recommended for continuation or for promotion to the next higher grade shall, unless sooner retired or discharged under another provision of law, be discharged upon the expiration of his or her period of continued service, or, if eligible, be retired. Any officer who would otherwise be discharged and is within two years of qualifying for retirement under some provision of law, shall be retained on active duty until he or she is qualified for retirement under that law, and then be retired.

Lieutenant Colonels, Colonels, Brigadier Generals, and Major Generals. An officer who holds the regular grade of lieutenant colonel, colonel, brigadier general, or major general, and who is subject to retirement for years of service may, subject to the needs of the service, have his or her retirement deferred and be continued on active duty if chosen by a selection board with the approval of the Secretary of the Air Force.

Above Major General. An officer subject to retirement for years of service who is serving in a grade above major general may, subject to the needs of the service, have his or her retirement deferred and be continued on active duty by the President.

Length of Deferral. Any deferrals of retirement and continuation on active duty under the Selective Continuation Program shall be for a period not to exceed five years, but the period may not extend beyond the date of the officer's 62nd birthday.

Declining Deferral. An officer who is selected for continuation on active duty but declines to continue on active duty shall be discharged, retired, or retained on active duty, as appropriate.

VOLUNTARY SEPARATION

The right of an officer to resign his commission or request release from active duty is subject to certain restrictions growing out of his military status. The acceptance of a resignation or request for release is an executive act which may be exercised by the President or the Secretary of the Air Force, as appropriate, through any proper office designated by him (AFR 36-12).

Normally, a tendered resignation or request for release from extended active duty will be approved. However, an application for separation may be disapproved when an officer is under investigation; under charges or awaiting result of trial; absent without leave; absent in hands of civil authorities; insane; in default with respect to public property or funds; is serving under a suspended sentence to dismissal; has an unfulfilled active duty service obligation or agreement; in time of war, or when war is imminent, or in a period of emergency declared by the President or Congress; in any other instance where the best interest of the Air Force requires retention.

A resignation or request for release must contain a complete statement of reasons and, when appropriate, will have attached thereto available documentary evidence to substantiate the reasons given.

Resignation for Hardship. An officer may tender his resignation when his or her retention in the service is causing undue hardship either to himself or to members of his or her family. In such instances documentary evidence must accompany the application.

Resignation as Conscientious Objector. A tender of resignation based on conscientious objection is handled on an individual basis with final determination made by the Secretary of the Air Force based on the facts and circumstances and the policies set forth in AFR 35-24. The officer must be conscientiously opposed to participation in war in any form; opposition must be founded on training and belief and must be sincere and deeply held.

Resignation in Lieu of Demotion or Elimination from Service. An officer under consideration for demotion or elimination may tender his resignation in lieu of further proceedings (AFR 36-2 or AFR 36-3).

Presumably, an officer faced with one of these situations who believes his or her case is worthy would elect to have the proceedings continue. But if he or she considers elimination or demotion almost a certainty, and is unwilling to accept demotion, he or she may choose to resign rather than submit to the embarrassment of having his or her short-comings aired before a group of fellow officers who constitute the board.

Resignation for the Good of the Service. A resignation "for the good of the service" is a serious matter. Such resignations are utilized under the following conditions: An officer whose conduct has rendered him or her triable by court-martial may tender a resignation for the good of the service if charges have not been preferred; or, tender a resignation for the good of the service in lieu of trial if formal charges have been preferred; or, if under suspended sentence.

It is not to be expected that such a resignation will be accepted when the nature of the offense or conduct is such that if placed before a court-martial a punishment more severe than dismissal would result. For example, an officer accused of fraud which might result in confinement in a penitentiary would not be permitted to resign for the good of the service. Historically, this form of termination of commission has been used in lieu of trial by voluntary action of an officer who is accused of serious transgressions of moral codes or other unofficerlike conduct which, if placed before a court-martial, would result in no greater sentence than dismissal.

> *"Tender consideration for worthless and incompetent officers is but another name for cruelty toward the brave men who fall sacrifices to these defects of their leaders."*—President Davis to the Confederate Congress, October 8, 1862.

FB-111 STRATEGIC BOMBER

14

Promotion

It is our goal to provide opportunity for everyone to rise to as high a level of responsibility as his talent and diligence will take him.

—John D. Ryan, Then Chief of Staff, USAF

The subject of promotion comes up for lively discussion whenever and wherever groups of officers gather. This is only natural since promotion is the avenue to increased responsibility and rank, including the pay and other emoluments that go with increased responsibility. Like the weather, promotion is discussed heatedly, but often without full consideration of just what the promotion system is supposed to accomplish, the methods available, or the problems involved in the operation of a career promotion system. To gain an appreciation of the promotion system—what it is, how it works, and some of the obstacles involved—a review of the laws and regulations governing promotion is essential.

DEFENSE OFFICER PERSONNEL MANAGEMENT ACT (DOPMA)

To an extent unparalleled by any other profession, the parameters of the careers of commissioned officers are established in law. This reflects the concern of the Congress and the Nation that the high quality of our officer force be maintained and the recognition that this quality is essential to national security. A landmark in the legislation governing officer personnel management was the enactment of the Defense Officer Personnel Management Act (DOPMA), which became effective on September 15, 1981.

DOPMA updated outmoded aspects of earlier law, introduced important new officer personnel management concepts, and completed the task, begun with the Officer Personnel Act of 1947, of unifying and standardizing the law as it applies to officers of the four armed services. While DOPMA deals in comprehensive fashion with all aspects of officer personnel management, the bulk of the legislation is concerned with the dimensions and characteristics of the career system for officers and with the specific rules governing promotion procedures. The changes from earlier systems and procedures were evolutionary, building on experience under prior laws to enhance the overall quality of the officer corps.

Two changes made by DOPMA are of particular significance to the Air Force. They are (1) permanent grade relief for the Air Force from the restrictive field grade authorizations contained in the Officer Grade Limitation Act of 1954, and (2) the creation of a single system of permanent promotions for all officers on the Active Duty List. In 1954 the Air Force, because its officer corps was young in terms of both age and length of service, was authorized relatively fewer officers in the grades of lieutenant colonel and colonel than were the other Services. Beginning in 1959, as the Air Force officer corps matured, these lower limits were increased on a temporary basis while a more comprehensive proposal for officer career management was developed and considered. The Air Force needed temporary grade relief on nine subsequent occasions before DOPMA provided the Air Force what it had long needed—adequate permanent authorizations in law to meet mission requirements and provide attractive career patterns.

The single system of permanent promotions for all officers on the Active Duty List introduced by DOPMA replaced a complex and sometimes confusing system of temporary and permanent promotions. Prior to DOPMA, officers on active duty typically changed insignia as a result of temporary promotion. Regular officers were also considered for permanent promotions which determined their tenure. A similar permanent promotion system for Reserve officers considered Reserve officers on active duty in competition with all other Reserve officers. With DOPMA, all Active Duty List officers are considered under the same rules governing promotion eligibility and consideration.

With DOPMA in force, the officer corps will continue to have a mix of Regular and Reserve officers on EAD, although all officers selected for promotion to the grade of major or above will be tendered Regular appointments. Most officers will continue to enter active service as Reservists. They will first be considered for a Regular appointment when they compete for selection to captain. Subsequent consideration points will be after five and seven years of service.

ELEMENTS OF THE PROMOTION SYSTEM

a. Promotion System Objectives. Air Force promotion plans for the future are based on these objectives.

(1) To fill Air Force positions. The primary purpose of the active duty promotion system is to select the best qualified officer available to fill Air Force positions; therefore, the system must be responsive to the needs of the Air Force. Promotion is not a reward for past service; it is based on demonstrated potential to serve in positions of greater responsibility.

(2) To provide reasonably stable, consistent, and visible career progression opportunities. The active duty promotion system must also provide incentive to the officer. It must make visible reasonable opportunity for promotion, to include accelerated promotion from below the promotion zone for officers possessing exceptional potential, to attract the officer toward an Air Force career. The active duty promotion system should be relatively stable, so as to provide each year group of officers opportunity comparable with that of other year groups.

(3) To maintain force vitality. To carry out its role in support of national objectives, the military service has a unique requirement to maintain a young and dynamic officer force that is capable of developing and managing a large combat

ready force and of assuming wartime leadership. The Air Force promotion system is designed to produce officers in sufficient quality and quantity while maintaining a well-balanced force in terms of age and experience.

b. Rate of Progression:

(1) A successful promotion program must contain a promotion plan as a basic element. The plan should provide for an officer to remain at a given level long enough to profit by experience, but not long enough to lose interest and initiative. This plan must provide the opportunity for an officer to advance to senior leadership positions while still young enough to meet the physical and mental demands of these positions. It should provide for the identification and accelerated advancement of the exceptionally well-qualified officer and for the elimination of the least qualified. It should be sufficiently flexible for recognition that an officer may reach peak capacity at a level below the top, yet can serve the organization well by being continued in a lower grade.

(2) Today, promotion planning provides the promotion phase points and promotion opportunity for both those outstanding officers promoted "on time" to each grade and for officers who have demonstrated exceptional ability and potential. For instance, the "due-course" officers in each year group are now considered for promotion to major based on 90 percent opportunity, with promotion occurring in the 11th year of service. The "fast burners" are selected earlier and are the resource for accelerated advancement to the top leadership positions in the Air Force. (See figure 1.)

Due-course Officer	*In-the-Promotion Zone*	*Quota*
1st Lieutenant	2 years	Fully Qualified
Captain	4 years	95%
Major	12 years	90%
Lt. Colonel	16 years	75%
Colonel	21 years	55%

Exceptional Officer	*Below-the-Promotion Zone (1)*	*Quota*
Major	9 to 11 years	5%
Lt. Colonel	11 to 16 years	7½%
Colonel	13 to 20 years	15%

(1) Below-the-promotion zone quota is expressed as a percentile of the in-the-promotion zone quota.

FIGURE 1. PROMOTION PHASE POINTS.

c. Promotion Opportunity:

(1) Of significant interest to the officer force is the opportunity for advancement in grade. Basically, promotion opportunity is determined by the percentage of each year group that can reasonably expect to be promoted to the next grade. While the Air Force strives to provide equal opportunities to each year group, the number that can be promoted each year is largely determined by Air Force re-

quirements. The promotion quota is a number equal to an established percentage of those officers in-the-promotion zone (IPZ) as indicated:

For Promotion to Grade of	
Captain	95% IPZ
Major	90% IPZ
Lieutenant Colonel	75% IPZ
Colonel	55% IPZ

(2) As a result of the quota being established as a function of those officers in-the-promotion zone (new eligibles), all officers selected from the below-the-promotion zone (BPZ) and from those above-the-promotion zone (APZ), i.e., previously considered but not selected, are at the expense of the IPZ. So even though officers previously failed for selection do not generate a quota, they compete equally, and if judged by the board to be better qualified may be selected. IPZ selection rates are reduced correspondingly since the quota limitation cannot be exceeded.

EXAMPLE OF QUOTA COMPUTATION

In-The-Promotion Zone (IPZ) Times Promotion Opportunity = Board Quota

Board Quota Times Below-The-Promotion Zone (BPZ) Opportunity = Max Below-The-Promotion Zone (BPZ) Quota

Board Quota Minus Below-The-Promotion Zone (BPZ) = In-The-Promotion Zone (IPZ) and Above-The-Promotion Zone (APZ) Max quota*

(Selections from Below-The-Promotion Zone, In-The-Promotion Zone, and Above-The-Promotion Zone equal the total board quota.)

Quota Computation (Mar. 1981 Major Board)

3064 (IPZ) X 90% = 2758 (Bd Quota)

2758 X 5% = 138 (BPZ Quota)

2758 – 138 = 2620 (IPZ and APZ Quota)*

Actual Select Rates

3064 (IPZ) + 1250 (APZ) = 4314 (IPZ + APZ Eligibles)

2620 (IPZ + APZ Quota) ÷ 4314 (Board Quota) = 61% (IPZ and APZ)

2290 (IPZ) ÷ 3064 (IPZ) = 75% (IPZ Select Rate)

*If full Below-The-Promotion Zone Quota is used.

(3) Because the Air Force cannot promote officers to the field grades unless vacancies exist in these grades, it may sometimes be necessary to adjust promotion opportunities downward. This might happen, for example, in a period when overall officer strength is being reduced. The outlook, particularly with the permanent field grade authorizations and management system provided by DOPMA, is

that the promotion opportunities and timing outlined above will be continued for the foreseeable future.

Over the past few years, we have seen senior officers who were with us in the post Korea "cold war" era retiring. This means that leadership is being transferred to the next generation of officers. The Air Force must continue to cultivate a promotion system that will select only those who best meet the challenges for future leadership.

PROMOTION BOARDS

So far we have discussed the new officer promotion system which DOPMA provides and the Air Force's promotion philosophy and procedures, but not how one officer gets promoted and another does not. No discussion of the subject is realistic that does not touch upon how one gets oneself promoted. There are those cynics who may tell you that the prerequisites are to be either a general officer's offspring, in-law or aide. One may reasonably doubt the truth of this, since there are too many evidences to the contrary in either direction. There are also others more sanguine who subscribe to the fatalistic philosophy that if they do a good job the system will take care of them, but this too is an insufficient explanation.

It might be more useful for the young officer, however, to discuss how a typical promotion board operates in order to understand better how the selection process—often shrouded in mystery and distorted by rumor—really works. First, to scratch some rumors, it does not help to have a friend on the board; job descriptions and word pictures *are* important as well as the blocks checked; and the process does not favor select groups.

A selection board uses the following officer records as instruments: all Officer Effectiveness Reports (OERs) and training reports, an official photograph, citations or orders for approved United States decorations, the officer selection brief, letters pertaining to nonattendance at/or ineligibility for professional military education, letters to the board from officers eligible in- or above-the-promotion zone, and specified unfavorable information such as court-martial orders, and Air Force forms reflecting the imposition of nonjudicial punishment.

The best-qualified method of selection is used by all selection boards. Since the board can only recommend a set maximum number it must arrange those eligible in an order-of-merit listing (best- to least-qualified), apply the quota to this listing and promote those officers above the line where the quota runs out. Officers are evaluated using the whole-person concept. Board members review carefully each officer's selection folder, especially in the areas of job performance and responsibility, leadership, breadth of experience, professional competence and academic and professional military education.

It is important to note that steps are taken to put the different OER systems into their proper historical context. The board is reminded that records contain evaluations under several different assignments. The ratee's job, unit of assignment, level of responsibility, and OER endorsements all play a part in evaluation. The selection folder reflects each officer's level of academic and professional military education, but the board judges to what extent these achievements enhance performance and potential to assume greater responsibility. Mere completion of such courses is not to be given disproportionate credit, nor are officers to be penalized for not having obtained advanced degrees or PME diplomas. Some officers, such as those assigned to engineering or scientific areas or to some particular rated specialty, may not receive an opportunity for career broadening and must not be penalized for that reason. The Air Force needs both specialists and generalists.

Boards are charged to promote the best officers they review for consideration, and the board members—highly qualified officers who have the experience and mature judgment to make accurate assessments of potential—must take an oath to perform their duties without prejudice or partiality, having in view the fitness of the officers considered and the efficiency of the USAF. Pains are taken to insure that board panels make selections on a uniform basis. Of course, in the last analysis one's view of the promotion selection process may depend upon whether or not one is selected, but the Air Force makes strenuous and conscientious efforts to insure that the selection basis is the best one possible.

HOW TO GET PROMOTED

This is the Sixty-Four Dollar Question, for which there is no perfect and unequivocal answer. However, some observations based on experience may be in order. First, hit the ground running when you go on active duty. Whether you go first to flight or technical training or get on-the-job training (OJT), master your *job* quickly and thoroughly and give it 110 percent effort. Although your seniors, peers, and subordinates should give you aid and comfort, don't assume that anyone is going to help you learn the ropes. After this, learn your *profession*, the profession of arms. It is not enough to be a specialist even in this age of specialization, to be just a pilot, navigator, or intelligence officer. The military profession is one of the oldest and most honorable professions. Learn it, understand it, respect it.

Along the way it may be useful to become an expert in some aspect of the military arts and sciences that few or no others know. If you are *the* expert, it may lead to some interesting experiences.

Strive for assignments at increasingly higher command and staff levels, including joint staff duties. When you can obtain advanced education and PME, go for it. Try to master at least one foreign language and it need not be an esoteric one. Thanks to a rather basic knowledge of French, one officer found himself posted to a sensitive joint staff position in Saigon at a crucial period of the Viet-

MISSILE LAUNCH CREW.

nam War, and later was picked to help initiate closer relations between Air University and France's École Superieure, École de Guerre Aerienne.

Since your fellow officer is at the same time your comrade, your peer, and your competitor for promotion, you must maintain a proper perspective about this business of promotion. Above all, keep a good sense of humor. Always be your own man or woman, never forgetting Polonius' advice to Laertes in *Hamlet*: "This above all, to thine own self be true. Thou canst not then be false to any man." After your career has ended and if you have soldiered like a professional, you will not only cherish the worth of it all, having followed the path of duty, honor and country, the rewards of professionalism and career development, the things you have done, the places you have been and the sights you have seen, but also you will cherish the camaraderie that was such an essential part of it.

15

Effectiveness Reports

War must be carried on systematically, and to do it, you must have good officers.
—George Washington

The measurement of an officer's effectiveness, with potentials, strengths, and weaknesses, is a matter of the gravest importance to the Air Force as well as to the individual. Effectiveness reports constitute the most important single source of information for selection and assignment of individuals to do the work of the Air Force. They are the means by which each individual may be placed where his or her work for a period of time will be of greatest value. Here also will be recorded the information which will determine selection for promotion, for favored assignment, for the higher schools, or for downgrading leading possibly to elimination or termination of service.

The junior officer is not so much concerned with why effectiveness reports are rendered, or even with the responsibility imposed upon the rating officers, as he or she is with how to obtain a favorable effectiveness report. This chapter is therefore divided into sections dealing with:

On what forms is an officer rated?

When is he rated?

By whom is an officer rated?

What conduct produces a higher rating?

Any officer who learns the answers to these questions, *and applies them,* will find himself in the position of rendering effectiveness reports on other officers sooner than he or she may now think. The basis of the following information is AFR 36-10.

ON WHAT FORMS IS AN OFFICER RATED?

All officers in grades lieutenant through colonel are evaluated on AF Form 707, "Officer Effectiveness Report," and AF Form 475, "Education/Training Report." (See sample forms in this chapter.) The AF Form 707 provides the normal means for evaluating and reporting on the performance and demonstrated potential of an officer in his or her assigned duties. Training reports are used to cover periods when an officer is in formal academic or technical training.

WHEN IS AN OFFICER RATED?

Determination of when officers on extended duty will be evaluated is explained on the chart taken from AFR 36-12 which is shown in this chapter.

BY WHOM IS AN OFFICER RATED?

The Rater is the person (Air Force officer, officer of another Armed Force of the United States, a member of a foreign military service, or a civilian) who is formally designated on an AF Form 2095, "Assignment/Personnel Action," as the officer's immediate supervisor. Most AF Forms 707 will be signed by three evaluators: the Rater, additional Rater, and Indorser. The additional Rater level can fluctuate from unit to unit, but it is normally the "rater's rater." Except for *Referral Reports*, three is the maximum number of evaluators who can complete the report. In general, a Referral Report is one containing a very low evaluation of the

WHEN WILL AF FORM 707 BE WRITTEN ON OFFICERS ON EXTENDED ACTIVE DUTY

RULE	A	B	C	D	E
	If ratee's grade is	and	and period of supervision has been at least	then report will be written with reason entered as	and GO TO table
1	colonel thru captain	1 year has passed since closeout date of last OER or TR	120 calendar days	annual	4-6.
2	lieutenant	6 months has passed since closeout date of last OER or TR		semiannual	
3	colonel thru lieutenant (see note 1)	rater changes for anticipated period of 120 calendar days or more, or attendance at formal training of 8 weeks or more is scheduled	120 calendar days (60 calendar days if more than 1 year has passed since the ratee's last report)	change of rater (CRO)	
4	colonel selected for brigadier general (see note 2)	promotion occurs	120 calendar days	promotion to general	
5	colonel thru lieutenant	determination of the appropriateness of actions under AFR 35-32, 36-2, 36-3, or 36-12 is needed, or ratee has performed in an unsatisfactory or marginal manner and a special report is appropriate	120 calendar days (90 calendar days if completing a period of observation on the control roster)	directed by (MAJCOM, wing, base, squadron, and so on) commander	
6		special report is directed by HQ USAF (see note 3)	as directed	directed by HQ USAF	
7		ratee has been declared missing in action (MIA), captured, or interned (see note 4)			

NOTES: 1. If an officer selected for promotion to brigadier general comes due for a report for any reason between the time selection is announced and assumption of the grade, the report is written.
2. No report is required under this rule, if the officer has previously received a report under rule 1 or 3 and note 1 above as a brigadier general selectee.
3. Authority to direct reports under this rule is retained by DCS/Personnel, HQ USAF. Special reports covering outstanding performance of duty are not permitted under this rule. If HQ USAF determines that special reports are required on certain officers for selection board use, HQ AFMPC/MPCAJB furnishes their names to the MAJCOMs, along with appropriate suspense dates, and directs that reports be submitted under this rule.
4. Reports for periods of missing in action, captured, or interned status of less than 15 calendar days are not prepared. If status of ratee remains in one of these categories for 15 calendar days or more, a report is rendered under this rule without regard to the number of days supervision. The report is closed on the day the ratee was placed in missing in action, captured, or interned status. Reports for periods during which the ratee was in missing in action, captured, or interned status are as directed by HQ AFMPC/MPCYO.

When is an officer rated?

officer being reported on. Reports made referral by the reviewer (the final evaluator) must be additionally endorsed. Referral reports will be provided to the officer reported on for his comments.

I. RATEE IDENTIFICATION DATA *(Read AFR 36–10 carefully before filling in any item)*

1. NAME *(Last, First, Middle Initial)*	2. SSAN *(Include Suffix)*	3. GRADE	4. DAFSC
HOFFMAN, MARIA B.	123-45-6789FR	Captain	8035

5. ORGANIZATION, COMMAND, LOCATION	6. PAS CODE
Air Force Human Resources Lab (AFSC) Lowry AFB, CO	PSSDJ9

7. PERIOD OF REPORT FROM:	THRU:	8. NO. DAYS OF SUPERVISION	9. REASON FOR REPORT
15 Feb 81	14 Feb 82	365	Annual

II. JOB DESCRIPTION 1. DUTY TITLE: Enter official duty title.

2. KEY DUTIES, TASKS AND RESPONSIBILITIES:

Concisely outline key duties, etc.

III. PERFORMANCE FACTORS

[1] *Specific example of performance required*

Factor	NOT OBSERVED OR NOT RELEVANT	FAR[1] BELOW STANDARD	BELOW[1] STANDARD	MEETS STANDARD	ABOVE[1] STANDARD	WELL[1] ABOVE STANDARD
1. JOB KNOWLEDGE *(Depth, currency, breadth)* What has the officer done to actually demonstrate depth, currency, or breadth of job knowledge in performance of duties? Consider both the quantity and quality of work.	O				X	
2. JUDGMENT AND DECISIONS *(Consistent, accurate, effective)* Does the officer think clearly and develop correct and logical conclusions? Report on how the officer grasps, analyzes, and presents workable solutions to problems.	O					X
3. PLAN AND ORGANIZE WORK *(Timely, creative)* Does this officer look beyond immediate job requirements? How well does she/he anticipate critical events?	O				X	
4. MANAGEMENT OF RESOURCES *(Manpower and material)* Does this officer manage to achieve optimum economy through effective utilization of personnel and material? Consider balance between minimum cost and false economy to the ultimate expense of the mission.	O				X	(initialed)
5. LEADERSHIP *(Initiative, accept responsibility)* How has the officer shown ability in initaiating ideas and action in organizing effort, obtaining cooperation and respect of others and in directing efforts of others?	O				X	
6. ADAPTABILITY TO STRESS *(Stable, flexible, dependable)* What is the effect of stress on the officer's performance? Does she/he work as well or better under adverse conditions? In difficult situations, heavy workloads and pressures, does her/his work deteriorate?	O				X	
7. ORAL COMMUNICATION *(Clear, concise, confident)* How well has this officer been able to present ideas orally?	O			X	(initialed)	
8. WRITTEN COMMUNICATION *(Clear, concise, organized)* How well has this officer been able to present ideas in written form?	O					X
9. PROFESSIONAL QUALITIES *(Attitude, dress, cooperation, bearing)* To what extent does this officer meet standards of bearing, dress, and courtesy and enhance the image of the Air Force officer?	O					X
10. HUMAN RELATIONS *(Equal opportunity participation, sensitivity)* How does this officer demonstrate support of the Air Force Equal Opportunity Program? Evaluation of this factor is mandatory.					X	

OFFICER EFFECTIVENESS REPORT, AF FORM 707 (front).

IV. ASSIGNMENT RECOMMENDATION: 1. STRONGEST QUALIFICATION: Based on your evaluation of the officer's
2. SUGGESTED JOB (Include AFSC): strongest qualification, recommend a future assignment.
3. ORGANIZATION LEVEL: Next step for this officer? 4. TIMING: Now? After a special school?

V. EVALUATION OF POTENTIAL:

Compare the ratee's capability to assume increased responsibility with that of other officers whom you know in the same grade. Indicate your rating by placing an "X" in the designated portion of the most appropriate block.

RATER	ADDN RATER	INDORSER	RATER	ADDN RATER	INDORSER	RATER	ADDN RATER	INDORSER	RATER	ADDN RATER	INDORSER

Lowest ← → Highest

VI. RATER COMMENTS

Organize the comments within the standards of good writing. Do not use headings; do not underline, indent, or capitalize merely to add emphasis. Include those comments required by paragraph 5-8b of AFR 36-10. Add any other comments not covered elsewhere as long as they're not excluded by paragraph 4-8 of AFR 36-10. The comments should serve to increase the value and meaning of the report. Amplify those positive aspects of the officer's performance deserving of special note.

NAME, GRADE, BR OF SVC, ORGN, COMD, LOCATION	DUTY TITLE	SSAN	SIGNATURE	DATE
JOSEPH P. DOAKES, Lt Col, USAF 380 Cmbt Spt Gp (SAC) Plattsburgh AFB, N.Y.	Director of Intelligence	123-45-4321FR	Joseph P. Doakes	15 Feb 82

VII. ADDITIONAL RATER COMMENTS ☐ CONCUR ☒ NONCONCUR

Review the ratings and comments of the rater for completeness and impartiality. Comments may be added if they add meaning to the report. If the Additional Rater qualifies as the Indorser, Section VIII below will be completed and Section VII will contain the statement "Additional Rater qualifies as Indorser." Additional Rater may indicate disagreement with ratings given in Section III by placing his/her initials to the right of the appropriate box. Significant disagreement requires justification.

NAME, GRADE, BR OF SVC, ORGN, COMD, LOCATION	DUTY TITLE	SSAN	SIGNATURE	DATE
ELLEN S. BROCK, Col, USAF 380 Cmbt Spt Gp (SAC) Plattsburgh AFB, N.Y.	Vice Commander	987-65-4321FR	Ellen S. Brock	16 Feb 82

VIII. INDORSER COMMENTS ☒ CONCUR ☐ NONCONCUR

When the Rater and the Additional Rater disagree, the Indorser must state with whom she/he agrees. Significant disagreement requires justification.

NAME, GRADE, BR OF SVC, ORGN, COMD, LOCATION	DUTY TITLE	SSAN	SIGNATURE	DATE
JOHN Q. PRIVATE, Col, USAF 380 Cmbt Spt Gp (SAC) Plattsburgh AFB, N.Y.	Commander	111-22-3333FR	John Q. Private	17 Feb 82

OFFICER EFFECTIVENESS REPORT, AF FORM 707 (reverse).

TRAINING REPORT

Training reports are usually submitted by the commandants of Air Force or other military service schools. The training report is rendered upon completion of the training course or at the end of the academic year when the course is of more than one year's duration. Training reports are also prepared at the completion of formal technical training courses whose prescribed length is eight weeks or longer. The training report contains such information as: (a) course

PERFORMANCE FACTORS	FAR BELOW STANDARD (Any Item)	BELOW STANDARD (Any Item)	MEETS STANDARD (Any Item)	ABOVE STANDARD (All Items)	WELL ABOVE STANDARD (All Items)
1. JOB KNOWLEDGE (Depth, currency, breadth)	• Has serious gaps in technical and professional knowledge • Knows only most rudimentary phases of job • Lack of knowledge affects productivity • Requires abnormal amount of checking	• Technical and professional knowledge is inadequate for the job • Must be assigned only routine duties and monitored regularly • Requires close supervision	1. • Demonstrates adequate technical and professional knowledge required for the job • Searches out facts and arrives at sound solutions to problems • Broad knowledge of related jobs and functions • Conversant with significant job-related developments	• Possesses keen insight and the ability to evolve it into practical solutions • Keeps informed of important developments in related fields • Can handle difficult situations effectively • Broad knowledge of related missions • Rarely requires guidance or assistance	• Possesses superb technical and professional knowledge • Sufficiently well versed in his or her job to discuss and implement improved methods resulting in savings in manpower or material • Maintains and increases professional and technical knowledge • Actively pursues new ideas and developments and their relation to the overall mission • Recognized authority in his or her field
2. JUDGMENT AND DECISIONS (Consistent, accurate, effective)	• Reluctant to make decisions on his or her own • Decisions are usually not reliable • Declines to accept responsibility for decisions	• Usually makes sound routine decisions • Tends to procrastinate on necessary decisions • Reluctant to evaluate factors before arriving at decisions	2. • Seeks out all available data before arriving at decisions • Consistently provides accurate decisions • Accepts responsibility for decisions and learns from incorrect judgments • Provides effective decisions by clear and logical thinking	• An exceptionally sound, logical thinker • Does not hesitate to make required decisions • Decisions are consistently correct • Opinions and judgment are often solicited by others	• Keen, analytical thinker • Makes accurate decisions under intense pressure • Extremely effective in exercising logic in broad areas of responsibility
3. PLAN AND ORGANIZE WORK (Timely and creative)	• Fails to plan ahead • Disorganized and usually unprepared • Objectives are not met on time	• Scheduling and organizational efforts normally fail • Encounters difficulty with tasks other than routine • Finished products are usually behind schedule	3. • Careful, effective planner • Anticipates and solves problems • Effectively balances resources • Finished products are consistently submitted on time	• Plans beyond requirements of present job • Plans coincide with related activities • Is flexible and able to adjust priorities • Frequently called on to organize complex tasks	• Able to anticipate critical events and makes prior provisions to deal with them • Plans encompass all feasible contingencies • Extremely effective in utilization of resources
4. MANAGEMENT OF RESOURCES (Manpower and material)	• Wastes or misuses resources • No system established for accounting of material • Causes delay for others by mismanagement	• Accomplishes conservation of material on a sporadic basis • Squanders resources to get job done	4. • Uses minimum material with good results • Establishes controls to ensure that manpower and material are accounted for and conserved • Develops and uses cost-effective methods	• Excellent results accomplished at minimum cost • Consistently suggests methods of conserving resources • Skillfully uses cost-effectiveness studies	• Extremely effective in use of material • Consistently seeks and projects ways of using existing equipment • Is often assigned to difficult and important projects where limited resources are a significant factor
5. LEADERSHIP (Initiative, acceptance of responsibility)	• Often weak. Fails to show initiative and accept responsibility • Lacks self-confidence • Inconsistent in dealing with subordinates	• Avoids responsibility • Displays confidence only when working with familiar subjects • Initiative and acceptance of responsibility adequate in most situations	5. • Accepts responsibility of assigned tasks • Consistently displays initiative • Commands respect of subordinates • Is fair and consistent in dealing with subordinates	• Demonstrates a high degree of initiative and acceptance of responsibility • Displays exceptional skill in directing others • Promotes enthusiasm by interest and sincerity • Acknowledged leader among his or her peers	• Consistently demonstrates outstanding initiative and acceptance of responsibility • Exhibits complete confidence in his or her ability to handle any task • Induces maximum effort from everyone • Is decisive in critical situations • Provides direction and guidance for broad areas of responsibility • Leadership not limited to subordinates and peers
6. ADAPTABILITY TO STRESS (Stable, flexible, dependable)	• Panics in new situation • Tendency to shirk difficult situations • Reaction is unpredictable	• Prefers to work on routine tasks • Jumps to erroneous conclusions in new situations • Hesitates to become involved in new situations	6. • Flexible and open to new ideas • Willingly seeks assistance in difficult situations • Provides reliable decisions under pressure • Consistently displays calm and controlled behavior	• Readily adapts to fluctuations and changing priorities • Consistently performs well in difficult situations • Anticipates changes and is prepared to react accordingly	• Responds quickly and effectively to crises • Systematically succeeds where others fail • Consistently provides outstanding leadership and guidance under difficult and stressful conditions
7. ORAL COMMUNICATION (Clear, concise, confident)	• Does not convey ideas clearly and concisely • Has limited vocabulary • Cannot express thoughts in a logical sequence	• Only occasionally able to verbally convey useful information • Briefings and discussions frequently exhibit a lack of confidence	7. • Gives direct and understandable responses to questions • Gives briefings which are organized and well presented	• Very articulate in a wide range of difficult communications situations • Puts extra effort into conversing well • Capable of persuading an audience	• Delivers concise, well-organized presentations • Is often called on to present and explain difficult and complex subjects • Can sway a hostile audience to his or her point of view
8. WRITTEN COMMUNICATION (Clear, concise, organized)	• Written communications are inadequate due to error in vocabulary, spelling, and grammar • Communications often raise doubt as to exact meaning • Others must continually seek clarification or correct errors	• Clarity of written communications is inconsistent • Only occasionally able to convey a cogent idea • Extensive editing and correcting is usually required before communications can be dispatched	8. • Writing is clear and concise • Written instructions and reports are readily understandable • Written communications are consistently well organized and grammatically correct	• Written reports can be easily followed by all readers • Communications are succinct and concise, containing only those words necessary to express an idea	• Able to describe complex or technical concepts so well that even the casual reader can readily comprehend the idea • Is consistently chosen for the most important and difficult writing assignments • Is frequently asked to edit the written correspondence of others
9. PROFESSIONAL QUALITIES (Attitude, cooperation, bearing)	• Displays a negative attitude toward the military • Bearing is slipshod and generally reflects carelessness • Totally unable to work with others • Does not accept or practice Air Force standards of bearing, behavior or grooming	• Does not accept AF standards and must be continually reminded to comply • Shows lack of enthusiasm with the success or failure of AF mission • Is aware of shortcomings but makes excuses for them • Frequently unable to work with others • Bearing, behavior and grooming create a very poor impression • Does not enforce Air Force standards of bearing, behavior or grooming	9. • Is aware of and follows Air Force policies and objectives • Remains current on developments and procedures • Cooperates fully with new ideas and policies • Volunteers for additional duties and promotes participation • Bearing, behavior and grooming create a good impression • Effective in working with peers and subordinates	• Practices and actively promotes AF policies among peers and subordinates • Pursues new developments and applies them to existing procedures • Actively promotes participation and willingly accepts jobs that others avoid • Bearing, behavior and grooming create a very favorable impression • Demonstrates exceptional skill in working with others and eliciting their cooperation	• Firm, fair, and uniform in enforcing Air Force policies on bearing, behavior, and grooming • Represents his or her organization and the AF for the most important events • Actively promotes organizational and AF objectives • Bearing, behavior and grooming are outstanding • Demonstrates clearly superior ability to work with others and to elicit their cooperation
10. HUMAN RELATIONS (Equal opportunity participation, sensitivity)	• Openly and knowingly practices discrimination • Uses racial epithets or sexual slurs maliciously • Is deliberately hostile to minorities or members of the opposite sex • Does not show any consideration or concern for other	• Displays very limited sensitivity to equal opportunity policies • Treats minorities or members of the opposite sex markedly different than other personnel • Employs inflammatory or derogatory terms toward minorities or members of the opposite sex • Tends to lack concern for peers and subordinates	10. • Treats all personnel fairly and equitably • Voluntarily participates in activities in support of equal opportunity • Shows concern and is sensitive to needs of others	• Establishes and enthusiastically maintains standards of equal opportunity • Encourages practice of equal opportunity and treatment in all activities • Displays a high degree of sensitivity and concern for others	• Actively demonstrates strong, visible, and credible support of equal opportunity • Is extremely knowledgeable in the area of equal opportunity and treatment • Displays extreme sensitivity and a deep concern in all dealings with peers and subordinates • Is extremely effective in solving human relations problems – solutions always reflect fair and equal treatment

Performance Standards. NOTE: These standards are not to be used or paraphrased as specific examples in section III of AF Form 707. They are simply standards by which the rater can judge which performance rating is supported by the specific example the rater is using. Use of general terms such as these in place of specific examples is inappropriate and is grounds for the report being returned to the rater for reaccomplishment.

Standards.

I. IDENTIFICATION DATA (Read AFR 36-10 carefully before filling in any item)			
1. NAME (Last, First, Middle Initial) Knutson, Charles R.	2. SSN (Include suffix) 000-00-0000FR	3. ACTIVE DUTY GRADE 1st Lt	4. PERMANENT GRADE 1st Lt
5. ORGANIZATION, COMMAND, LOCATION AND PAS CODE 657th Personnel Research Lab Lackland Air Force Base TX (AFSC)	6. ACADEMIC PERIOD 36 days	8. PERIOD OF REPORT FROM 8 Sep 79	THROUGH 14 Jun 80
	7. LENGTH OF COURSE 39 weeks	9. REASON FOR REPORT ☒ FINAL ☐ ANNUAL ☐ DIRECTED	
10. NAME AND LOCATION OF SCHOOL OR INSTITUTION Trinity University, San Antonio, Texas			
11. NAME OR TITLE OF COURSE Graduate Program in Business (final year under AFM 213-1)			12. DUTY AFSC MA

II. REPORT DATA (Complete as applicable)				
1. COURSE HOURS COMPLETED 30	2. COURSE HOURS FAILED 0	3. AFSC AWARDED	4. AERO RATING AWARDED	5. DEGREE AWARDED MA
6. COURSE SUCCESSFULLY COMPLETED (Final report only) ☒ YES ☐ NO (If "NO" give reason)				
7. TITLE OF THESIS Job Analysis - Key to Better Management			8. ACADEMIC FIELD Business Administration	
9. DISTINGUISHED GRADUATE (Final report only) ☒ YES ☐ NO DISTINCTION MADE FOR THIS COURSE				

III. COMMENTS (If applicable or appropriate)

ACADEMIC/TRAINING ACCOMPLISHMENTS (Special achievements related to curriculum, research, communication, etc.)

Use this report in such management areas as promotions, school selections, assignments, and career planning. Use the same care and attention in preparing this report as you do in preparing AF Form 707 (Officer Effectiveness Report). Include in this area a word picture of each student that accurately and completely portrays his or her skills or abilities. Include those comments required by paragraph 7-3.

PROFESSIONAL QUALITIES (Bearing, appearance, conduct)

OTHER COMMENTS

Add any other comments not covered elsewhere. Enter explanation if paragraph 7-2e(9) is used in section II of report.

IV. REPORTING OFFICIAL			
NAME, GRADE, SVC, ORGN, LOCATION Joseph P. Antonucce, Col,USAF 6570th Personnel Research Lab Lackland AFB TX (AFSC)	DUTY TITLE Commander	SIGNATURE Joseph P. Antonucce	
		SSN (Include suffix) 000-00-0000FR	DATE 14 Jun 1980

EDUCATION/TRAINING REPORT, AF FORM 475 (front).

hours completed, (b) course hours failed, (c) any AFSC or Aero rating awarded, (d) degree awarded, (e) was course completed, and (f) comments.

Unfavorable Information Files (UIFs), AFR 35-32. UIFs provide commanders with a repository of substantiated derogatory information concerning the member's personal conduct and duty performance which may form the basis for administrative, personnel or judicial actions. These files are "For Official Use Only" and are closely controlled at all levels to preclude unauthorized disclosure. UIFs may include but are not limited to: AF Form 1137, Unfavorable Information File Summary, documentation concerning placement on the Control Roster, failure to discharge a just financial obligation, misconduct or substandard performance resulting from emotional instability, drug or alcohol abuse or similar deviations from accepted norms of behavior. UIFs are maintained by the CBPO/DPMQA and contain only that unfavorable information which has been verified for file by the individual's commander or higher authority. In addition, a copy of the AF Form 1137 on officers is maintained at MAJCOM level. Exact disposition instructions for the UIF and its contents are contained in Table 35-5, AFM 12-50. Except for documentation concerning court-martial convictions, certain (specified) civil court convictions, drug abuse, records of certain (specified) Article 15 UCMJ actions mandatory for file and Control Roster actions, the documentation must be referred to the member for comment. The notification of "intent to file" optional information in the UIF will be sent by the commander to the member for his or her comments.

WHAT CONDUCT PRODUCES A MORE FAVORABLE RATING?

Of obvious importance to the officer who seeks to obtain good effectiveness reports—as he should—are the factors weighed by the rating official. These become self-evident with an examination of the effectiveness report forms. If, with objectivity, you cannot rate yourself in the more desirable columns, you'd better start a regime of self-improvement quickly.

An Aid to Receipt of Good Effectiveness Reports. Good effectiveness reports which lead to selection for promotion and preferred assignments, must be earned.

Forbes Magazine, a publication for business executives, included the following helpful principles as developed by Rogers & Slade, Management Consultants. Observation of these simple truths will help any officer in the execution of his responsibilities and in doing so will assist him in getting good effectiveness reports.

In my relations with those who supervise me, I will—

(1) Accept my full share of responsibility.
(2) Make sure I know what is expected of me.
(3) Do what is requested in the best manner I know.
(4) Be agreeable when asked to do something difficult or unpleasant.
(5) Be honest—and not try to cover up my errors.
(6) Stand up for my decisions—but admit it promptly if they are shown to be wrong.
(7) Accept criticism in good spirit and not let a few rebukes "get me down."
(8) Point to needed improvements but make sure my ideas are well-thought-out and clearly presented.

In my relations with those under my supervision, I will—

(1) "Sell" the job to be done, and not "pull my rank."
(2) Be firm, but reasonable.
(3) Treat those working under me as human beings—consider their feelings.
(4) Accept proffered suggestions—or explain why they should not be used.
(5) Give full job instructions and not "pass the buck."

(6) Be friendly but not intimate—not play favorites.
(7) Set a good working example—not break rules I expect others to follow.
(8) Back up my people when they are right and give credit when and where due.

An Aid to Avoidance of Bad Effectiveness Reports. Those who have had the opportunity of evaluating thousands of effectiveness reports incident to selections for promotion or elimination have commented upon the patterns which are followed in very many instances by officers whose records are below standard.

Summary. Officers will be helped to avoid bad reports by eliminating any tendency or display of unprofessional conduct including intemperate use of liquor, transgressions with the opposite sex, obscenity or vulgarity; by avoiding the generation of ill-will among seniors, associates, or juniors; by avoiding a reputation of mediocre or undependable performance of duty; and by avoiding a reputation of being a poor credit risk.

An officer may earn splendid reports which will lead to a successful and envied career by establishing a reputation of finding always a way of "getting things done"; by winning the good will of those with whom he or she is surrounded, seniors and juniors alike; and by preparing himself today and tonight for the discharge of the tasks of tomorrow or next week.

HOW TO WORK FOR AN "S.O.B."

In times past there has been much conversation in the service about the difficulties of working under certain officers. The editorial staff considered the matter and decided to attempt to offer something constructive about it. Surely here was a field in which great good was possible. An officer was invited to write his views. His answer is reproduced below.

Dear Sir:

You asked if I'd write a piece on how to work for an S.O.B. The mere thought raised wells of sympathy within me for those unfortunates, who as I did in times gone by, now suffer under a hair-shirt boss who is intolerable. How to live with the Grouchy, the Unreasonably Impatient, the Unfair, the Mean and Vicious—the summation of all the S.O.B.'s I've served under? My enthusiasm was quickened by the thought that I might be able to help those who had fallen on such evil times as to inherit an S.O.B. as unit commander. Ah, the troubles I've known from S.O.B.'s!

Well, I took pen in hand and began to cast about among the multitude of my S.O.B. superiors to select the most horrendous one as my opening illustration, so as better to explain how I survived my painful ordeal while retaining my sanity. With elation upon discovering my perfect example, I began to describe old General Blank, the biggest S.O.B., surely, in all the world. As I wrote, however, I began to recall how, after I grew to know General Blank, I learned of his nerve-shattering war experiences in Asia, of his being finally relieved of his command and sent home more or less in physical collapse. I recalled his singular touchiness about his wife, a nosey young lady who caused no end of trouble within the base, and I wondered if the great differential in ages between Blank and his young wife—25 years—made him thus sensitive. These and other things about Blank occurred to me. On second thought, I decided he was not my candidate for Senior S.O.B.

When I remembered Major Dumguard, though, I knew I had my prize S.O.B.

Dumguard's extreme impatience, his growled answers, his sudden violent angers over nothing, his indifference toward my problems—these and many other characteristics of a bonded 100-proof, aged in wood S.O.B. came to my mind. Yet my resolve faded when I remembered Dumguard's young son hanging so long between life and oblivion with spinal meningitis; the deep shock of his wife's death; the fact that he lagged in promotion far behind his colleagues.

In fact, I must admit it, I can tell no one how to work for an S.O.B., because I've never worked for one. I have worked for men who were suffering from illnesses physical or mental, and who vented symptoms of these on me occasionally. I have worked for men who were bewildered, discouraged, tired, hurt, nervous, miserable, afraid. These too, made my life unpleasant. But I see now that these men were not true S.O.B.'s. They were human beings in some sort of trouble, men in pain, whether from real or fancied ills. Therefore I must turn back to you unmarked pages on S.O.B.'s. I can't recall a one. Is it possible there aren't any?

Very truly yours,

ACTION TO VOID, CORRECT, OR ATTACH SUPPLEMENTAL INFORMATION TO OFFICER EFFECTIVENESS REPORTS AND TRAINING REPORTS

For the vast majority of officers devotion to duty, the application of a reasonable amount of common sense, and the desire to succeed, will result in effectiveness reports which accurately portray actual performance of duty. However, there will be occasions when an individual feels that he or she has been unjustly, unfairly, or inaccurately evaluated. Therefore, in an effort to take every possible precaution to protect the individual, the Air Force has established the Officer Personnel Records Review Board. Any Air Force officer who believes that any part of his or her personnel records is inaccurate, unjust, or unfairly prejudicial to his career may apply to the Board for review. Valid requests by individuals can result in—

(1) Voiding of an effectiveness report or indorsements thereto.

(2) Deletion of any portion of the comments of the rating or indorsing officials on effectiveness reports.

(3) Correction of minor errors such as incorrect dates, AFSC, duty title, and period of supervision of an effectiveness report.

(4) Attachment of a letter to an effectiveness report.

The details of applying for review by the Board are in AFR 31-11.

A LEGENDARY CLASSIFICATION SYSTEM

To a distinguished military leader of a past era of the German Army is attributed classification of officers which applies, we had best admit—even to the Air Force. According to this legend there are only four classes of officers. *First,* the brilliant and industrious. They make the best staff officers, for their talents provide maximum service to commanders. *Second,* the brilliant and lazy. They are the most valuable and constitute the commanders. Their tendency to avoid troublesome and time-consuming detail enables them to retain the perspective which is necessary in the art of making decisions. Their plans tend to the simple, the direct, the most promising for easy success. *Third,* the stupid and lazy. While this group will add little to military luster they can be used on small tasks which are necessary to be accomplished. At least, they will do no great harm. They can be retained and used. *Fourth,* the stupid and industrious. Great damage may result from their actions. Attacking the ill-advised with zeal and energy, they may induce a disaster. They are the most dangerous. They must be eliminated!

EFFECTIVENESS REPORTS—VINTAGE 1813

Extracted from the Adjutant General's School Bulletin, April 1942, and reprinted below are excerpts from an effectiveness report which has been gathering dust these many years. Names of the officers have been changed; and any similarity to persons living or dead is coincidental.

"Lower Seneca Town, August 15th, 1813.

Sir:

I forward a list of the officers of the —th Regt. of Infty. arranged agreeable to rank. Annexed thereto you will find all the observations I deem necessary to make.

Respectfully, I am, Sir,

Yo. Obt. Sevt.,
Lewis Cass"

—th Regt. Infantry

Alexander Brown—Lt. Col., Comdg.—A good natured man.
Clark Crowell—first Major—A good man, but no officer.
Jess B. Wadsworth—2nd Major—An excellent officer.
Captain Shaw—A man of whom all unite in speaking ill—A knave despised by all.
Captain Thomas Lord—Indifferent, but promises well.
Captain Rockwell—An officer of capacity, but imprudent and a man of violent passions.

Captain Dan I. Ware Captain Parker	Strangers but little known in the regiment.
1st Lt. Jas. Kearns 1st Lt. Thomas Dearfoot	Merely good—nothing promising.
1st Lt. Wm. Herring 1st Lt. Danl. Land 1st Lt. Jas. I. Bryan 1st Lt. Robert McKewell	Low, vulgar men, with the exception of Herring. From the meanest walks of life—possessing nothing of the character of officers and gentlemen.

1st Lt. Robert Cross—Willing enough—has much to learn—with small capacity.
2nd Lt. Nicholas Farmer—A good officer, but drinks hard and disgraces himself and the Service.
2nd Lt. Stewart Berry—An ignorant unoffending fellow.
2nd Lt. Darrow—Just joined the Regiment—of fine appearance.

2nd Lt. Pierce 2nd Lt. Thos. G. Slicer 2nd Lt. Oliver Warren	Raised from the ranks, but all behave well and promise to make excellent officers.
2nd Lt. Royal Gore 2nd Lt. Means 2nd Lt. Clew 2nd Lt. McLear	All promoted from the ranks, low, vulgar men, without one qualification to recommend them—more fit to carry the hod than the epaulette.
2nd Lt. John Sheaffer 2nd Lt. Francis T. Whelan	Promoted from the ranks. Behave well and will make good officers.

Ensign Behan—The very dregs of the earth. Unfit for anything under heaven. God only knows how the poor thing got an appointment.

Ensign John Breen Ensign Byor	Promoted from the ranks—men of no manner and no promise.

Ensign North—From the ranks. A good young man who does well.

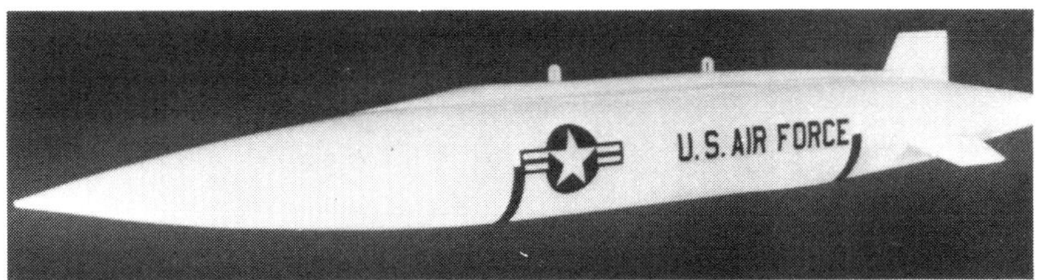

SRAM ATTACK MISSILE

AGM-62A WALLEYE TV-GUIDED MISSILE

16

Educational System of the USAF

Untutored courage is useless in the face of educated bullets.

—George S. Patton, Jr.

A successful Air Force must have well-educated leaders. Consequently, the Air Force has adopted far-reaching programs of military education. These programs apply to all officers of the Air Force, both regulars and members of the Air National Guard and Air Force Reserve.

SERVICE SCHOOLS

Resident Officer Courses. Selected active duty and nonextended active duty officers are eligible to attend the USAF professional service schools. Limited quotas for nonextended active duty officers are reserved annually to the Air Force Reserve and the Air National Guard (See AFR 53-8 for eligibility criteria).

Seminar and Correspondence Courses. Correspondence courses are offered by the USAF professional schools and the Industrial College of the Armed Forces. The Air War College and the Air Command and Staff College offer seminars at many bases in the United States and overseas. These nonresident programs parallel in so far as practicable the resident courses. All eligible active duty and civilian personnel may enroll in these courses (See AFR 53-8 for eligibility criteria).

Service Schools under the Joint Chiefs of Staff. Selected active duty Air Force officers are eligible to attend the regular course of the Joint Service Schools (Armed Forces Staff College, National War College, and Industrial College of the Armed Forces).

CORRESPONDENCE COURSES

University Correspondence Courses. University correspondence courses are offered to military personnel through the cooperation of certain colleges and universities; courses available are listed in catalogs prepared by the Defense Activity for Non-Traditional Education

Support (DANTES). (See AFR 213-1.) This opportunity should be embraced by those officers who lack sufficient credits to obtain a degree and others who wish to obtain college training in selected subjects. Education Services Officers can supply full information.

Service School Extension Courses. The mission of extension courses is to provide a progressive non-resident course of military instruction for personnel of all components (AFR 50-12). In general, these courses consist of a course in each of the several military specialties. Extension courses are also provided for Air War College, Air Command and Staff College, and Squadron Officers School.

Ambitious officers need not await a detail to a school; they may acquire much of the training they would receive by successful completion of these courses.

SPECIAL TRAINING

Education with Industry. Provision has been made for a hands-on opportunity for selected officers to study industry's organization, management, and technology, so that they may apply this knowledge to their Air Force specialty. For ten months the officer will be assigned to a civilian company. Students selected must be volunteers. The civilian industrial organizations in which this training will be accomplished are determined by the Air Force Institute of Technology. (See AFR 50-5.)

Short Courses in Weapons Training. Special training provides experienced officers and airmen with additional, short-course instruction in new weapons systems. It is usually one to two weeks long, and may be taught either by a civilian plant manufacturing weapons (Factory Special Course) or by Air Training Command. Experienced personnel already on active duty can find further details for possible attendance at such courses in AFR 50-9.

Training "On the Job." Utmost importance must be attached to training "on the job," whatever assignment is given to an officer. Here he or she will have best opportunity to apply school training as well as to develop his or her capabilities under his or her own steam. The Department and the major commands will strive to provide for each officer a well-rounded series of assignments which will facilitate progress.

Many assignments will be especially sought by the officer because clearly they fit into the scheme of his professional advancement. But other assignments may be quite unwelcome, initially, because they seem to depart from a preferred course. Here the officer must watch his or her step. The work of the Air Force must be done. It comes first—before the desires or charted plan for individual advancement. Incidentally, duty in such circumstances as Vietnam is the very best "on tho job" training for the young officer.

THE NATIONAL DEFENSE UNIVERSITY

The National Defense University includes the National War College, the Industrial College of the Armed Forces, and the Armed Forces Staff College, military educational institutions operated from the level of the Joint Chiefs of Staff. They are concerned with joint task force and theater operations, national and international plans and policies. The National War College and the Industrial College of the Armed Forces are of coordinate rank with the other Senior Service Schools, and the Armed Forces Staff College with the Intermediate Service Schools.

National War College—Fort Lesley J. McNair. *Mission:* To help prepare selected personnel of the Armed Forces and the State Department for the performance of joint and combined high level policy functions, command and staff functions, and strategic planning duties. For this purpose, it conducts a course of study of Government agencies and the military, economic, scientific, political, psychological, and social factors of power potential which are essential parts of national security.

Prerequisites: Selection of Air Force active duty line officers to attend is made by USAF Senior Service School Designation Boards. Nominations for school attendance are made immediately following selection for promotion to colonel and lieutenant colonel. Nominees compete for selection up to four years after nomination provided maximum PLS/TAFCS for attendance is not exceeded. Attendees must satisfy the following constraints:

a. Have less than 21 years PLS/TAFCS as of 1 January of the fiscal year of class graduation.

b. Have more than 3 years of intervening service since last PCS education assignment as of 1 July of the year of class entry.

c. Have enough retainability to serve at least 3 years following school attendance as of the day of class graduation.

d. Have served at least 2 years on station as of the projected departure date (except officers on short tours who must be within 90 days of completing current tour.)

e. Must not have attended another senior service school. For information concerning the criteria for attendance by nonactive duty AF Reserve and Air National Guard officers, see AFR 53-8.

MAXWELL AIR FORCE BASE, ALABAMA.

f. Be a pinned on Lieutenant Colonel prior to class entry.
Frequency of classes: one per year.

Industrial College of the Armed Forces—Fort Lesley J. McNair. *Mission:* Conduct a course of study in joint and combined organization, planning and operations, and in related aspects of national and international security, in order to enhance the preparation of selected officers for duty in all echelons of joint and combined commands.

Prerequisites: Prerequisites for attendance are the same as those shown for the National War College.

Special Courses. Specially scheduled short courses are conducted at the College or in cities throughout the country for officers of the National Guard, Reserve officers of the Army, the Navy, and the Air Force, and selected executives of industry, educators, and prominent citizens.

Armed Forces Staff College—Norfolk, Virginia. *Mission:* Conduct a course of study in joint and combined organization, planning and operations, and in related aspects of national and international security, to enhance the preparation of selected officers for duty in all echelons of joint and combined commands.

Prerequisites: Selection of Air Force active duty line officers to attend the Armed Forces Staff College is made by USAF Intermediate Service School Designation Boards. Nominations for school attendance are made immediately following selection for promotion to major. Nominees compete for selection up to two years after nomination, provided maximum PLS/TAFCS for attendance is not exceeded. Attendees must satisfy the following constraints:
a. Have less than 15 years PLS/TAFCS as of 1 January of the year of class graduation.
b. Have had at least 3 years of intervening service since last PCS education assignment on the date of class entry.
c. Have enough retainability to serve at least 3 years following school entrance.
d. Have not been selected for promotion to lieutenant colonel.
e. Must not have attended one of the intermediate service schools.
f. TOP SECRET clearance is required.
g. Be a pinned on major prior to class entry.
Frequency of classes: two per year.

SCHOOLS OF THE AIR UNIVERSITY

General. A major effort of all military forces is directed at improving the skills and professional competency of its personnel. Efficient utilization of technical weapons and a perpetual growth of leadership in the profession of arms must be attained. In the Air Force, as in other professions, it has been found that there is no substitute for formal instruction at functional schools.

One must differentiate between the "training" and "professional education" aspects of these Air Force schools. *Training* is defined as instruction in which emphasis is placed upon practice to gain technical skills, endurance, or facility. *Professional education* is instruction and study which has the broader purpose of transmitting principles, concepts, and aspirations. Courses which provide "training" have been assigned to the Air Training Command. Schools designed to further the "professional education" of officers have been assigned to the Air University, as agent of the Air Training Command. The professional education of

Air Force officers takes place in successive steps during their careers in coordination with assignments to broader and more responsible duties.

The Air University employs one unified system to provide selected Air Force officers with the knowledge and abilities essential for increasingly important command and staff positions. To fulfill its far-reaching educational mission, Air University (AU) operates colleges, schools, institutes, centers, and supporting agencies. Its headquarters and most of its major activities are located at Maxwell AFB, Ala., historic site of an early Wright Brothers' flying school. The Air Force Institute of Technology operates at Wright-Patterson AFB, Ohio, and the Extension Course Institute functions at Gunter AFS, Ala. Air Force ROTC detachments are at 141 universities and colleges throughout the United States and Puerto Rico.

Air War College. This is the Air Force's senior service school. It is attended by USAF, Army, Navy, Marine, and foreign officers, plus selected officials from the State Department and other U.S. Federal governmental agencies. Air Force officers selected to attend this 10-month course are chosen from Colonel to Lieutenant Colonel nominees. The school mission is to prepare senior officers for high command and staff positions. Lectures and seminars familiarize students with problem situations, and eminent guest lecturers talk on related subjects. Group problem solving is encouraged through analytical reasoning, reflective thinking, and group discussion, and each student undertakes individual and group research projects. The college also directs the Aerospace Research Institute and the Combined Air Warfare Course.

Air Command and Staff College. This is the middle-rung professional school, lasting ten months, to be taken as early as practicable. Some officers of sister and allied services are chosen, but the great majority are Air Force officers serving in the grade of major. The school mission is to prepare selected officers for the command and staff duties of majors and lieutenant colonels. The school curriculum focuses on command and staff functions, problems of national security, and employment of military forces in support of national security policies. Instruction emphasizes staff communication skills necessary for the graduate to be an effective staff officer. The college stresses free expression of ideas and thought and provides the environment for independent and creative thinking and research.

Squadron Officer School. This is the first rung of the Air Force's professional education ladder, lasts 8½ weeks, and is attended by first lieutenants and captains who have between two and eight years active commissioned service. The school mission is to prepare lieutenants and captains for command and staff tasks required by the Air Force, while providing a foundation for further professional development. The core curriculum focuses on communication skills, leadership, management, and force employment with major emphasis on leadership through student-centered activities. The competitive atmosphere at Squadron Officer School provides the environment for the officer student to analyze and improve his leadership skills through practical application. The total environment of the school encourages the free expression of ideas and creative approaches to problem solving.

Academic Instructor and Foreign Officer School. The mission of the AIFOS is to increase the effectiveness of selected instructor personnel of the Air Force, to raise the Air Force level of communicative skills, and to prepare international officers for assignments elsewhere in Air University. The school conducts the five-week Academic Instructor Course (AIC) several times a year. The Foreign Officer School is in charge of the preparation of international officers for further Professional Military Education Schooling.

Leadership and Management Development Center. The mission of LMDC is to provide instruction in special education courses and to conduct research and to provide consultant services in subject areas for which the Institute has special competence. In carrying out its mission, the LMDC conducts curriculum evaluation and documentary research and offers short courses in the Judge Advocate School and in the Chaplain School. It also conducts a Professional Military Comptroller School, a Professional Personnel Management School, a Unit Historian Development Course, and the USAF Commander's Seminar. The LMDC also has administrative control of the USAF Chaplain Resource Board.

Air Force Institute of Technology. AFIT conducts scientific, technological, and other specialized education to satisfy USAF requirements. Programs at Wright-Patterson AFB include those conducted by the Civilian Institutions Directorate, the School of Engineering, the School of Systems and Logistics, and the School of Civil Engineering. Officers administered by the Civilian Institutions Directorate take on-campus college and university courses under Air Force sponsorship to meet the needs of the service in a wide variety of graduate and undergraduate programs. The Airmen Education and Commissioning Program (AECP) provides an avenue for outstanding airmen to complete their college education and go on to earn commissions at Officer Training School (OTS).

Air Force ROTC. The Air Force ROTC program, directed by Air University, makes available qualified, selected cadets as potential USAF officers, and is the major source of Air Force officers. The AFROTC Airman Scholarship and Commissioning Program and the AFR 39-10 Early Release Program provide opportunities for outstanding airmen to complete their college education and earn commissions. More information concerning the Air Force ROTC Program can be found later in this chapter.

Extension Course Institute. ECI is the Air Force's correspondence school for military subjects, in contrast to the Defense Activity for Non-Traditional Education Support (DANTES, at Pensacola, Florida), which provides catalog listings of high school, technical, and college academic correspondence courses available

HEADQUARTERS BUILDING OF THE COMMAND AND STAFF COLLEGE.

through certain colleges and universities. ECI offers more than 400 courses covering Air Force career areas, and five general courses based on the curricula of resident Air Force general service schools. Members of the Regular Air Force, Air Force Reserve, Air National Guard, Civil Air Patrol, members of other services, and civilians employed by the Air Force comprise the student body of more than 274,000. ECI has awarded more than 4,750,000 diplomas.

Community College of the Air Force. The Community College of the Air Force, with 115,000 registrations through 1980, offers means for individuals to acquire associate in applied science degrees. By the end of 1980, more than 11,000 degrees had been awarded through the CCAF.

Officer Training School. Officer Training School (OTS) is a short lead-time officer procurement program conducted by Air Training Command at the Lackland Training Annex, San Antonio, Texas. Its mission is to procure and train qualified college graduates to meet specific Air Force officer requirements. Sixteen twelve-week classes are conducted each year. The curriculum provides the basic knowledge and skills required of newly commissioned officers. OTS provides commissioning opportunities to graduates of all accredited colleges and to enlisted personnel qualified for the Airman Education and Commissioning Program. Pilot candidates undergo an additional three weeks of flight instruction to determine their qualification for Air Force pilot training.

Air University Library. The AU Library provides complete bibliographic and reference service for students, faculty, and staff personnel. Books and bound journals number about 375,000, and military documents more than 500,000 items, including complete files of regulations, manuals, and directives in the authority section. There is an extensive Audio-Visual Center, a modern Reading Laboratory, and a comprehensive Cartographic Information Division with 580,000 maps.

THE UNITED STATES AIR FORCE ACADEMY

The United States Air Force Academy was authorized by Act of Congress and approved by the President early in 1954, to provide a professional academy to supply a nucleus of officers trained intensively and specifically for lifetime careers in the Air Force.

The official name is the United States Air Force Academy. (It may be called the Air Force Academy but do *not* call it the Air Academy.) Those admitted to the Air Force Academy are called Air Force Cadets.

The Air Force Academy is located 8 miles north of Colorado Springs, Colorado. The Air Force Academy is built on 17,900 acres of ranchland in El Paso County, just east of Pike National Forest and 60 miles south of Denver, in the southeastern part of Colorado. Climate of the area is semi-arid with cool summers and moderate winters. Located in the north part of the site, at 7,100 foot altitude, is the main complex of the Academy where the cadets live and attend classes. Included in this area are the chapel, cadet dining hall, academic building and library, science laboratories, cadet social center and dormitories, gymnasium, parade and drill fields, and athletic fields. Nearby is the Air Force Academy Prep School.

The first class of cadets graduated in June 1959. Upon graduation from the four-year course, cadets receive a bachelor of science degree and commissions as second lieutenants in the Regular Air Force.

UNITED STATES AIR FORCE ACADEMY.

Eligibility Qualifications. There are five specific eligibility requirements for admission to the Air Force Academy. They are:

1. Must be a citizen of the United States.
2. Must be of good moral character.
3. Must have attained the age of 17 years and must not have passed the age of 22 years as of 1 July of the year admitted to the academy.
4. Must be unmarried and have no dependent children. Any Air Force Cadet who marries will be discharged from the academy.
5. Must be in good physical condition.

Educational Qualifications. Any candidate who has graduated from a secondary school in the top 40 percent of his class and has attained good grades, or will have so graduated by the date of his or her admission to the Air Force Academy, has a reasonable chance of qualifying academically. Completion of either the College Board Admissions Testing Program (ATP) or the American College Testing Program (ACT) is required. As a general guide, an individual seeking admission to the Air Force Academy should have as a scholastic background a minimum of 15 units credit including four units each of math and English. (A unit of credit is defined as the satisfactory completion of a year's study in any subject in a secondary school.) In addition, a demonstrated leadership potential is required which includes participation in athletic and nonathletic extracurricular activities.

AIR FORCE FLYING SCHOOLS

The Air Training Command. All undergraduate flying training is conducted by the Air Training Command. Pilot and navigator training is conducted entirely under military auspices at Air Force bases under the supervision of Headquarters, Air Training Command, Randolph AFB, Texas. Currently, undergraduate pilot training is conducted at Columbus AFB, Miss.; Laughlin AFB, Texas; Reese

AFB, Texas; Vance AFB, Okla.; Williams AFB, Ariz.; and Sheppard AFB, Texas. Navigators are trained at Mather AFB, Calif.

AIR FORCE TECHNICAL SCHOOLS

As the Air Force, along with the world at large, grows increasingly technical in its operations, the need for technically trained officers increases also. The Air Training Command operates a number of Technical Training Schools to which officers may be assigned. Courses are designed to advance the officer's proficiency in his or her career field whether jet aircraft maintenance, electronics, missile maintenance and operation, intelligence, or hundreds of other such technological and support field careers. The main centers of such training are at Chanute AFB, Keesler AFB, Sheppard AFB, and Lowry AFB. Courses vary in length.

Aerospace Medical Division. The Air Force School of Aerospace Medicine had its origin in "The Medical Research Laboratory and School of Flight Surgeons" at Hazelhurst Field, Long Island, New York, where it was established in May 1919. Seven years later it was moved to Brooks Field, Texas, and designated as "The School of Aviation Medicine." When Randolph Field was completed it was moved to that base.

This school, with headquarters now at Brooks Air Force Base, instructs all flight surgeons for the Air Force, all flight nurses for the Air Force and Navy, and other technicians of the Air Force medical team. Certain courses are open to airmen as well. Students have come to this school from 31 nations.

Instruction is only part of the school's mission. It also conducts an extensive research program in aerospace medicine, seeking methods of prevention and cure of ailments brought about by our high speed invasion of the air, and of space farther and farther from the earth. Research into the selection standards for flying personnel, aiding and improving innate abilities for flight, improvement of flight equipment, and air evacuation procedures are additional fields of study.

AIR FORCE TEST PILOT SCHOOL

The Air Force has established at Edwards Air Force Base, California, a Test Pilot School for the training of flight test officers. The training program, one year long, will turn out 24 test pilots, 6 test navigators, and 10 test engineers per year. To qualify for admission at the school, officers must hold bachelor degrees in engineering, one of the physical sciences, or mathematics. Rated officers must have flying experience including the following: 1,000 hours in supersonic or jet aircraft or in helicopters or one year as aircraft commander in a multi-engine aircraft. The maximum age is 32 years. Two classes per year (July and January) will be conducted, each comprising 25 students. Graduates will go into flight test duties. Application to attend the school may be made to AFMPC/RPF, Randolph AFB, Texas 78150.

VETERANS EDUCATIONAL ASSISTANCE PROGRAM (VEAP)

VEAP was established in 1977 as a replacement for the Vietnam Era GI Bill. Participants in VEAP have every dollar they contribute matched by the Veterans Administration on a 2 for 1 basis. Participants can contribute monthly amounts of from $25 to $100 in $5 increments. The total member contribution is limited to $2700 with the VA paying $5400 for a maximum of $8100. The program can be used after 6 years or the completion of an initial tour of service.

VIETNAM ERA GI BILL

Personnel serving on active duty prior to 1 January 1977 are entitled to educational benefits under the Vietnam Era GI Bill. Under this program individuals may receive up to $342 per month for full-time study at VA approved institutions. Additional amounts are provided for each dependent of an eligible member. The program may be used in-service. Current legislation terminates these benefits on 31 December 1989.

THE AIR FORCE RESERVE OFFICERS TRAINING CORPS (AFR 45-48)

Curriculum. The Air Force ROTC education program provides pre-professional preparation for future Air Force officers. It is designed to develop men and women who can apply their education to their initial active duty assignments as Air Force commissioned officers. In order to receive a commission, AFROTC cadets must complete all requirements for a degree in accordance with university rules and regulations as well as completing certain courses specified by the Air Force. AFROTC courses are normally taken for academic credit as part of a student's electives. The amount of credit given toward a degree for AFROTC academic work varies with different colleges and universities. The two major phases of the curriculum are the General Military Course and the Professional Officer Course.

The General Military Course (GMC). The GMC is a two-year course normally taken during the freshman and sophomore years. The course covers two main themes—the development of air power, and the contemporary Air Force in the context of United States military organization.

General qualifications for entry into Air Force ROTC GMC are: (1) must be a full-time student at a college which offers Air Force ROTC; (2) must be a United States citizen (for scholarship appointment); (3) must be in good physical condition; (4) must be of good moral character; (5) must be at least 14 years of age for GMC entry; and, (6) must be at least 17 years of age for scholarship appointment.

The Professional Officer Course (POC). The Professional Officer Course is a two-year course of instruction in Aerospace Studies, normally taken during the student's junior and senior years. The curriculum covers Air Force management and leadership and American defense policy.

Qualifications for admittance to the Professional Officer Course are: (1) United States citizenship; (2) good moral character; (3) at least 17 years of age with parent/legal guardian consent; (4) two academic years remaining (undergraduate, graduate, or combination of the two); (4) qualification on the Air Force medical examination; (6) interview and selection by a board of Air Force officers; (7) successful completion of a four-week field training course if a Four-Year Program cadet, or a six-week field training course if a Two-Year Program applicant; (8) qualification on the Air Force Officer Qualifying Test; and, (9) if pilot or navigator candidate, able to fulfill all commissioning requirements prior to age 26½. Scholarship recipients must fulfill commissioning requirements before reaching age 25 (a waiver of previous military service up to age 29) on June 30 in the estimated year of commissioning. Nonscholarship students must fulfill all commissioning requirements prior to age 30.

Programs. Two routes to an Air Force commission are available to college students in the AFROTC. Entering students may enroll in the Four-Year Program, and students with at least two academic years remaining in college may apply for the Two-Year Program.

Four-Year Program. College students may enroll in the Four-Year Program on campuses where it is offered and on those campuses where cross-enrollment agreements are in effect which allow students not on the main campus of the hosting university to enroll in their AFROTC programs. Students may enroll in the General Military Course in the same manner and at the same time as other courses. Enrolling in the GMC is with *no military obligation.*

During their freshman and sophomore years, students enroll in the GMC. They then may compete for entry into the Professional Officer Course which is normally taken during the last two years of college. Selection into the POC is highly competitive and is based upon qualification on an Air Force medical examination, scores achieved on the Scholastic Aptitude Test (SAT) or American College Test (ACT), scores achieved on the Air Force Officer Qualifying Test (AFOQT), college major, grade-point average, successful completion of a four-week field training course at an Air Force base, and the recommendation of the Professor of Aerospace Studies.

Two-Year Program. The Two-Year Program consists of the Professional Officer Course, the last two years of the Four-Year Program. It is designed to provide greater flexibility to meet the needs of students desiring Air Force opportunities. The basic requirement is that applicants have two academic years remaining at either the undergraduate or graduate levels, or a combination of both.

After being nominated by a Professor of Aerospace Studies, applicants seeking enrollment in the Two-Year Program are evaluated on scores achieved on the SAT/ACT, scores achieved on the AFOQT, the Air Force medical exam, and a personal interview by a board of Air Force officers.

Applicants must successfully complete a paid six-week field training course at an Air Force base during the summer. Those meeting all requirements may then enroll in the Professional Officer Course. When applicants are enrolled in the POC, they enlist in the Air Reserve which enables them to receive a $100 nontaxable allowance each month.

Flight Instruction Program. Air Force ROTC pilot candidates not yet certified as private pilots participate in the Flight Instruction Program (FIP). The program is conducted during the last year of AFROTC, and it is at Air Force expense. Flying lessons are taken at a Federal Aviation Administration (FAA) approved civilian flying school near the respective campus.

Air Force ROTC Nurse Program. Student nurses pursuing at least a baccalaureate degree at a National League of Nursing-accredited school of nursing or certain state-approved schools of nursing are eligible to enroll in either the Four-Year or Two-Year Programs. They may also compete for two-year scholarships which will finance the last two years of college.

Air Force ROTC nurses also compete for entry into the nurse/supervisor/internship program as their initial assignment. They enjoy a marked advantage when competing for USAF sponsored programs such as nurse anesthetist, flight nurse, midwifery, and ob-gyn practitioner as well as graduate education programs.

Enrollment procedures are the same as for entry into the Four-Year and Two-Year Programs. Student nurse cadets or student nurse applicants must meet a minimum grade-point average of at least 2.50 or higher on a 4.00 scale.

Cadet Pay and Other Benefits. An AFROTC cadet is entitled to many of the benefits received by active Air Force personnel. Social and other extracurricular activities, together with leadership and academic training, are intrinsic to Air Force ROTC. But there are also more tangible benefits.

The Air Force provides all AFROTC uniforms and textbooks for on-campus courses and field training. Room and board as well as a salary are paid during the summer field training. All cadets enrolled in the Professional Officer Course receive a nontaxable subsistence allowance of $100 each month during the school year.

All scholarship cadets receive the nontaxable subsistence allowance in addition to tuition, laboratory and incidental fees, and reimbursement for textbooks as long as they remain on scholarship status. Cadets in the Professional Officer Course may travel free on military aircraft on a space-available basis. Cadets in the Flight Instruction Program receive up to 25 hours for free flight training while in school.

Credit for Previous Military Training. The Professor of Aerospace Studies may grant credit toward completion of the General Military Course for previous military experiences. (See AFR 35-48.)

College Scholarship Program. AFROTC Scholarships are available to qualified applicants in both the Four- and Two-Year Programs. Each scholarship provides full tuition, laboratory and incidental fees, and full reimbursement for curriculum-required textbooks. In addition, scholarship cadets receive a nontaxable $100 subsistence each month during the school year while on scholarship status. This subsistence does not begin until enlistment in the Air Force Reserve and enrollment in the AFROTC College Scholarship Program.

High School Students. Competitive four-year scholarships are available primarily to high school seniors/graduates who will pursue specific scientific/technical degrees (e.g., engineering, mathematics, physics, etc.); limited numbers of scholarships are available to those who will enroll in certain nontechnical degree programs (e.g., business administration, accounting, biology, chemistry, etc.).

Applicants for four-year scholarships are evaluated on the basis of: (1) achievement on the SAT or ACT; (2) high school academic record; (3) recommendation from a high school official; (4) record of extracurricular activities; and, (5) personal interview by a representative of AFROTC.

All scholarship finalists are scheduled for, and must successfully pass a medical examination. Final requirement for activation of a scholarship is acceptance at a college or university which offers AFROTC and enrollment in the program.

College Students. Other scholarship opportunities exist, on a competitive basis, for students already enrolled in college. These programs may vary slightly each year so interested students should contact the Professor of Aerospace Studies at a college offering AFROTC for current and specific information. Scholarship applications are made directly to the PAS.

Applicants for AFROTC scholarships are selected on the basis of the "whole-person" concept that includes both objective (e.g., grade-point average, etc.) and subjective (e.g., interview evaluation, etc.) factors.

Air Force ROTC Pre-Health Professions Programs (HPP)—Premedical and Preosteopath. An AFROTC pre-Health Professions Program in selected medical areas is offered to encourage students to earn commissions through Air Force ROTC and go on to acquire doctorates in health career fields. Men and women college students and active duty Air Force airmen who are pursuing a medical or osteopathic degree are eligible to compete in the pre-Health Professions Program which includes scholarships. Pre-HPP members will be commissioned second lieutenants in the Air Force upon completion of AFROTC and baccalaureate degree requirements.

Additional tuition assistance for graduate level health schooling expenses is guaranteed under the auspices of the Armed Forces Health Professions Scholarship Program for pre-Health Professions Program graduates upon acceptance to their appropriate graduate level school. This scholarship will sponsor the remaining health professions schooling.

Air Force ROTC Airman Scholarship and Commissioning Program (ASCP). Enlisted members of the Air Force may apply for college scholarships available to them through the Air Force ROTC Airman Scholarship and Commissioning Program. The scholarships cover costs of tuition, laboratory and other incidental fees, and books and provide a subsistence allowance of $100 a month while the cadet is enrolled in Air Force ROTC and pursuing an academic curriculum leading to a degree and second lieutenant's commission.

Scholarship application and eligibility requirements are contained in AFR 53-20. Base education offices process applications and have the latest information on the program which culminates in a discharge of the airman from active duty, enrolling in AFROTC on campus, and earning a commission.

Veterans. Honorably discharged veterans who meet the same criteria as other students may also enter the Four-Year Program at the start of their junior year. The required four-week summer camp may be taken later in the program. Veterans may qualify for the G.I. Bill or VEAP in addition to scholarship money, the tax-free allowance, and other financial aid to which they are entitled.

Early Release Program for Active Duty Air Force Airmen. Active duty Air Force airmen may be released from active duty early for the purpose of enrollment in the Air Force ROTC Professional Officer Course and ultimately receive an Air Force commission. Airmen who can complete degree requirements in no more than two years and are accepted by an institution offering AFROTC, may apply for early release following award of an AFROTC enrollment allocation.

Application and eligibility requirements are contained in AFR 39-10. Air Force ROTC detachments process applications and insure that all eligibility requirements can be met. Interested airmen should contact the nearest detachment in order to obtain the latest information on early release to enroll in the Air Force ROTC POC.

Accepted applicants are eligible for all AFROTC benefits. Upon completion of all AFROTC and degree requirements, cadets are commissioned as second lieutenants in the Air Force.

Women in Air Force ROTC. Air Force ROTC has become a prime source of commissioned Air Force women officers. Women compete equally for scholarships. If selected for a scholarship, they receive $100 each month, tuition costs, laboratory fees, and reimbursement for textbooks.

Women cadets complete the same courses on campus as men cadets and have the same opportunities for cadet leadership. A limited number of pilot, navigator, and missile positions are made available on an annual basis.

Military Obligation. AFROTC cadets who complete the program are appointed second lieutenants in the United States Air Force Reserve. As officers, they incur active duty obligations as follows:

(1) Nonflying officers serve four years;

(2) Pilots serve six years after completion of flying training;

(3) Navigators serve five years after completion of navigator training;

(4) Pre-Health Professions Program members must accept an Armed Forces Health Professions Scholarship, if offered, and will normally incur an additional commitment depending upon the time required to complete the appropriate health professions school. Graduates who fail to gain entrance into the appropriate health professions school by their graduation date will be called to active duty for four years; and,

(5) Nurse graduates must agree to accept a commission in the Air Force Nurse Corps. Two consecutive licensing exams are permitted, if required. Nonscholarship nurse cadets who twice fail the licensing exam will be afforded the option of requesting active duty as a line officer in lieu of mandatory release. Scholarship nurse cadets who twice fail the licensing exam will serve four years as a line officer, the same as any other nonflying AFROTC graduate.

Marriage. Cadets may be married while enrolled in the program. All applicants must agree to accept any world-wide active duty assignment and in some cases these may be "unaccompanied tours"; however, the Join Spouse program is designed to assign both husband and wife in the same area if both are military members.

Educational Opportunities. Certain qualified AFROTC cadets may request an educational delay in reporting to active duty in order to complete graduate work or a professional school at their own expense.

Minuteman Educational Program. Cadets selected for duty as missile launch officers in the Minuteman Missile Program will have the opportunity to participate in the Minuteman Education Program which can lead to a master's degree. This program pays full tuition and no additional service commitment is incurred.

Tuition Assistance. Most Air Force bases have extensive on-base graduate college programs. Officers may take classes on their own time with the Air Force paying 75 percent of the tuition costs. Any service commitment runs concurrent with other obligations.

Air Force Institute of Technology (AFIT). Graduate study for officers in master's as well as doctorate programs may be completed through AFIT. Selected outstanding AFROTC graduates majoring in engineering, scientific, and technical fields may enter through this AFIT program.

Extension Course Institute. A wide range of correspondence courses are available through the Extension Course Institute to all Air Force personnel. Courses available include both professional military and technical programs.

How to Apply for Air Force ROTC. Students applying for the Four-Year Program register for Air Force ROTC in the same manner and at the same time as they register for other college courses.

Students interested in applying for the Two-Year Program must contact the Professor of Aerospace Studies at a campus where AFROTC is offered. Application must be made early in the academic year preceding field training. The student must have two academic years of study remaining as an undergraduate or graduate student after completion of the six-week field training.

For more information on the AFROTC program at a specific college or university, write to the Professor of Aerospace Studies or the Air Force ROTC Admissions Counselor at that school.

SELF-EDUCATION

Self-improvement and study must be a part of the program for any officer who aspires to high success. Attendance at schools and performing the required duties of an assignment are not enough. The working day is short, eight hours or less in many instances. The profitable use of leisure time is an important factor and healthful, interesting recreation is certainly part of it. The development of skills through selected hobbies may be invaluable. The development of real enjoyment in planned reading will be of great aid to the ambitious officer. Current fiction, biography, and serious books of the day constitute a part of it. The officer must keep abreast of events and be able to talk intelligently with intelligent people, both military and civilian. The study of military history is a most important part of a wise program. An amazing number of officers who have achieved high position have been thorough students of history, with emphasis upon military history. Study of contemporary world history should be required. Self-study of selected subjects to enhance or broaden a professional career should be included. The *Air University Suggested Professional Reading Guide* provides a systematic program of selective reading.

Study of a selected foreign language is becoming necessary, since the Chief of Staff has stated the requirement for each officer to become proficient in reading, writing, and speaking at least one foreign language. The officer who chooses to plan the use of some small part of his leisure time, not overlooking the need for recreation and enjoyment of family, will gain a material advantage over his or her peers who plan and act less wisely. There are small things in which officers should be proficient which can be mastered quickly. Ability to type letters and reports so that they have a professional makeup and appearance will get over the frequently encountered obstacle of the lack of competent clerical help, especially in the field. Knowledge of the elements of bookkeeping and accounting will serve all officers well on all manner of assignments.

There are other types of knowledge becoming necessary for the officer in this modern technological age. In studying career field requirements the Air Force Educational Requirements Board has recommended certain disciplines that officers should master, either in college or afterwards. They include oral and written communication, logic, basic statistics, basic college mathematics, probability theory, introduction to electronic data processing, basic personnel management, and human relations. In most cases the officer will have to pursue these subjects on his own time in an off-duty education program.

The Air Force authorizes attendance for participation at meetings of technical, scientific, professional, or similar organizations when attendance contributes to Air Force missions or programs, and provides a medium for exchanging technical data (AFR 30-9). Other provisions authorize officers to accept graduate fellowships offered by recognized foundations.

PROFESSIONAL ASSOCIATIONS

Listed below are the names of associations and societies of possible interest to Air Force officers. The list is by no means inclusive.

Aerospace Medical Association, Washington National Airport, Washington, DC 20001.
Air Force Aid Society (AFAS), 1117 N. 19th St., Arlington, VA 22209.
Air Force Association (AFA), 1750 Pennsylvania Ave. NW, Washington, DC 20006, $13 per year (includes subscription to *Air Force Magazine*).
Air Force Historical Foundation, (AFHF), Bldg. 361, Bolling AFB, DC 20332, $12.50 per year for junior officers (includes subscription to *Aerospace Historian*).
American Military Institute, (AMI), Eisenhower Hall, Kansas State University, Manhattan, KS 66506, $20 per year (includes subscription to *Military Affairs*).
American National Red Cross (ARC), 17th and D Sts. NW, Washington, DC 20006.
Armed Forces Communications and Electronics Association (AFCEA), 5205 Leesburg Pike, Falls Church, VA 22041.
Company of Military Historians (CMH), N. Main St., Westbrook, CT 06948 (includes subscription to *Military Collector and Historian*).
National Defense Transportation Association (NDTA), 910 17th St. NW, Suite 301, Washington, DC 20006, $25 per year (includes subscription to *Defense Transportation Journal*).
National Guard Association of the United States, One Massachusetts Ave, NW, Washington, DC 20001 (includes subscription to *The Guardsman*).
Reserve Officers Association (ROA), One Constitution Ave. NE, Washington, DC 20002 (includes subscription to *The Officer*).
Society of American Military Engineers (SAME), 607 Prince St., Alexandria, VA 22314, $20 per year (includes subscription to *Military Engineer*).
The Retired Officers Association (TROA), 201 N. Washington Ave., Alexandria, VA 22314, $15 per year (includes subscription to *The Retired Officer*).

PROFESSIONAL PERIODICALS AND JOURNALS

Following are some of the service and other journals of possible interest to Air Force officers.

Air Force Magazine (monthly), The Air Force Association, 1750 Pennsylvania Ave. NW, Washington, DC 20006, $13 per year with membership in AFA.
Air Force Times (weekly), 475 School St., Washington, DC 20024, $32.50 per year.
Airman (monthly), USGPO, Washington, DC 20402, official USAF periodical distributed within the Air Force.
Air University Review, USGPO, Washington, DC 20402, distributed within the Air Force.
Armed Forces Comptroller, Box 91, Mt. Vernon, VA 22121, $7 per year.
Armed Forces Journal (monthly), 1414 22nd St. NW, Suite 303, Washington, DC 20037, $18 per year.
Aviation Week and Space Technology, 1221 Avenue of the Americas, New York, NY 10020, $39 per year.

17

Career Development

A career of service to the Nation as an officer of the United States Air Force is an unexcelled opportunity for worthwhile achievement, contribution, and recognition.
—General Robert J. Dixon, USAF (Ret.).

The career development of an Air Force officer expresses itself through assignments, classifications, promotions, and education. These four factors constitute and control career progression in every field of Air Force activity. As an Air Force officer, you should understand these factors in order to discharge effectively your part of the responsibility for intelligent management of your career. (The subjects of promotion and education are dealt with in Chapters 14 and 16, respectively.) Each of the four controlling factors is dominated by one over-riding consideration, the needs of the Air Force. The Air Force wants your career to develop logically and efficiently, but cannot support this objective at the expense of failing to meet military requirements. Remember that the Air Force's only product is national security and unlike other businesses, we cannot afford to have a bad year.

CLASSIFICATION SYSTEM

The Air Force Classification System provides a means of determining manpower requirements in terms of skills and of identifying personnel qualifications.

Purpose of Classification. A classification system is essential to the management of personnel resources. When reviewing the manpower pool to program personnel to meet position requirements, one finds a group of individuals each of whom is different to some degree in basic aptitudes and ability. Because of these differences, it is necessary to identify 570,000 positions in the Air Force, as well as to identify the individuals qualified to serve in them. This identifying process, known as the Air Force Specialty Coding System, thus provides the Air Force with the capability to determine requirements, both present and future; and enables the system to adjust to changing conditions and requirements. Thus, the basic purpose of classification is to provide personnel managers and users with the guidance needed to place the *right person in the right job at the time required.*

Air Force Specialties. A high degree of similarity exists in many of the established Air Force positions. Actually there are almost as many positions as there are people. It becomes apparent that the labeling or coding system cannot provide identification for so many positions if the system is to be an effective management tool. Closer examination of the positions reveals that, while many are almost identical in scope, others—though related functionally— are similar to a lesser degree. It is evident, therefore, that a single label can identify those positions which require the same basic abilities as well as the personnel possessing the abilities to perform satisfactorily in one or more of the duties. These labels, which have been designated as Air Force Specialties ("AFSs"), are given numerical codes, Air Force Specialty Codes ("AFSCs"). Narrative descriptions are prepared which describe the duties and responsibilities, knowledge, education, training, and other factors required to attain full qualification in the specialties. The concept of grouping together as many positions as practicable forms the basis of our *concept of practical specialization.*

The "Shredout." Other positions, which are basically related but to a lesser degree, may also be grouped together under one AFS, but these positions may require additional coding to identify differences in specific equipment or functions. This identification is provided by "shredouts"—the adding of letter identifiers to the specialty codes. This procedure reduces the unmanageable number of positions to a number that can be readily managed and is more meaningful to the personnel manager.

Officer Codes. The officer codes consist of four digits. The first two digits identify the utilization field in which the specialty is located; the third digit in combination with the first two identifies the specific Air Force Specialty; and the fourth digit identifies the level of qualification of the individual. When this digit is "1," it indicates that the officer is qualified at the entry level of the AFS. If it is a "4," "5," or "6," it indicates that the officer is fully qualified in the AFS. Unlike airman requirements, which are identified in manning documents by use of the specific skill levels, officer requirements are always shown at the fully qualified level of the AFS. Similar AFSs are grouped and referred to as utilization fields. Officers who do not possess the AFS at the fully qualified level are assigned duty at the entry level until they become eligible for award of the intermediate or fully qualified level. This provides the Air Force with an inventory of requirements and available resources by levels of qualification.

Airman Codes. In the airman area, almost 477,000 positions are reduced to approximately 226 specialties to which are affixed approximately 200 additional codes or shredouts, to provide specific equipment or functional identity.

Career Fields. To establish further inter-relationship, the specialties are grouped functionally based on similarity of the skills required. These groups, referred to as Career Fields, are arranged in an overall structure which includes all the skills required by the Air Force in the accomplishment of its current or future missions.

The *concept of functional grouping* recognizes the versatility of our manpower pool, and since we are interested in placing people in jobs they can do, we establish functional groupings based on the common qualifications and aptitudes people must possess to perform successfully in a group of jobs. Under this concept we emphasize that it is not necessarily important that the jobs look alike or are closely structured. It is important, however, that people placed in the jobs have similar qualifications.

What an "Air Force Specialty Code" Stands For. In the application of the principles of our concepts in establishing a specific specialty description in the officer area, the following steps are necessary:

First, the labeling or coding process provides a descriptive title and four-digit code under which are grouped closely related positions. Why are numerical codes established in addition to the descriptive title? The four digit codes for officers, and the five-digit ones for airmen, can be used more readily in statistical reporting, in manning documents, for assignment actions, and more accurately identify the career areas. A review of the codes, AFSCs, in the Communications-Electronics Utilization Field will illustrate an application of the coding system for officer specialties. Take, for example, the code 3024. The first two digits, "30," identify the utilization field in which the five different specialties are located. A third digit, "2," in combination with the first two, identifies the specific specialty—Communications Electronics Systems Officer. A fourth digit, "4," tells the reviewer the level of qualification of the individuals who possess the specialty. When the fourth digit is a "1," we know that the individuals possess potential for full or partial qualification. A "4" indicates that the officer is fully qualified in the operating level AFSC. When a "6" is the fourth digit the individual is fully qualified at the staff level and is qualified to supervise any of the subordinate operating levels. The fourth digit also enables personnel managers to review the personnel inventory and determine the Air Force capability of meeting its requirements for qualified personnel.

Second, the remainder of a specialty description consists of four, or five, parts. First, a brief summarization of the scope of the specialty which is logically entitled the Specialty Summary. Next is the Duties and Responsibilities Section. Here are grouped the broad duties of the specialty and the specific tasks or jobs which make up each of the listed duties. The third section of the description, the Specialty Qualification Section, outlines the mandatory or desirable qualifications of the specialty in terms of knowledge, education, experience, training, and other factors indicating the skills and ability to perform and progress in the specialty. The fourth section of the specialty, the Specialty Data Section, indicates the grade spread of the specialty and a group code from the DOD Occupational Conversion Table, a listing of related military and civil service jobs. As mentioned previously, shredouts or suffixes to the codes are established to identify specific equipment, functional variations, or positions associated with a common specialty. When suffixes are appropriate to a specialty, they are included in a fifth part of the description of the specialty. The suffix becomes an integral part of the AFSC and the specialty description covers the duties and qualifications associated with each shredout of the specialty.

Use of the Classification System. Up to now the philosophy, practices, concepts, and their application in the management and operation of the classification system have been discussed. Now the procedural policies and controls

related to the system should be examined. In establishing each specialty, recognition must be given to career progression opportunities afforded the individuals as well as mission accomplishment. As mentioned previously, the specialty qualification section of a description indicates the qualifications required to adequately perform the duties described. In the case of officers, minimum periods of experience in the duties have also been established; for airmen, minimum periods of demonstrated proficiency have been established prior to progression to each higher skill level. In addition to the minimum periods of experience and demonstration of proficiency, mandatory or desirable schooling has been established as a prerequisite for award of AFSCs.

It should be apparent that the Air Force Personnel Classification System, its concepts and techniques, are based on sound management principles. Contrary to the belief of many Air Force personnel, classification practices and policies do not result from "off the cuff decisions" or changes made merely for the sake of change. The overall system, its concepts, the placement of responsibility, the controls and tools used to evaluate the qualifications of our personnel inventory, have been conceived from the opinions of the major commands and are fundamentally based on sound ground rules. The system possesses the adaptability to respond and adjust readily to meet immediate and long-range requirements and thus provide a workable framework for the management of the most valuable of our assets, the people who insure the accomplishment of the Air Force mission.

More detailed information may be found in AFR 36-1, "Officer Classification Regulations, and AFR 35-1, "Military Personnel Classification Policy."

CAREER AREA	CODE	UTILIZATION FIELD TITLE	AFSC	AFS TITLE
INTL POLTL-MIL AFFAIRS	02	INTL POLITI-COMILITARY AFFAIRS	0216	International Politico-military Affairs Off
DIS PREP	05	DISASTER PRE-PAREDNESS	0516	Disaster Preparedness Staff Officer
			0524	Disaster Prep Off
			1025	Pilot, Helicopter
			D	T/UH-1F/P
			E	CH-3
			F	CH-53
			G	UH-1N
			H	HH-1H
			J	HH-3
			K	HH-53
			Z	Other
			1035	Pilot, Search-Rescue
			B	HC-130
			Z	Other
			1045	Pilot, Transport
			F	C-131
			H	C-135, C-137
			J	C-130
			K	C-140
			L	C-141
			M	C-9
			N	C-5
			Q	T-39
			R	T-43
			Z	Other (includes U-3)
			1055	Pilot, Tactical Airlift
			A	C-123
			B	C-130
			E	C-7
			G	AC-130A/E
			H	C-130 (Cmbt Talon)
			Z	Other
			1065	Pilot, Tanker
			C	KC-135
			D	KC-10
			Z	Other
			1115	Pilot, Ftr
			A	F-101
			C	F-106
			D	F-104
			E	F-105
			F	F-4
			G	F-111
			H	F-5
			K	A-7
			L	A-37
			M	F-15
			N	A-10
			P	T-38B
			Q	F-16
			R	Plt, Sys Opr, F-4
			S	Plt, Sys Opr, F-111
			Z	Other (includes T-33, T-37, T-38, MPQM/QF-100, MPQM/QF-102)
			1135	Pilot, Spec Tactics Ftr
			A	F-105
			B	F-4
			C	EF-111
			Z	Other
			1145	Forward Air Controller
			B	Tac Ftr Qual, OV-10A Plt
			C	Tac Ftr Qual, O-2A Plt
			E	OV-10 Pilot
			F	O-2A Pilot
			H	OA-37 Pilot
			Z	Other

CAREER AREA	CODE	UTILIZATION FIELD TITLE	AFSC	AFS TITLE
			1475	Air Ops Off, Plt, Recon/EW/Abn Comd & Con
			A	TR-1/U-2
			B	RF-4
			C	RB-57
			D	EB-57
			F	SR-71
			K	WC-130
			L	DC-130
			M	EC-135
			N	RC-135
			P	WC-135
			T	E-3A
			U	E-4A/B
			Y	General
			1485	Air Ops Off, Plt, Helicopter/Search-Rescue
			B	HH-1N, HH-1H, HH-43
	14	PILOT (Continued)	C	UH-1H, UH-1N, T/UH-1F
			E	CH-3, CH-53
			F	HH-3, HH-34, HH-53
			N	HC-130
			P	Helicopter S&R
			Q	Fixed Wing S&R
			Y	General
			1495	Air Ops Off, Plt, Air Ops, General
			A	Physiological Support
			B	Aircrew Life Support
			C	Acft Performance Off
			D	Undergrad Pilot Tng
			Z	Other
			1525	Nav-Bombardier, Strat
			A	B-52 (ASQ-38) Nav
			B	B-52 (ASQ-48) Nav
			C	B-52 (ASQ-38) RN
			D	B-52 (ASQ-48) RN
			E	FB-111
			Z	Other
			1535	Navigator, General
			A	T-43
			G	KC-135
			K	HC-130
			Z	Other
			1545	Navigator, Airlift
			A	C-5
			E	C-123
			G	C-130
			H	C-130E (AWADS)
			J	C-135
			K	VC-137
			L	C-141
			Z	Other
			1555	Weapon Systems Officer
			A	F-101
			C	F-4
			D	RF-4
			E	F-111
			F	SR-71
			Z	Other
			1565	Navigator, Recon/Abn Comd & Control
			G	DC-130
			J	WC-130
			K	EC-135
			L	RC-135
(Continued)			M	WC-135
			T	E-3A

OFFICER CLASSIFICATION STRUCTURE CHART.

SCIENTIFIC & DEVELOPMENT	27	ACQUISITION PRGM MGT	2716 2724	Acquisition Mgt Off Acquisition Proj Off
	28	DEVELOPMENT ENGINEERING	2816 2825 2835 2845 2855 2865 A B E Z 2875 A B Z 2895	Staff Dev Engrg Mgr Electronic Engineer Mechanical Engineer Astronautical Engr Aeronautical Engineer Experimental Test Pilot Alft/Tanker/Strat Bmbr Ftr/Interceptor/Recon Helicopter/VSTOL Other Experimental Test Nav Alft/Tanker/Strat Bmbr Ftr/Interceptor/Recon Other Project Engineer
PRGM MGT	29	PROGRAM MGT	2996	Program Manager
COMM-ELECT	30	COMMUNICATIONS-ELECTRONICS	3016 3024 3034 3055 3096	Comm-Elect Staff Off Comm-Elect Systems Off Comm-Elect Maint Off Comm-Elect Engineer Comm-Elect Director
LOGISTICS	31	MISSILE MAINTENANCE	3116 3124 C F G Z 3196	Msl Maint Staff Off Msl Maint Officer BGM-109 LGM-25 WS-133A/M, WS-133B Other Msl Maint Director
	40	AIRCRAFT MAINTENANCE & MUNITIONS	4016 4024 4054 A B 4096	Maintenance Staff Off Acft Maint Officer Munitions Officer Munitions EOD Aerosp Maint Director
COMPUTER SYSTEMS	51	COMPUTER SYSTEMS	5116 5135 A B C D E 5155 5164 5176	Computer Sys Staff Off Computer Sys Dev Off Basic Software (Non-functional) Applications Software (Functional) Data Base Admin Comp Math Techniques Comp Performance Eval Computer Ops Officer Comp Sys Plans & Prgms Officer Computer Sys Manager
CIVIL ENGRG	55	CIVIL ENGINEERING	5516 5525 A C D E F G 5596	Civil Engrg Staff Off Civil Engrg Officer Architect/Architectural Engineering Civil Engineering Industrial Engineering Electrical Engineering Mechanical Engineering General Engineer Civil Engineer Director
CARTOG-RAPHY/GEODESY	57	CARTOGRAPHY/GEODESY	5716 5734	Cartographic/Geodetic Staff Officer Cartographic/Geodetic Officer
LOGISTICS (continued)	60	TRANSPORTATION	6016 6054	Transportation Staff Off Transportation Officer
	62	SERVICES	6216 6224 6234 6244	Services Staff Off Services Ops Off Services Sales Off Food Service Officer

SCTY POLICE	81	SECURITY POLICE	8116 8124	Scty Police Staff Off Security Police Off
SP INVES	82	SPECIAL INVESTIGATIONS	8216 8224	Spec Inves Staff Off Spec Inves Officer
BAND	87	BAND	8716 8724	Band Staff Officer Band Officer
LEGAL	88	LEGAL	8816 8824	Judge Advocate, Staff Judge Advocate
CHAP-LAIN	89	CHAPLAIN	8916 8924	Staff Chaplain Chaplain
MEDICAL	90	HEALTH SVCS MANAGEMENT	9016 9025	Health Svcs Administrator, Staff Health Svcs Administrator
	91 92	BIOMEDICAL SCIENCES	9116 9124 A B C D E F G 9136 9146 9156 A B C G H 9166 9176 *9186 B 9196 9216 9226 9236 9246 9256 9266 B C D 9276 9286	Bioenvmtl Engr, Staff Bioenvironmental Engr General Industrial Hygiene Medical Constr Sanitary Architecture Biomed Engr Other Medical Entomologist Staff Biomed Scientist Biomed Lab Officer Biomed Lab Science Microbiology Chemistry Blood Bank Other Aerosp Physiologist Health Physicist Clinical Psychologist Clinical Neuropsychologist Clinical Social Worker Dietitian Occupational Therapist Physical Therapist Pharmacist Optometrist Biomedical Specialist Audiologist Speech Other Podiatrist Physician Assistant
	93	PHYSICIAN	9316 *9346 A *9356 A B C D *9366 A B C E F G H J L M N 9376 *9386 A B C E G H J K L N R 9396	Staff Clinician Family Physician Family Practice Spec Aerosp Med Physician Aerosp Med Spec Preventive Medicine Occupational Medicine Family Practice Spec Pediatrician Allergy Adolescent Medicine Cardiology Endocrinology Neonatology Gastroenterology Hematology Neurology Infectious Diseases Medical Genetics Nephrology Physical Med Physician Internist Allergy Oncology Cardiology Endocrinology Gastroenterology Hematology Rheumatology Pulmonary Diseases Infectious Diseases Nephrology Nuclear Medicine Emergency Physician

CAREER AREA	CODE	UTILIZATION FIELD TITLE	AFSC	AFS TITLE
	94 95	PHYSICIAN (continued)	*9416	Surgeon
			A	Thoracic Surgery
			B	Colon & Rectal Surgery
			C	Cardiac Surgery
			D	Pediatric Surgery
			E	Peripheral Vascular Surgery
			F	Neurological Surgery
			G	Plastic Surgery
			9426	Urologist
			9436	Ophthalmologist
			9446	Otorhinolaryngologist
			*9486	Orthopedic Surgeon
			A	Hand Surgery
			B	Pediatrics
			*9496	Obstetrician/Gynecologist
			A	Endocrinology
			B	Oncology
			C	Pathology
			D	Maternal-Fetal Medicine
			*9526	Pathologist
			F	Neuropathology
			*9536	Diagnostic Radiologist
			B	Neuroradiology
			C	Nuclear Medicine
			E	Special Procedures
			9556	Dermatologist
			9566	Anesthesiologist
			9576	Neurologist
			*9586	Psychiatrist
			A	Child Psychiatry
			9596	Radiotherapist
MEDICAL (continued)	97	NURSE	9716	Nursing Administrator
			*9726	Mental Health Nurse
			A	Mental Health Nurse Specialist
			9736	Operating Room Nurse
			9746	Nurse Anesthetist
			*9756	Clinical Nurse
			A	OB-Gyn Nurse Practitioner
			B	Pediatric Nurse Practitioner
			C	Prim Care Nurse Prac
			D	Educ Coordinator
			9766	Flight Nurse
			9776	Nurse-Midwife
			9786	Envmtl Health Nurse
	98	DENTAL	9816	Dental Staff Officer
			9826	Dental Officer, Gen
			9836	Oral Surgeon
			9846	Periodontist
			*9856	Prosthodontist
			A	Fixed Prosthodontics
			B	Removable Prosthodontics
			9866	Orthodontist
			9876	Oral Pathologist
			9886	Endodontist
			9896	Pedodontist
	99	BIOMEDICAL SCIENCES (Continued)	9916	Staff Veterinarian
			9925	Veterinarian
			*9936	Veterinary Scientist
			B	Toxicology/Pharmacology
			C	Radiology/Biophysics
			D	Psychology
			E	Physiology/Biochem
			H	Microbiology
			9946	Veterinary Clinical Spec
			A	Surgery
			B	Internal Medicine
			C	Radiology
			D	Pathology
			G	Laboratory Animal
			9956	Veterinary Health Spec
			B	Public Health
			D	Ecology/Environmentology

COMMANDER AND DIRECTOR SPECIALTIES

AFSC	TITLE
0016	Director of Personnel Management
0026	Organization Commander
0036	Director of Operations
0046	Director of Logistics
0056	Comptroller
0066	Air Commander
0076	Planning & Programming Officer
0086	Missile Commander
0096	Director of Resource Management

SPECIAL DUTY IDENTIFIERS

IDENTIFIER	TITLE
0900	Commander, Cadet Sq, USAF Academy
0910	Air Attache
0920	Recruiting Service Officer
0930	Historical Officer
0940	Instructor
0950	Training Commander, Officer Training School
0960	Computer Systems Program Director

PREFIXES

PREFIX	TITLE
A	Commander
B	Radio Frequency Allocation Officer
C	Automated Functional Applications Analyst
D	Automated Systems Program Designer
F	Aircraft Systems Flight Evaluation
G	Nonrated Officer Aircrew Duty
J	Medical Service Resident
K	Instr Plt/Nav/Msl Lch Off/Air Wpns Con/Space Ops Officer
L	Area Specialist
M	Stdn/Flt Examiner/Msl Evaluator/Air Wpns Con Evaluator/Space Ops Evaluator
N	Squadron Operations Officer
P	Parachutist
Q	Advanced Medical Service Resident
R	Medical Service Specialist
S	Fighter/Interceptor Weapons Instructor
T	Technical Instructor
U	Rated Expertise
V	Operational Flying
W	Military Consultant to the Surgeon General
X	Safety
Y	Analytical Studies Officer

REPORTING IDENTIFIERS

IDENTIFIER	TITLE
0001	Patient
0002	General Officer
0003	Student Officer Authorization
0004	Unallotted
0005	Nuclear Weapons Custodian
0006	Pilot Trainee
0007	Navigator Trainee
0101	Senior Medical Student
0102	Medical Service Student (Other)
0103	Medical Education Program Student
0104	Dental Education Program Student
0105	Veterinary Education Program Student
0106	Uniformed Services University of Health Sciences (USUHS) Student
0107	HPSP Medical Student
0108	HPSP Dental Student
0109	HPSP Veterinary Student
0110	Excess Leave Law Student
0111	Funded Legal Educ Prgm Law Student
0112	Nondesignated Lawyer
0113	HPSP Biomedical Science Student
0114	AFROTC Educational Delay Law Student

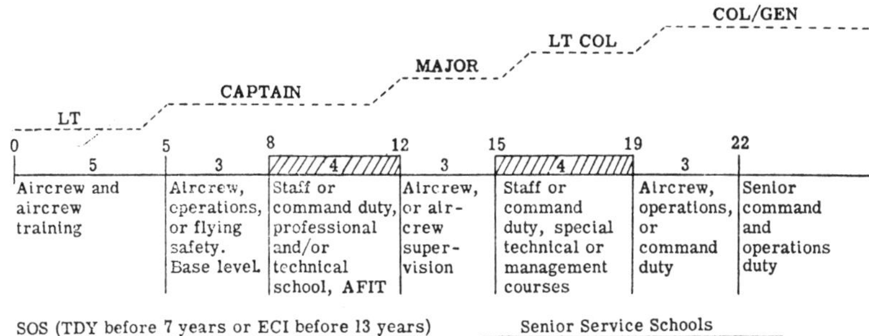

Career Pattern for Rated Officers.
(Fourteen years aircrew/aircrew related duty out of 22 years rated)

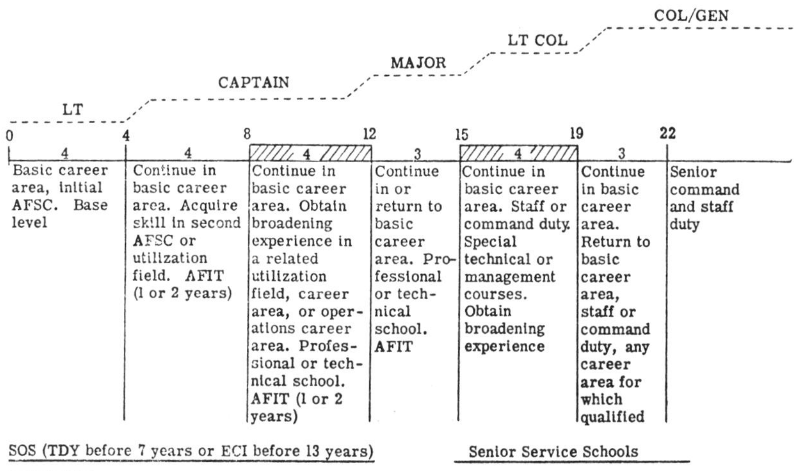

CAREER PATTERNS FOR RATED AND NON-RATED OFFICERS.

ASSIGNMENTS

Assignments are a central point of interest for all officers. Ideally, the needs of the service, professional development, and personal desires are all fulfilled in the assignment process. Personnel support of operational requirements, the attainment of career objectives, and the satisfaction received from one's profession are determined by its operation. Unless it is soundly conceived, understood, and applied by all concerned, there can be little hope of satisfying each of these requirements. This important function concerns you as an individual officer.

Basic Policies. While the assignment of officers in the Air Force is governed by many separate policies concerning a wide range of situations, the following fundamental considerations form the cornerstones and the unifying force of the process.

First among these fundamental considerations is the need for officers of specific grades and skills within Air Force activities. Officers must be ordered to duties where their talents are required for effective mission accomplishment.

The junior officer of today will assume positions of great importance and responsibility in the future. To equip him or her for these responsibilities as he or she progresses in service and rank, the officer is assigned to varied types of duties and schools for experience, professional growth, and technical competence, an understanding of the complexities of the Air Force, leadership ability, and command and executive experience.

The Air Force will accomplish its mission most effectively with officers having high morale. The satisfaction that comes from successfully meeting a challenging assignment and doing the things that one wants to do leads to high morale. For this reason, much emphasis is placed on the desires of the officer when making an assignment decision. Careful and impartial consideration is given to the officer's stated assignment preferences, his or her request for training, location of assignment, etc. Marital and dependent status and family considerations play some part in the final determination of each assignment. Every effort is made to match the *individual's desires* with his *professional needs* and the *needs of the Air Force.*

Other Factors Affecting Assignment. Assignment decisions are influenced by many other policies that stem from the foregoing considerations. Those believed to be of most interest to you are discussed below.

Oversea Service. Assignments outside the continental limits of the United States are a normal part of military service. The general policy is that officers will be selected for oversea duty in such a manner as to assure an equal distribution of such service.

The length of an oversea assignment varies according to location and whether or not you are accompanied by your dependents. At some stations, for example, you are required to serve in an unaccompanied status, i.e., without your dependents. In Germany, the tour is 3 years based on the fact that your dependents are allowed to join you, and services such as commissary, exchange, and hospital are available to support them. You may, however, elect to serve in Germany without your dependents. In this case, the duration of your assignment would be two years.

Another aspect of foreign service of interest to you is the frequency of oversea assignments. Basically, the frequency of oversea assignments is determined by the ratio of requirements between oversea units and those located in the United States. This ratio varies from negligible oversea requirements in some skills to the opposite extreme of all requirements for other skills being in oversea units. Whatever the ratio happens to be for your particular skills, however, you can expect to be assigned overseas in turn; i.e., all other factors being equal, the individual who has been in the United States for the longest period of time goes first.

This sequence is followed unless you volunteer for an oversea assignment. Volunteers normally move to the top of the oversea availability list and are considered for the next oversea requirement for which they are qualified. You may volunteer for oversea duty if you have sufficient service retainability to complete the prescribed oversea tour.

Departmental/Joint Activities Assignments. Certain assignments, by virtue of the nature of work involved and the mission of the organization, are placed in a special category. These positions include the Air Staff; Air Attache; Military Advisory Assistance Groups; Missions; Exchange Program; Joint Service Schools and organizations; other government agencies; joint and combined staffs, and the Secretary of Defense Staff. Only those officers demonstrating high personal qualities and professional competence are selected for assignment to positions in those organizations. These positions usually require officers in the grade of captain or higher. The tour of duty varies with the organization. You may express your desire for such an assignment in your military record (see below).

Education and Training. Education and training opportunities are available throughout your career in the Air Force (see Chapter 16). These opportunities occur at various intervals in your career and are integrated in your assignment pattern based on individual need and your personal desire. Typical career assignment patterns shown elsewhere in this chapter provide for such education.

The principal education and training programs available to you are summarized in AFM 50-5. This manual includes a brief description of the course, its duration, and the prerequisites for attendance. Training courses offered by other services to Air Force officers are also included in this prospectus.

Selection of individuals to attend courses of training varies with each course and is made on a competitive basis. Generally speaking, except for service colleges, selections are made from officers who apply or indicate a desire to attend a particular course in their assignment preference statement. Application procedures are explained in the manual. The policy of selecting volunteers for the various training courses does not preclude the selection of non-volunteers. Frequently, the number of qualified applicants is not sufficient to meet Air Force training requirements. When this occurs, you may be selected to attend a course of training for which you did not apply. In all cases, the needs of the service and the professional development of the individual are determining factors. Selection of officers to attend service colleges, Command and Staff College and higher, is made by a board of officers convened periodically at Headquarters AFMPC. All officers who meet the eligibility requirements are considered, and if selected, are required to attend the course. Officers to attend the Squadron Officers School are selected by the commanders of major air commands based on criteria and quotas received from Headquarters USAF.

Upon completion of flying, technical, or special training you will normally be assigned to duties that utilize the training received. Officers completing postgraduate education normally receive one or more tours in duties closely allied with the field of postgraduate study. Following a course of instruction at a service college, you will resume your normal assignment pattern.

Flying Status. Officers possessing an aeronautical rating are normally assigned to positions involving the operation, command, and supervision of manned aerospace systems. Periodically, selected officers are assigned to other duties in order to broaden their experience. Officers assigned to positions which do not require an aeronautical rating normally will not have an opportunity to fly.

Career Status. Officers in the Air Force may have career status as regular officers (Reg AF) or Reserve officers in indefinite reserve status (IRS), both serving on an indefinite period of active duty. All other officers are Reserve officers with established dates of separation (DOS). Because of the limited period of active duty, an officer in non-career status with a DOS is ineligible for certain assignments, such as service schools.

Active Duty Service Obligations and Commitments. An important aspect of your military service is "active duty service commitments." These operate to determine the length of the active duty in the Air Force you are required to serve.

An active duty service commitment is a period of active military service, effective upon commissioning, which you are required to serve as a result of pre-commission training. The length of this period of active service varies according to the source of commission. For officers who are commissioned through the AFROTC program and are not programmed for flying training, the period of required service is four years.

The active duty service commitment may also result from a change of station assignment or government-sponsored training received as a commissioned officer while on active duty, i.e., after you have received your commission. Duration of the service commitment as a result of training varies according to the length and type of training completed.

Performance of Duty. As in any organization, progression in your career to positions of greater responsibility is influenced to a great extent by the manner in which you perform the duties to which you are assigned. Periodically, your supervisor will prepare an Officer Effectiveness Report (OER) which reflects his or her appraisal of your duty performance (see Chapter 15). It also includes his or her evaluation of your ability to accept and perform duties of greater responsibility. These reports are used in determining your assignments, including selection for postgraduate and service colleges, and your promotion to higher military rank.

Physical Fitness. If you are hospitalized for medical treatment, your assignments subsequent to such hospitalization will be dependent upon your physical condition. Each case is considered individually, and, if physically capable, you will resume your normal assignment pattern.

Humanitarian Reassignment or Deferment. Reassignment solely for humanitarian reasons is not authorized in the military service. The determining factor in the approval of a request for transfer is the need of the service. A request for reassignment to a particular station for humanitarian reasons may warrant giving preference to a person making such request over other qualified persons eligible for reassignment to meet the requirements of the service. The fact that the transfer will increase the morale and effectiveness of the individual are considered in authorizing changes of station in these circumstances, but transfer based on this fact alone is not authorized.

You may request a reassignment, or deferment from reassignment, when conditions warrant such action. Approval is based on an evaluation of the facts and circumstances contained in each request and the availability of a valid

military requirement. How to apply and the information required is explained in AFR 36-20.

Assignment Preference Statement. Each officer has the opportunity to record his or her career objectives in the Career Objective Statement (AF Form 90). This statement, together with your performance record, duty history, and professional and educational background, provides officials with the information necessary to make a decision concerning your assignments. Although requirements of the Air Force take precedence, every effort is made to honor personal preferences. Accordingly, you are urged to take advantage of this opportunity.

In addition, you may apply for special duty assignments to one of the following programs: AFROTC, Civil Air Patrol, Advisors to Air National Guard and AF Reserve units, Office of Special Investigations (OSI), Recruiting Service, Air Force Academy, and Auditor General. Information concerning the type of duty involved, location of assignments, and how to apply is contained in AFR 36-20, Chapter 8.

Counseling. Unit personnel officers are responsible for explaining and making available information on the officer distribution system, officer career management program, and opportunities for technical training and professional education. In addition, your immediate supervisor and your unit commander are available to advise you concerning your career in the Air Force, answer any question about assignments and assist in solving any problem you may have. These officers are your primary contacts and you are expected to use them.

Operation of the System. Assignment of support officers in the Air Force is centralized. AFMPC extracts officers from commands for oversea duty. Information concerning assignment to a specific unit and station and to a particular position is contained in AFR 36-20.

Bibliography. The following publications contain more detail about the subjects outlined above and are available in all organizations.

AFR 35-1. This regulation establishes military classification policies and procedures.

AFR 36-1. This regulation describes the officer classification system and lists the various occupational specialties used in the Air Force.

AFM 50-5. This manual contains information on professional, flying, and technical training available to military personnel, and includes a brief description of each course and eligibility requirements.

AFR 36-12. Release from EAD.

AFR 36-20. Officer Assignments.

AFR 36-23. This regulation provides guidance for utilization and career development of commissioned officers in the active Air Force.

AFR 36-41. USAF Exchange Program with air forces of other governments.

AFR 36-6, Vol. I and Vol. II. Officer Assignment Information Directory.

AFR 36-51. This regulation establishes active duty service commitments for officers of the Regular Force and EAD service commitments for career officers of the Reserve components.

AFR 36-94. Specified Period of Time Contracts.

AFR 53-11. *Air Force Institute of Technology.*
This outlines the role of the Institute of Technology in Air Force educational and training programs, general eligibility criteria for participation, and procedures for admission.

OFFICER CAREER OBJECTIVE STATEMENT (Lt Col and below)
(THIS FORM IS AFFECTED BY THE PRIVACY ACT OF 1974 - See reverse)

Use reverse for remarks.

(1) GRADE	(2) SSAN	(3) NAME (Last, First, Middle Initial)	(4) DATE (Yr & Mo)
04	123-45-6789	SMITH, JOHN A.	1 Sep 79

I. IMMEDIATE ASSIGNMENT OBJECTIVES

A. CONUS	DAFSC	BASE	DUTY TITLE OR POSITION AND LEVEL
FIRST PREFERENCE	(5) 7316	(6) Randolph	7. Base DP
SECOND PREFERENCE	(8) 7316	(9) Lackland	10. Chief CBPO
THIRD PREFERENCE	(11) 7316	(12) Mather	13. Base DP

B. OVERSEA	DAFSC	COUNTRY OR AREA	TOUR LENGTH
FIRST PREFERENCE	14. 7316	(15) Belgium	(16) 36
SECOND PREFERENCE	17. 7316	(18) Germany	(19) 36

(20) OVERSEA VOLUNTEER STATUS *(Check only one)*
[X] NONVOLUNTEER [ ] VOLUNTEER [ ] COT
[ ] EXTENDED TOUR [ ] ANY SHORT TOUR
[ ] ANY LONG TOUR [ ] WORLDWIDE
[ ] NON-CONUS RESIDENT

C. MAJCOM/GEOGRAPHIC PREFERENCES SHOULD YOUR BASE OR OVERSEA VOLUNTEER PREFERENCES BE UNAVAILABLE

(21) CONUS STATE	(22) 1ST CONUS AREA	(23) 2ND CONUS AREA	MAJOR COMMAND (24) 1ST	(25) 2ND	(26) 3RD
California	5	6	ATC	MAC	AFSC

II. SPECIAL ASSIGNMENT PREFERENCES

A. CAREER BROADENING *(Complete only if you desire this assignment next)*

ASSIGNMENT	DAFSC	DUTY TITLE OR POSITION AND LEVEL
FIRST PREFERENCE	(27) 7356	28. MAJCOM Special Services Staff Officer
SECOND PREFERENCE	(29) 7364	30. Base Chief Social Actions

B. RETRAINING *(Complete if you want to leave your current AFSC permanently)*

DAFSC	(31) FIRST PREFERENCE	(32) SECOND PREFERENCE	(33) THIRD PREFERENCE

C. RATED SUPPLEMENT: THIS INDICATES PREFERENCE ONLY AND DOES NOT CONSTITUTE A VOLUNTEER STATEMENT. TO VOLUNTEER COMPLETE SECTION IIA ABOVE.

ASSIGNMENT	DAFSC	DUTY TITLE OR POSITION AND LEVEL
FIRST PREFERENCE	(34)	35.
SECOND PREFERENCE	(36)	37.

D. SPECIAL DUTY APPLICATION: DO NOT COMPLETE THESE BLOCKS UNLESS YOU ARE A VOLUNTEER FOR A SDA ON YOUR NEXT ASSIGNMENT AND MEET THE ELIGIBILITY CRITERIA IN AFR 36-20, CHAPTER 8.

(38) TYPE	(39) DATE DESIRED	40. LOCATION/POSITION
ROTC	June 1980	Det PAS

III. LONG RANGE OBJECTIVES

A. NEXT ASSIGNMENT PLUS ONE

41. DAFSC	42. DUTY TITLE OR POSITION	43. MAJCOM/LEVEL/LOCATION ETC.
0026	Base Commander	Southeast/west

B. ACADEMIC AND PROFESSIONAL MILITARY OBJECTIVES

	ACADEMIC MAJOR OR COURSE	DEGREE	STUDY METHOD	ENTRY DATE	COMPLETION DATE
ACADEMIC	44. Counseling	45. MA	46. Off duty	47. Sep 79	48. May 81
PME	49. National War College		50. Residence	51. Jul 81	52. Jul 82

IV. CURRENT ASSIGNMENT INFORMATION

VOLUNTEER STATUS FOR PCS MOVE WITHIN CONUS

[ ] VOLUNTEER [ ] NO PREFERENCE [X] DESIRE TO REMAIN AT CURRENT STATION

CURRENT	BASE	MAJCOM	DAFSC	AUTOVON AND EXT	DEROS
	Randolph	AFMPC	7316	487-4053	

DATE ARRIVED STATION	DUTY TITLE	SIGNATURE
14 Sep 78	Chief, Career Mgt Section	(Signed)

SAMPLE AF FORM 90, OFFICER CAREER OBJECTIVE STATEMENT (front).

AUTHORITY: 10 U.S.C. 8012; 44 U.S.C. 3101; and Executive Order 9397.
PRINCIPAL PURPOSES: To allow an officer to indicate assignment preference within primary Air Force Specialty and two choices outside primary utilization field. Gives opportunity to indicate if retraining is desired and three choices for new career areas. Permits officer to volunteer for Special Duty assignment. Lets rated officer indicate rated supplement duty preferred. Gives officer three choices for MAJCOM and geographic location of assignment and two choices for oversea area of assignment. Permits officer to indicate long range objective and academic and professional military education desired.
ROUTINE USES: Allows MAJCOM and AFMPC Career Managers to learn the type of assignments training, and long range progression pattern the officer desires. The SSAN is used for identification and records.
DISCLOSURE IS VOLUNTARY. If information is not provided, including SSAN, career managers will be unable to properly assign officers in line with the officers' progression desires.

REMARKS

My first choice for my next assignment is ROTC Det PAS anywhere in the SE or SW. If this is not available, I desire assignment to ATC as a Special Services officer or as a Base/MAJCOM Social Actions officer. If these are not possible I'd like to be considered for a DP or CBPO position preferably in ATC.

SAMPLE AF FORM 90, OFFICER CAREER OBJECTIVE STATEMENT (reverse).

CAREER GUIDES (AFR 36-23)

The Air Force maintains career progression guides for all career fields. The appropriate progression guide is set forth in AFR 36-23 for each career area. A typical example is provided in this chapter to illustrate the method.

CAREER PROGRESSION GUIDE—SCIENTIFIC AND DEVELOPMENT ENGINEERING

General. A brief functional summary of the S&DE Utilization Fields associated with R&D functions is given below. More detailed specialty descriptions are given in AFT 36-23 and in AFR 36-1.

Utilization Fields. Within the S&DE Career Area, officers perform similar activities related to the four utilization fields, which are:

(1) AFSCs 26xx—Scientific. (See chart).
(2) AFSCs 27xx—Acquisition Program Management.
(3) AFSCs 28xx—Development Engineering.
(4) AFSCs 2996—System Program Director.

In general, some officers may be narrowly utilized in the functions they perform while others may be more broadly utilized. The preferred patterns of utilization include activities such as the following.

Scientific Functional Specialization. This field is composed of many different scientific disciplines grouped together because of the commonality of function —namely, to accomplish basic or applied research under controlled and defined conditions in order to produce results that will add to the pool of existing scientific knowledge. Also included are the functions of:

(1) teaching college-level courses in subjects that encompass the academic disciplines that relate to the specialties in this field.

(2) providing specialized technical services in areas that are not directly related to R&D.

Acquisition Program Management Functional Specialization. This field is composed of officers who work the diverse elements associated with a new system's entry into the operational environment. A key element of this process is the delegation by the Air Staff of its executive role of managing all the participating commands and industrial organizations to the system program manager (2996) who exercises management through a system program office. These elements include development, test and evaluation, financial management, configuration management, procurement, production, logistical support, training, and site activation. Officers from the Scientific, Development Engineering, and other areas enter this field by lateral transfer at the captain, major, and early lieutenant colonel grades.

Development Engineering Functional Specialization. This field includes the engineering specialties concerned with the designing, developing, installing, modifying, testing, and analyzing of materials, techniques, methods, systems, or processes. Also included is teaching college-level courses in academic disciplines that relate to developing engineering specialties, intelligence collection, and data handling. A significant portion of the activities accomplished in this field may be described as test and evaluation and are primarily concentrated in Systems Command's test centers, test ranges, and laboratories. Because this

CAREER PROGRESSION GUIDE--SCIENTIFIC

YEAR	PHASE	GRADE	PME	TRAINING	EDUCATION
--29	EXECUTIVE/LEADER	Projected average authorizations	Refer to AFR 53-8 for residence eligibility; AFR 50-12 for correspondence eligibility (see Base Education Officer for more information)	See AFM 50-5 for appropriate current courses	Graduate degree requirements in utilization field. For AFIT eligibility see Base Education Officer for current AFIT program quota.
--28					
--27					
--26		Colonel 39% of authorizations			
--25					
--24				Attendance of Professional Meetings and Symposia.	
--23					
--22					
--21	STAFF	Lieutenant Colonel 13% of authorizations			
--20					Doctorate
--19			National War College, Industrial College of the Armed Forces, Air War College		
--18					
--17				Locally available courses in specialty.	
--16	ADVANCED DEVELOPMENT	Major 22% of authorizations		Research & Engineering Technology (Education with Industry-Lt-Lt Col eligible)	
--15					Master's
--14					Academic Disciplines Include:
--13			Armed Forces Staff College	Armament Development (Education With Industry--Lt-Lt Col eligible)	Graduate opportunities in many Scientific, Engineering, and Management areas. Appropriate Academic Degrees Required for Specialty Qualification Are:
--12			Air Command and Staff College	Short courses as required by specialty.	Masters: Human Factors, Psychology
--11	INTERMEDIATE DEVELOPMENT				
--10					
-- 9		Captain 50% of authorizations			
-- 8					
-- 7					
-- 6					
-- 5			Squadron Officer School		Bachelor's Degree mandatory in appropriate academic discipline, i.e.,
-- 4	INITIAL				Science, Engineering, Mathematics, Physics, Chemistry, Metallurgy
-- 3					
-- 2		Lieutenant 12% of authorizations			
-- 1					

CAREER PROGRESSION GUIDE--SCIENTIFIC (Continued)

ASSIGNMENTS	OPTIMUM PHASE POINTS	YEAR
-Occupy key managerial positions at all echelons of command. -Formulate broad objectives for basic/applied research. -Major program, laboratory management, direction of staff agencies of field units. -R & D senior command and staff. -Command level liaison with other military departments of Government research agencies. -Continue independent laboratory research in special field.	Transition limited number rated career specialists back from operations area.	29-- 28-- 27-- 26-- 25-- 24-- 23--
-Technical management of research programs within field of specialization. -Initiate, manage, provide technical direction for basic/applied research performed under contract with universities or industry. -Technical staff duty at HQ USAF, Department of Defense. -Independent laboratory research. -Technical liaison duty with other Military Departments or Government Research Agencies. -Director of Academic Program at the Air Force Academy in discipline associated with specialty.	Transition limited number of rated officers to operational commands for R&D liaison. Return limited number of rated career specialists to operations area.	22-- 21-- 20-- 19-- 18--
-Technical management of research programs within field of specialization. -Initiate, manage, provide technical direction for basic/applied research performed under contract with universities or industry. -Technical duty in service engineering support activity. -Staff duties in R&D Technical Service engineering support at HQ USAF and MAJCOMs. -Direct/teach academic programs in disciplines associated with scientific specialty. Alternatives at Phase Completion: -Continue in similar work activities. -Continue independent laboratory research in special field. -Transition to Acquisition Program Management Utilization Field (27XX).	Return rated nonspecialists to operations area upon completion of DDA. Scientific officers may transition into the Acquisition Program Management Utilization Field at the Major grade.	17-- 16-- 15-- 14-- 13-- 12-- 11--
-Conduct basic/applied research at laboratory level. -Initiate, manage, provide technical direction for basic/applied research performed under contract with universities or industry. -Instruct at Air Force Academy, Air University in college-level courses in disciplines associated with scientific specialty. -Duty with technical service engineering support functions. Alternatives at Phase Completion: -Those mentioned below: -Transition to Acquisition Program Management Utilization Field (27XX) -Transition of officers into field from the Operations Career Area who possess exceptional qualifications in science via AFIT programs at the MS or PhD levels. -Return of Scientific officers attending AFIT at completion of Development Phase. BROADENING ASSIGNMENT	Transition rated officers into field via AFIT, technical training. Career specialists enter Advanced Degree programs between 2-14 year points. Earlier or later entry based on Air Force requirements.	10-- 9-- 8-- 7-- 6-- 5--
-Work with experienced research scientist at the laboratory level in basic/applied research organizations; system Command's Laboratories, or Aerospace Med. Div. -A number of officers will be assigned to service engineering technical support functions. Entry Requirements: -Bachelor's degree in the appropriate scientific specialty is required (see AFM 36-1). Alternatives at Phase Completion: -Continue in similar work activities. Attend AFIT Graduate Program on a selected basis. Seek parallel assignment such as Technical Intelligence Officer or Air Force Academy Instructor.	Basic AFSC technical course earliest date.	4-- 3-- 2-- 1--

activity continues through the life cycle of a system, some test and evaluation also is done in other commands.

System Program Director Functional Specialization. The System Program Director (Program Manager) directs and provides the executive management supervision for major system acquisitions programs, as delineated by AFR 800-2, Program Management, and attachments through all phases of the acquisition life cycle. This particular AFSC is the prime focus of the system acquisition business and requires extremely well-qualified and well-versed officers.

S&DE Career Progression Charts. A career progression chart has been prepared for each of the four S&DE Utilization Fields. Each career phase is identified on these charts and is correlated with expected military grade. However, in practice there is no sharp distinction in duties between one grade and another. In fact, some lieutenants may manage the acquisition of major items of equipment while a few lieutenant colonels are occupied in bench-oriented rather than management activities. The small number of authorized full colonel positions in the Scientific and Development Engineering Fields may be explained by their transition into the more highly management-oriented Acquisition Program Management Field and the System Program Director Field.

Career Phasing. The normal pattern for the career development of most S&DE officers is one of academic specialization, increased graduate education, and broadening technical experience with increasing responsibility for program supervision and management. Officers are expected to develop executive and managerial skills and abilities. These include both understanding the principles of good management and the many procedures that are specific to managing R&D in an Air Force-industrial-scientific complex. Qualified officers who assume supervisory and managerial responsibilities, in most cases, have an opportunity to attend courses in technical and systems management. Few officers will have identical career patterns. Individual plans must be flexible to permit deviations for technical or specialty training as required.

In general, S&DE careers may be divided chronologically into five phases.

Initial Phase (0 through 3 years). *Assignments.* The initial assignment of a young officer who enters one of the S&DE utilization fields ordinarily will be to an office where he or she will assist more experienced officers or civilians. An effort will be made to provide an opportunity for orientation in the administration or management of technical activities, as well as in the technical activities themselves, so the young officer will be able to choose and to compete for appropriate graduate training. The emphasis in this phase is to develop individual skills by working in the specialty.

Education and Training. Officers are encouraged to apply for admission to the AFIT graduate educational programs to increase their knowledge in their planned career areas. Advanced graduate training is considered essential in many S&DE Utilization Fields due to the uniqueness of the tasks accomplished; therefore, officers in these fields are encouraged to advance their graduate education. Such officers will compete for selection for AFIT on a best-qualified basis.

Officers are encouraged to attend the technical courses that are available for their development and that are directed to their individual talents and abilities. Career officers are encouraged to complete Squadron Officer School.

Intermediate Development Phase (4 through 10 years). *Assignments.* During this phase most career broadening should occur for officers who began their careers in the R&D functions. They will transfer either into S&DE authorizations that are not directly related to systems acquisition or into other utilization fields. S&DE officers who were first assigned to non-R&D functions will transfer into R&D organizations. Officers with appropriate academic qualifications may transfer into the field from other career fields by a duty AFSC change. Others will come in through AFIT programs. Some pilots and navigators will enter this field during this phase for career-broadening tours.

A limited number of officers will be assigned to HQ USAF, combined staffs, and other activities to duties that demand a high degree of knowledge in S&DE fields. Some officers will be assigned to Air University and to the Air Force Academy to teach college-level courses.

Education and Training. In this phase, every effort will be made to give officers opportunities to complete their advanced graduate education according to Air Force requirements and applicant qualifications. Locally available specialty courses and education-with-industry opportunities should be considered in terms of the individual officer's career plan. Those with exceptionally outstanding performance may be selected to attend the Program Management Course at the Defense Systems Management College. All officers are encouraged to complete Squadron Officer School early in this phase, either in residence or by correspondence. Later in this phase, all officers are encouraged to complete the ACSC course by correspondence/seminar. Officers are expected to attend and take part in professional society meetings and symposia presented in their specialty.

Advanced Development Phase (11 through 16 years). *Assignments.* Officers in this phase perform advanced R&D work in their special fields. They supervise junior colleagues who work at the task level. They also supervise relatively large-scale R&D projects performed under contracts with universities or industry; design and manage the performance of selected programs; and serve as staff officers at various headquarters levels.

A large number of requirements for S&DE officers in functions other than R&D continue to exist during this phase. Lateral transfers encourage an exchange between the operational and the R&D communities. A limited number of officers may transit into the S&DE Utilization Fields during the first half of this phase, in most cases qualifying though AFIT advanced degree programs. A limited number of pilots and navigators who served previous career-broadening tours in this field will return for another tour.

Education and Training. In most cases, this phase will be the last opportunity for qualified officers to enroll in and complete AFIT programs at the master's and doctorate levels. Officers are encouraged to reevaluate these opportunities in light of current qualifications and current AFIT selection criteria. Officers may elect to attend education-with-industry programs if appropriate to career plans. Officers in the Acquisition Program Management Field should attend specialized training courses conducted by AFIT or pursue such training on their own time. Those with exceptionally outstanding performance will be selected to attend the Program Management Course at the Defense Systems Management College.

Majors not selected to attend an intermediate level course should complete ACSC by correspondence. Selected officers will attend a senior-level professional school. Others are encouraged to complete Air War College or Industrial College of the Armed Forces by correspondence. Officers are expected to attend and take part in professional society meetings and symposia presented in their specialty.

Staff Phase (17 through 21 years). *Assignments.* Officers in this phase will assume increased responsibilities for technical management in their special fields at the program or system level. At this level they formulate policies and plans for large-scale R&D activities; initiate, manage, and provide technical supervision for R&D work performed under contract; furnish liaison with other military departments and government agencies involved in R&D work. A limited number of the highest qualified officers will continue to conduct specialized research or engineering efforts independently. A limited number of pilots and navigators who previously served career-broadening tours in this field will be selected for return to R&D duties for the latter phases of their careers.

Education and Training. Usually, qualifying graduate education will have been completed before this phase. However, the pace of technological advance demands continuous upgrading to ensure contact with advances as they occur. Officers in R&DE fields are encouraged to enroll in professional short courses and to take graduate training as required. Opportunities to participate in education-with-industry programs are available as required by the officer's career plan. Officers in the Acquisition Program Management Field are encouraged to attend specialized training programs conducted by AFIT.

Officers may be selected to attend either NWC or AWC. Others are encouraged to complete the AWC course through seminar or correspondence. ICAF also may be completed by correspondence (check locally for current criteria). Officers are expected to attend and take part in professional society meetings and symposia presented in their specialty.

Executive or Leader Phase (22 years plus). *Assignments.* Officers in this category are expected to direct, monitor, and supervise Air Force R&D programs as program or laboratory directors or in management positions at intermediate or higher headquarters. A limited number of officers may continue to perform important R&D functions related to their specialized knowledge and experience.

Education and Training. Attendance in special and selected professional short courses is encouraged. Officers are expected to attend and take part in professional society meetings and symposia as appropriate.

PROFESSIONALISM

The ultimate objective of career development is a quality called "professionalism," brought to its highest standard of excellence. All of us readily recognize the meaning of professionalism when we apply the term to a doctor, a lawyer, or an engineer. What does it mean when applied to an Air Force officer? To assess this, it is first necessary to appreciate the unique scope of an Air Force officer's professionalism—the vast field of knowledge, skill, and comprehension that is encompassed in an Air Force career. The military forces of the United States constitute, incomparably, the most stupendous effort of the American people. Fully twenty percent of the federal budget is expended on the military forces: over 2 million people are in the armed services, more than the total workers in the three greatest corporations of U.S. industry. The complicated, expensive tools of our military forces rest on the entire range of U.S. technology and industry, involving deeply other professional areas such as medicine, engineering, law, education, and all forms of science. Moreover, the military forces are an integral, crucial part of U.S. diplomacy and international relations: separation of "peacetime" foreign policy from the existence and strength of U.S. armed forces is permanently outmoded. Finally, the armed forces exert, through their diverse requirements and activities, a major impact on the U.S. economy. Plainly, it is not enough that an Air Force officer be expert in a particular field of Air Force operations: he or she is a part of a vast complex which touches all aspects of American life; must seek to achieve a professionalism which matches the uniquely multiple strands of responsibility and knowledge which characterize his or her profession above all others.

Professionalism begins at less lofty levels than the ultimate responsibilities of the top officers of the Air Force. It begins with *sure* skill and knowledge relative to the particular task assigned to an Air Force officer. It begins when an Air Force officer knows how to perform his or her own job at maximum efficiency, routinely, prudently, smoothly. The reliable sign of the true professional is the calm, seemingly easy performance of tasks under all circumstances, without noise or heroics, without unnecessary risk or extravagant use of resources.

When we looked at the performance of Air Force officers in Vietnam we saw professionalism on every hand. War is no place for amateurs; it is the professional who counts. The Secretary of the Air Force once asked an Air Force captain in Vietnam to name the single greatest asset of Air Force units operating there. The Secretary had expected the answer to be: "superior equipment." But the captain answered: "The professionalism of our personnel; they know their jobs and perform them efficiently, expertly."

Professionalism is expanded and exploited when an Air Force officer knows how to create professionalism in airmen. This means teaching, careful supervision, discipline, morale. It is the mark of the officer "pro" that he or she has working with him airmen who are also "pros."

Because of the vast field of concerns, the Air Force officer must steadily increase professional competence through expanding his or her own knowledge. This means taking every opportunity to enhance educational levels, up-grading basic educational degrees to master or doctorate, attending military schools, reading widely. In particular, the Air Force professional should come to understand the very serious business management aspects of the Air Force, since the net capabilities of the Air Force will always be an expression of the efficiency with which Air Force officers use the resources made available by the Congress. Sixty billion dollars a year—the largest single-purpose expenditure of the nation —is the business responsibility of Air Force officers. This money is utilized in a thousand ways; as pay, for procurement of equipment and supplies, for construction, for maintenance and operations, for research and development. The business end of this problem is not only the more than 100,000 contracts with U.S. industry, but also the accurate determination of requirements, the formulation of realistic plans, the efficiency of use of available resources. Each Air Force officer, participates in this vast business of the Air Force. The Air Force can obtain the most national security from its huge annual budget only when each Air Force officer obtains the most efficient results from his or her use of his or her part of the total.

In his field of career development each Air Force officer should constantly

C-141.

seek to heighten and broaden his competence so that at each succeeding level of his career he will fully deserve the accolade: a professional.

> *The true measure of a professional is his ability to adapt to new situations.*
>
> Harold Brown
> Secretary of the Air Force
> (1966)

18

The United States Air Force as a Lifetime Career

Today, as always, if a nation is to keep its freedom it must be prepared to risk war.
—President L. B. Johnson, 1964.

The relative attractiveness of a military career is the subject of continuing national debate. It should be. The regular officer corps of our United States armed forces constitutes the most important single bulwark of the security of the republic. Those great groups of American professionals—the engineers, the doctors, the lawyers, the business executives—contribute signally to the advancement of American civilization. Even such, however, cannot be compared with the corps of professional military officers in the light of danger. Danger is with us—continuing national danger in an age of monstrous weapons. The arduous war in Vietnam is not likely to be the last such limited conflict involving threats of escalation to larger war. For these reasons the corps of professional military officers must be of the highest quality obtainable. Only *you* know whether you might enhance that quality. Furthermore, only *you* know whether a military career appeals to you. If you are undecided, perhaps the following considerations would help in reaching a conclusion.

Whatever career a person may choose, certain criteria seem to apply to his choice. Among these are the following:

Pay and Emoluments
Security
Advancement Possibilities
Living Conditions
Prestige of the Career
Challenge of the Career
Associations
Retirement Benefits

Rarely has a young person deliberately chosen a career in which, in his or her mind, most of the above criteria seemed unfavorable. Yet, few career fields offer maximum favors in *all* of the criteria cited. The engineer who is building a railroad in the upper reaches of the Amazon would not cite *living conditions* as the basic attractiveness of the work. The wildcat oil prospector would not cite *retirement benefits* as the forte of that profession. The priest would not cite *pay*. Thus each career emphasizes some and slights some of these criteria. Just so does the military career. Let's look at it objectively.

PAY AND EMOLUMENTS

Officers of our armed services receive comparatively good pay and emoluments.

Medical Expenses are normally a minor item to most officers and their families. The Air Force provides medical attention for the officer and his dependents. Thus the financial hazard of serious illness is for the most part avoided. Though an officer may suffer an extended illness his pay continues in full, except for flying pay.

Insurance costs are mitigated, due to the availability of government life insurance and certain survivor benefits.

Commissaries and Base Exchanges offer food and sundries at costs lower than encountered in the average civilian establishment.

Travel and Transportation expenses are partially met by government allowances. It must be admitted, however, that the average officer can suffer a net loss in this regard. (See Chapter 5.)

Leave With Pay is available to each officer on the basis of 30 days per year. (See Chapter 9.)

Travel to foreign countries and all about the United States is an experience that officers take as a matter of course, but which would cost a shocking amount of money if undertaken privately. Our civilian friends spend many hard-earned dollars globetrotting for pleasure and education. (See Chapter 23.)

In summary, officers receive pay and emoluments at a relatively good standard.

SECURITY

The officer enjoys a relatively high degree of security. He or she cannot be separated from the Service except by due process of applicable laws. Physical disability incurred in the first eight years of service results in separation without retirement benefits, although important Veterans Administration benefits accrue. After eight years of service, the officer who becomes physically disabled receives retirement status and pay, the latter depending on the degree of disability. For all practical purposes, a Regular officer has a lifetime security, barring only serious misconduct and incompetence.

Security may seem a pale and lifeless quality to young people contemplating a vigorous career. Yet, it would be well to remember that, with the addition of a family, even the young must give due concern to the sureness of their living income. Officers never grow wealthy on their military pay, but they can enjoy the peace of mind which comes from an assured source of livelihood for themselves and their dependents. A long, stable future may be planned with confidence. It is perhaps asking the impossible to urge that a young man or woman give realistic consideration not only to the merits of security in his or her present job, but also to the needs of old age, and even security for his or her yet unborn children. If one can do this, however, one sees that security is a career criterion of sterling quality. (See Chapter 12.)

ADVANCEMENT

For the Regular officer, advancement in rank is available at stipulated periods. The opportunity to advance more rapidly is almost unlimited. There is no need to marry the daughter of the boss, or to seek help through special influences. Merit alone will take an officer upward. Of course the competition is keen, and grows sharper as higher ranks are reached. Yet a person may aspire to any military position to which personal endowments can carry him or her.

On the other hand, not every officer can be a general; mathematics points this out inexorably. To the contrary, most officers will end even full careers as lieutenant colonels. Therefore, if you correctly assess yourself as no better than average, you cannot reasonably expect to wear stars. The stars are there, however, waiting for the officer with the ability and the will to win them.

Advancement in the Air Force is more than a matter of promotion; it includes also advancement in responsibility. Whereas in some professions advancement takes the form of a steadily increasing standard of income, often these increases of income are not associated with increased responsibility and authority. A man or woman may labor at essentially the same level of responsibility for 12 to 15 years, doing work which grows dull and burdensome. Not so in the Air Force.

LIVING CONDITIONS

The living conditions of an Air Force officer and his family range across a wide spectrum. Not all married officers can expect to be provided quarters at air bases. There simply are not that many quarters yet in existence. Consequently, officers will find themselves occupying rental housing, commensurate with their ability to pay, in civilian communities adjacent to air bases.

Although housed off the air base, an officer and his family will nonetheless find the air base the center of their activities. The various facilities of the base, the Exchange, the Officers' Club, the swimming pools, golf courses, social gatherings—all of these aspects of the air base will have their appeal.

The officer's children may attend schools on the air base, or, more probably, will enroll in schools of the adjacent civilian community. Your children's schooling may be marked by numerous shifts between schools in various parts of the world.

Perhaps the most notable "living condition" of an Air Force career is the frequent changes of station. Air Force officers live like nomads; here today, Japan tomorrow, Germany yesteryear. These uprootings are inherent in the life of a career officer. Often they mean temporary (and not so very temporary) separation of the family. This is a consideration which should be weighed well before choosing the Air Force as a career, for it arises often.

Hazard is part of Air Force life. As a member of the Air Force, each officer is subject to being dispatched suddenly into danger. The war in Southeast Asia was stark evidence of this fact. On the other hand, in these days of nuclear-armed intercontinental ballistics missiles, perhaps danger is simply where you find it. The fact should not be overlooked, however, that hazard is a greater factor in the prospects of a career Air Force officer than it is in almost any civilian vocation.

PRESTIGE OF THE CAREER

Officers of the Air Force are generally accorded a place of high respect in the minds of Americans. The career officer is a member of an honorable profession the importance of which is beyond question. The officer is accepted in any social group, is considered an excellent credit risk, is assumed to be an intelligent and responsible citizen. These are the basic attributes of prestige in its actual application. Public opinion polls continue to reflect the public's high rating of the military officer in America.

CHALLENGE OF HIS CAREER

Few, if any, career fields offer greater challenge to the individual than does the Air Force. Never before in the history of mankind has so much power been assembled in organized fashion as is represented by the capabilities of the United States Air Force. Making ready, or employing if need be, this almost incredible power is a vast job of endless ramifications. The task embraces fully the entire fields of engineering, law, medicine, political science, personnel management, industrial production, research, science, even advertising. There is hardly an aspect of American genius that does not reflect itself in the operation of our Air Force. At the same time, there is not a single "approved solution" in the Air Force. Nothing is static. Events and the progress of science do not permit nor tolerate stereotyped procedures. The problems are many, difficult, and urgently in need of solution. If a person seeks a challenge to his or her own abilities, he or she need look no further than an Air Force career.

CAREER ASSOCIATIONS

Probably the most satisfactory aspect of an Air Force career lies in the associations it offers with a peerless company of fine men and women. The officers of the Air Force and their families are a group whose qualities are unexcelled by any professional corps in the nation. They are honorable people who adhere to a high code of conduct. They are loyal people, who serve the nation without stint. They are able people, who meet the great tasks of the Air Force and the difficulties of Air Force life with sure competence. They are interesting people, who contribute to the knowledge and culture of the nation from their worldwide experiences. They are, on the whole, good people whom it is good to know. There is a comradeship of the Air Force that has no parallel in civilian professions and to which no adequate value can be assigned. This is the truly sparkling attraction of an Air Force career.

Not only does the Air Force career offer fine associations with the people of the Air Force, it also affords a matchless opportunity to meet and know men and women of every walk of life in this and many other nations. Few civilian Americans compare with the average Air Force officer in the cosmopolitan nature of their friendships. Almost any Air Force officer of 20 years service can name personal friends in every state of the union and in a large number of foreign countries.

RETIREMENT BENEFITS

Regular officers of the Air Force enjoy excellent prospects relative to retirement. Subject to the approval of the Department of the Air Force, an officer may retire after 20 years or, more probably, 30 years' service. (See Chapter 12.) Retired officers receive a substantial percentage of their active duty pay, and retain privileges of access to air base clubs, commissaries, exchanges, and athletic and medical facilities. Taken together with a reasonable insurance program, these provisions permit a retired officer to live comfortably and in dignity, though not luxuriously.

The Air Force officer may look forward to retirement at an age early enough to permit him to enjoy leisure or embark on another career. So many of our civilian friends must remain at their posts of work until death or crippling disease fells them. On the other hand, an Air Force officer may retire after 20 years' service at approximately 45 years of age or at any time thereafter, subject to current fiscal policies of the Congress.

SUMMARY

The foregoing constitutes a reasonably accurate appraisal of the cited criteria as they apply to an Air Force career. They are the major pros and cons involved. In these days of harsh materialism and bitter cynicism, one sometimes hesitates to utter a sentence resting on idealism. Yet, one last word needs saying. Every decent man or woman will seek, *must* seek, a way of life which permits them to respect their own work. There is an innate urge common to all honorable people to be useful. The Air Force, standing as it does today between freedom and its implacable enemies, holds open wide a door to you who would feel that your career is one *you* can respect.

19

The Code of the United States Air Force

It is not the critic who counts, nor the man who points out how the strong man stumbled, or where the doer of deeds could have done them better. The credit belongs to the man who is actually in the arena; whose face is marred by dust and sweat and blood; who strives valiantly; who errs and comes short again and again; who knows the great enthusiasms, the great devotions, and spends himself in a worthy cause; who, at the best, knows in the end the triumph of high achievement; and who, at the worst, if he fails, at least fails while daring greatly, so that his place shall never be with those cold and timid souls who know neither victory nor defeat.—Theodore Roosevelt.

The code of the Air Force is a standard of life. It is a set of principles at which all officers aim, both in their official duties and in their personal concerns, in actions seen and known by the world, and in the deepest privacies of the mind. These things, these ideals, are not to be found in regulations. There is no checklist, no approved solution for each individual's every situation. Rather, the code of the Air Force is an attitude of the spirit, rarely arrived at consciously, but strongly governing the reactions of Air Force officers to all the many sides of Air Force life. Insofar as this spirit of the Air Force runs strong in the veins of its officers, the nation will have a military air arm of indomitable quality, and freedom will have a champion more able than any it has ever before known.

PRINCIPLES OF THE AIR FORCE CODE

The code of the Air Force cannot be captured in its entirety in the narrow lifelessness of the printed word, for this code is a living reality that must be experienced to be fully understood. Yet, some of the deep foundations upon which our code rests can be identified. These are criteria by which we have lived and fought and some of us have died; these are values we have sought to uphold, the true hallmarks of the United States Air Force.

Patriotism. Patriotism as practiced in the Air Force is an intelligent devotion to the interests of the United States *above every other consideration.* It is based upon the conviction that preservation of the American system of life, with its noble traditions, free institutions, and infinite promise is at once the highest practical morality and the most enlightened self-interest. Patriotism springs from deep knowledge that this land, this people, are in truth the hope of the earth. Such patriotism is not jingoism, blindly unaware of faults in our American system, stupidly touting it as perfect. Rather, such patriotism works to remove faults that do exist, while remaining confident that here in America the fallible hands and limited minds of humans have done their best work; that ours is the most promising road yet found toward a goal of maximum human welfare; towards freedom, dignity, and justice for each individual; and peace for the world. We know we have not yet reached this goal, but we believe the United States is man's longest stride in that direction, and we mean to see to it that the ground thus far gained is not lost, nor the path onward blocked.

Insofar as this purpose requires sacrifices of an Air Force officer, that sacrifice is made. Not only in the high drama of war does patriotism invoke a readiness to make personal sacrifices, but also in the humdrum of routine duties, during "off-duty" hours, in the thousands of small choices with which daily life confronts us. Patriotism invests the duties assigned to an Air Force officer with an importance far beyond their apparent measure. Thus, patriotism in the Air Force is intelligent devotion to the interests of the United States day by day, at work or at leisure, in war or peace, in life or, if need be, in death.

Honor. Honor is the highest form of self-respect. A person must live with himself alone for a very substantial part of his or her life. It is imperative to peace of mind and peace of soul that one respect his or her own character. Thus, honor is that code of conduct which springs not only from a "do unto others as you would have done unto you" philosophy, but rather from a determination to do only those things with which one's inner thoughts are serene. A person of honor does not lie, steal, cheat, or take undue advantage of another. The question here is not whether you would be seriously hurt if someone did such things to you. The question is whether you yourself could be proud of such practices on your own part. To be known as a man or woman of honor has many practical advantages, but these are incidental. The greatest reward of honorable people is that they respect themselves. Somehow such self-respect shows, and others recognize it as the sign of an honorable human being. We of the Air Force seek to be counted as such.

Courage. Courage is ascendancy over fear. Note that where there is courage there must be fear. In this sense a lion is not courageous, for it does not know fear. The Air Force wants no officers who are unafraid. The perils which threaten the United States in this nuclear era, in this time of limited wars that could escalate to Armageddon, are such that a person who is without fear is either a fool or ignorant. The Air Force *does* want officers who can conquer their fears, suppress them in the interests of the nation, and courageously carry on despite fear. Such officers are dangerous to the enemies of the United States, whereas a rash fool is not. Nor is the Air Force satisfied with officers whose courage is a thing of a moment, a flash of strength invoked for a brief instant only. Instead, officers are needed whose courage is as steady and long-enduring as the threat to the nation, who can steel themselves for the long pull over the years.

Loyalty. Loyalty is the quality of sincere confidence in and support for the purposes, methods, and capabilities of one's associates, whether superiors or subordinates. Loyalty is a quality which precludes sneering comment on the faults of your squadron commander, or bleating complaints about the errors of your new crew chief. It is basically an attitude of warm friendship toward your comrades. Perhaps an illustration makes this point more plain. Two little lads were discussing their dogs. The one pointed out that his dog was a thoroughbred of impressive lineage, including champions. The other lad, gazing at his mongrel pup, said: "Spotty is the best little black and brown dog with one ear half gone that there is in the whole world." That is loyalty, and it was probably richly deserved. If you look carefully, you will find that many more of your associates than you may imagine well deserve your loyalty.

Discipline. Discipline is the cement that binds together any military force. Without it, an Air Force is a mob. Obedience of orders in letter—and, more important, in spirit—is the heart of discipline. Air Force officers who grudgingly, complainingly, or unenthusiastically obey orders are poorly disciplined officers who encourage poor discipline in the airmen they supervise. Such are not the officers who build and maintain an effective fighting force. Almost everyone agrees that discipline is a fine thing—for everyone else. Too many of us are tempted to believe that orders and regulations were meant for the masses, but not for the Great Me. Yet we all should be aware that unless we practice discipline in our routine conduct, the stress of combat, of emergency, and of violent uncertainties can easily force us from the iron path of obedience to orders, with results disastrous to our cause and to ourselves. One thing seems certain: We shall have disciplined officers of our Air Force, or we will one day have no Air Force, no nation, and no freedom.

A low-flying RF-101 of 2d Air Division, Vietnam, casts its shadow over the missing spans of the My Duc highway bridge in North Vietnam.

Readiness. One of the most striking qualities of Air Force officers is their relative readiness to meet whatever tasks arise. The Far East Air Forces transformed themselves overnight from a slow-paced occupation force to the combat sword of the United Nations over Korea. Yet, two days prior to the entry of United States air forces into combat in Korea, not an officer of the Far East Air Forces had an inkling such a task would arise. Eight days later they had totally destroyed the North Korean air force of over 300 aircraft. Once again, the Air Force was called upon for quick reaction to the combat situation in Vietnam. Thousands of Air Force men deployed into battle in Southeast Asia to fight a bitter, difficult war. Our officers can be ready. They can be well-trained, well-informed, clear-headed, in good physical condition, masters of their duties. Such readiness must be maintained, for some day a bigger gong than Korea or Vietnam may suddenly ring out a dread alarm, calling Air Force officers to the trial of the ages. For this, each of us must stand prepared.

Frugality. We of the Air Force represent the largest single item in the United States budget. We are an expensive force, utilizing fabulously costly equipment and supplies. In fairness to all American taxpayers (including ourselves) we ought to hold this expense to the minimum consistent with national safety. There is more to this quality of frugality, however. Air Force units in Vietnam sometimes found to their sorrow that the cost of wastefulness was ineffectiveness, and that the price of ineffectiveness was a longer war with more lives lost. One of the dark hopes of the Communist world is that the United States will bankrupt itself through its far-flung national security programs. Let's not contribute to that enemy scheme.

Caution. Caution has a proper place in the code of the Air Force. The great strength we hold poised is the chief defense of the United States. Our defense is threatened by the most massive power for evil ever assembled. The situation is dangerous and explosive. A mistake can lead to total war. Was there ever a condition of affairs which more strongly recommended caution? Moreover, Air Force officers operate equipment noted for its great speed and awesome effects. Recklessness is often paid for in lives; it is certainly paid for in millions of dollars annually. In these circumstances, no Air Force officer is justified in abandoning caution. A famous Air Force combat ace once said: "When I get in an airplane, I slow down." It would be well if all Air Force officers accepted this advice: When you get in an Air Force uniform, slow down and proceed with caution.

Sense of Responsibility. The quality known as responsibility is one of the most valued in an officer. Its most frequent evidence is the execution of work that should be done whether or not one is directly charged with that work. How many times can you hear the phrase from the substandard officer: "Oh, well, it's not my headache." But it *is* your headache, if it hurts the Air Force. The sense of responsibility found in the best of Air Force officers will not be satisfied with a job merely well done. The question is: Has the job been done to the best of my ability? If not, it isn't finished. On an air base in the Far East, a very high ranking Air Force general walked across a runway. As he hurried towards the operations building, the general looked back over his shoulder, hesitated, finally said: "Damn it, I've got to go back. I saw a bolt on that runway. Should have picked it up. They ruin tires, you know." Here was a busy officer, not too busy to ignore that deep-rooted sense of responsibility.

Teamwork. Teamwork is the method by which officers of the Air Force do the impossible more rapidly. Teamwork makes football champions of eleven men who, if they all insisted on being ball carriers, would gain nothing. Teamwork makes every officer's task easier, and yet the overall result obtained is greater success. It is the nearest thing in life to getting something for nothing. The officer who insists on playing a lone hand, who always wants to carry the ball, who feels that associates should do their own work without bothering him or her, will find that the path is rough and leads not very far. Cooperation with others is a first essential. The Air Force is a team organization, and teamwork is the oil of the machine.

Ambition. To be ambitious is the mark of a superior officer. Every officer has a right to be ambitious and will be a better officer for it. People tend to assess one another according to the evaluation placed by the individual upon himself. If you exert yourself to become qualified for higher positions and greater responsibilities, you will eventually become convinced that you are ready for advancement. Thus, ambition based on qualification will give you confidence and aggressiveness, qualities indispensable to superior officers. Stonewall Jackson said: "You can be what you determine to be." When Jackson used the word *determine,* he meant willingness to do the work, gain the qualifications, and seek the opportunity necessary for advancement to your goal. Ambition can be a driving force for self-improvement, and should be. Only one caution is relevant: Set the sights of your ambition no higher than the level of your willingness to work, for the two are linked as with iron.

Adaptability. One of the qualities most employed by Air Force officers is adaptability, meaning the ability to make the best of any situation. Life in the Air Force is a rough-and-tumble of wildly varying conditions. Officers serve in all parts of the world, in an almost infinite number of differing conditions. Aircraft change endlessly, new maintenance problems arise constantly, all things are subject to frequent alteration. Each officer must be prepared to adapt to the conditions he or she may find in the assignment. Living accommodations, offices, operational facilities, the quality of superiors and subordinates, all may differ radically from the officer's previous assignment. Yet it is with these new factors that the officer must work. It is the officer, not they, who must adapt. Thule cannot compare with Eglin for surf bathing. Adaptability and flexibility are essential to meet such diametrically different situations as are the routine experiences of Air Force officers.

Ascendancy of the Civil Power. Every officer of the Air Force takes an oath to defend the Constitution of the United States against all enemies, foreign and domestic. Fundamental in our Constitution is the concept of ascendancy of the civil power over the military. Concrete expression of this intent is found in our civilian Secretary of the Air Force, our civilian Secretary of Defense. The Air Force expects every officer to support always the continuation of this system of civilian control of the military, for it is an indispensable part of the American way. Officers should seek every means to avoid collisions between the civilian and military authority at any level. Rather, we should find ways to coordinate these two forms of authority for the good of the nation. It was thus that America has been built to its present eminence; it must be thus that America continues its climb to the stars.

Relations with Civilians. One of the most important accomplishments expected of an Air Force officer is the maintenance of favorable relations with civilians. This fact becomes doubly important when one realizes that about 90 percent of John Q. Citizen's attitude toward the Air Force is a direct reflection of the impression individual Air Force officers make upon him. It is imperative to the continued strength of the Air Force that American citizens hold a favorable view of the Air Force. The future of the Service, in a very large measure, turns on the relations of Air Force personnel with their civilian neighbors.

Every Air Force officer is a public relations officer, a source of information about the Air Force to civilians with whom he comes in contact. Such information is not necessarily conveyed by words. If you, an Air Force officer, are irresponsible toward your just debts, conduct yourself in an obnoxious manner in your civilian neighborhood, display yourself in a drunken condition, ignore the laws and the customs of the community—how much confidence would you expect the civilian observer of these malpractices to place in the Air Force you are a part of? On the contrary, if you conduct yourself as a sober, responsible citizen of the civilian community, your civilian acquaintances will tend to feel that the Air Force is an organization in which confidence can be safely reposed. Thus, by your *pattern of conduct,* you form important impressions regarding the Air Force.

Your civilian friends are interested in their Air Force. They have a right to be. It is their best hope for averting total war and their most effective instrument with which to win victory in limited war. The Air Force is close to the American citizen's vital interests for it is the force which stands between his home and the bombs of the enemy. Incidentally, the Air Force is one of the citizen's greatest single reason for taxes. John Q. Citizen has a lively and legitimate interest in the Air Force, and it is to the individual Air Force officer that he most frequently turns for information. The manner in which the citizen's interest is met by you will have a decided effect on the future of the Air Force. Accordingly, each officer should *welcome* the opportunity to provide factual, proportioned information about the Air Force to his civilian friends, taking care always, of course, to avoid classified matters. If your civilian friend points with alarm to a report that a C5A aircraft has blown up, explain how such things happen, *and how infrequently they happen.* If a civilian asks whether your quarters allowance is tax exempt, tell him that this is true. Also explain to him the *overall* financial status of Air Force personnel. In other words, so far as security considerations permit, give your civilian friends the *whole* truth. Throughout, remember that the United States Air Force is the property of the American citizen. He has built it, manned it, paid for it. He has an awesome lot of blue chips riding on its wings.

Air Force officers should seek to take a full part in the life of the civilian community. Be sociable with your neighbors. Become a member of their clubs, societies, and churches. Take your share in community enterprises. In short, be a good neighbor, a respected and desired neighbor. While doing so, you will not only perform well a duty expected of you as an Air Force officer, you will also add greatly to your own enjoyment of life.

ON HEALTH AND HAPPINESS

Many lists of attributes considered essential to a successful Air Force officer omit two critically important items: Health and Happiness. The traditional picture of an officer of our armed services is that of a man or woman so dedicated to duty, so disciplined, that they "carry on" despite illness of the body, sickness of the soul, or agonized turmoil of the mind. Baloney! An officer in such condition may indeed "carry on," but is a menace to all. He or she is incapable of doing an adequate job in assigned tasks. He or she is almost as dangerous as though he were drunk or insane. The Air Force has issued an order *requiring* all Air Force officers to take action to guard their own physical health and mental freshness. Thus the Air Force recognizes a fact: officers cannot do their best work unless they are healthy and happy. Let's consider the matter.

Mental health—we'll call it happiness—is a natural state. We must concern ourselves, therefore, with those things that disturb happiness, those troubles that beset us with such intensity or persistence as to destroy our natural serenity.

Physical Troubles. The most common, and perhaps the most serious disturber of our inner peace, is bad health. An officer who is suffering from ulcers, or heart trouble, is a worried officer, anxious about the future and that of his or her family. He or she may well be an uncomfortable officer, suffering often or constantly from the illness. The next physical examination may end his or her career in the Air Force, leaving the serious problem of finding a new way to earn a living of sufficiently high standard for his or her dependents. This officer is in trouble. How did it happen?

A young, vigorous officer has some difficulty imagining himself other than healthy when he or she *is* healthy. Still, it might be well to look ahead a bit. The Air Force is a tense organization, bearing the most grave and urgent responsibilities ever assigned to a military force. Officers of the Air Force lead lives that are characterized by steady pressure, sudden change, taut occasions. Unless an Air Force officer is more than usually careful of his or her health, the long years of tension will finally make their mark, grievously.

Fast Lunches. Being specific, consider the matter of eating. How many times does the Air Force officer "grab a bite of lunch"? In twenty years of service, this can happen over 5000 times! What does this procedure of eating rapidly and without regard to a balanced diet do to health? Any doctor can give answers ranging over a wide variety of physical troubles. Which one would finally hit *you* depends on your particular makeup. One of them certainly would.

Liquor. Now, liquor. One of the most certain though expensive methods of breaking down good health is the excessive use of liquor. All right, what's excessive? Unfortunately there's no pat answer. Life would be far simpler for most people if there *were* a pat answer—if it could be said that two drinks of an evening are *not* dangerous, three drinks are dangerous. Not so. It all depends on you. There is one certainty, however: if a person drinks to the point of being "high" each evening, or even a few evenings a week, something is going to give. Perhaps it will be the stomach, perhaps the heart, perhaps the liver. Sooner or later.

If you prefer, you need not drink at all. No one will think the less of you.

Drugs. Do not use them, *not ever!*

Exercise. Exercise, taken moderately, is good health insurance. Probably the best form of exercise for Air Force officers is golf. When one is young and bubbling with energy, such games as tennis, squash, or basketball may seem more attractive. Bear in mind, however, that such forms of exercise only rarely can be continued through middle age. Golf is moderate enough physically to be played throughout life, and offers enough exercise to give one all the benefits exercise has to offer. The Air Force's weight control program and the physical fitness program underline the responsibility of the officer to keep trim and fit so as to be able to carry out the often arduous duties.

Weight. Watch the weight. Almost all people tend to pick up weight as they grow older. The trouble with this is, the more weight one carries, the more work must be done by a tiring heart. Thus the normal strain on the heart, due to age, is doubled by the greater amount of tissue it must support. Keeping the weight down will *not* prevent a heart attack nor cure heart trouble. Excessive weight *will* bring on such troubles earlier, aggravate them, and cause them to be more persistent.

Worry. Most doctors look on worry as the chief debilitating agent of modern times. Anxiety is their high-priced word for it. Worrying is probably the most expensive of all pastimes, because it not only prevents effective action to remove the cause of worry but also degrades the physical and mental strength of the worrier, thus rendering him less capable of taking any action. Worrying, however, is more a symptom than a disease. This thought brings us to the area of mental health—happiness.

Happiness. Mental health—serenity, happiness—this state of mind is usually firm and stable when there is no cause for concern about health, family, career, or the finances. It is true that an entire spectrum of special situations that do not fall within these four categories could be described. It is also true that about 99 percent of all unhappiness stems from the fatal four already mentioned. The question of health as it applies to the officer has already been mentioned. What about the health of others?

Dependents' Health. An officer should never allow his official duties of the moment to prevent his taking proper steps to assure the health of his loved ones. This may necessitate absenting oneself from one's place of duty for a few hours to baby-sit while the spouse goes to the doctor for a needed check up. Do so. Your commander will approve, especially if you ask before hand. The point is this: if an officer fails to see that his spouse or child gets proper medical attention at the time it can do the most good, sooner or later we may find the officer's spouse or child seriously ill. Then we have a situation by which the officer is deeply disturbed. The Air Force loses his most effective services, not for a few hours but for weeks or months.

Family Problems. Turning more generally to the matter of the officer's family, no person enjoys the best mental health when his or her family relationships are abnormal. The foremost consideration in this area is the relationship between husband and wife. Air Force officers ought to realize that they cannot live a daily home life full of strain and turmoil and at the same time do a top notch job at work. If there is persistent trouble between husband and wife, it should not be allowed to fester. Rather, the two should agree to discuss their problem with the chaplain or the doctor, as they choose. Thresh it out, bring it up into the light of day. False pride is poor justification for suppressing a situation that can wreck lives, careers, and the health of all involved. Officers *must* realize that marital problems are the proper concern of the Air Force, since these strains can and very often do result in poor health, poor efficiency, or both. If an officer is suffering from appendicitis he or she does not hesitate to lay this problem before a doctor. He or she should as quickly seek competent advice on family problems, since these can be, in the long run, more damaging than most bodily ailments.

Career Dissatisfaction. Concern about one's career can become a source of worry, of bitterness, of frustration amounting to a major menace to mental health. Perhaps you feel you have been forced into a career channel to which you are unsuited. Perhaps you believe you have been mistreated in the matter of promotion. These thoughts can lead to a state of mind which is incompatible with serenity and will markedly degrade the performance of an officer in his assignment. Should such a problem develop, the officer is well-advised to talk it over with his commander in complete frankness. The commander may not agree with you, but at least he will know your state of mind. By sharing your problem with him, you lighten it somewhat. Better still, the commander may be able to do something about it. What can be lost by such action?

Finances. Finances are perhaps the greatest single source of worry among officers. Finances are also the greatest single source of career and marital troubles. Every officer should spend ample time *thinking*—not worrying—about his finances. Trite as it may sound, there is no good substitute for a budget. A budget is simply a financial plan, a program for matching obligations and income. Any officer can draw up a budget for himself. The trick is to stick to it.

Easy Credit. Foremost among the financial traps awaiting officers is the easy credit of these times. Officers are considered excellent credit risks, since they are required to meet their just debts. They can buy on credit—on "terms" —almost anything they are willing to sign for. There is a great temptation, therefore, to go for the high-priced car, the deluxe television set, the quality furniture. The best help in resisting these temptations is the cold logic of a personal budget. No amount of glitter from the shiny chrome on the Cadillac will make the budget think its income is greater than it is.

If, however, an officer slips into a welter of debts, the best action is to consult with a reputable bank. Seek to consolidate a number of debts into one large one, to be paid off in a manner consistent with one's income. An oversea assignment in a low-cost-of-living area may help to solve a temporary financial problem.

Other than by such notably stupid practices as spending too much money on liquor or gambling, the chief methods of encountering financial troubles are in connection with renting or buying a house, buying a car, and purchasing furniture or appliances.

Renting a House. Renting a house is an experience that often befalls Air Force officers. In connection therewith, it is probably sufficient merely to note certain salient aspects. First of all, a lease on a house is binding on an Air Force officer unless moved out of the general area by competent orders. To obviate argument, this contingency should be expressly stated in the lease. Second, the officer should be very clear as to what the rent covers. How about utilities? Water? Heat? Who pays for these? Third, if an extra month's rent in advance is required by the landlord, exactly what are the conditions under which this advance is *not* repaid to the officer? Is fair wear and tear chargeable to the officer? Before renting, ask to see a few utility bills for the *winter* months. These are, in effect, additional rent. Such matters are best determined firmly before the lease is signed.

Buying Houses. If an officer must buy a house, very great care is indicated. Servicemen may make use of the FHA loan authority, so that a house may be purchased with a down payment of less than 10 percent of the sales price—*provided* the sales price is no greater than the FHA appraised value of the house. An officer is usually ill-advised to pay more than the FHA appraisal. The best deal can be made directly with the seller. Real estate agents' fees are in the area of 6 percent of the sales price. So far as the buyer is concerned, this fee is simply an addition to the cost of the house. Buying a house is probably the most serious financial move an officer will ever make. Thorough investigation of all angles is imperative.

Buying Cars. The primary mistake of most officers in buying an automobile is to sign up for a Cadillac on a Chevrolet budget. When financing charges are included, any full equipped automobile is a very expensive item. If a thousand or more dollars are added to jump from the "low-priced" three to the middle or higher priced makes, we are talking about an amount almost equal to a junior officer's total annual pay. Further, the higher the initial price, the higher the upkeep cost. To operate a car that gets 15 miles to the gallon will cost about $200 a year more than to operate a car that gets 20 miles to the gallon. When it comes to spare parts, another factor sets in with a vengeance. Compare the cost of a new muffler for a Cadillac with the same item for a Ford. Generators, starters, water pumps, tires—these, too must be replaced when they wear out. In this area one finds a significant differential. Finally, compare the percentage of turn-in allowance to original cost. Whatever your final decision, at least take these matters into account.

Buying Furniture. An officer can easily go overboard in purchasing furniture and appliances. You will be overseas about one-third of your service. Repeated moves or storage will reduce furniture to a shambles. Buy only the items of furniture you really must have. Stay away from very expensive, delicate items. Don't try to match the Joneses. Maybe the Joneses do not live a third of their lives in faraway places.

Bad Checks. Monitor zealously your check balance, and indeed your entire credit rating. This includes checks written by your spouse, as well. Impress upon him or her the need for responsibility, and if you share a checking account, be sure that each party knows the extent to which he or she may use it. Keep an amount in your account as a cushion above your monthly expenditures. If possible, arrange with your bank either to notify you if there is a slip-up before "bouncing" a check, or authorize them to cover the amount by drawing against your savings account.

Summary. The foregoing comments have touched only lightly on some of the more common sources of trouble. Whatever may be your specific situation, if you have a problem that menaces your health of mind, of body, or of soul, seek competent advice, and seek it early. Remember that your health is a matter of proper and sincere concern to your superiors in the Air Force, to your doctor, and to your chaplain. Don't ignore the help they can give you, and will gladly give if you ask. You are far more important to the Air Force than any piece of machinery, any weapon, any gadget. You, the Air Force officer, are the kingpin of a weapons system that guards the survival of civilization. Take care of yourself.

Moderation—a sense of proportion—in the areas discussed will keep most officers out of trouble. If, nonetheless, troubles arise, don't just worry. Do something!

Part Three • Your Family

20

Social Life in the Air Force

The value of tradition to the social body is immense. The veneration for practices, or for authority consecrated by long acceptance, has a reserve of strength which cannot be obtained by any novel device. —Alfred Thayer Mahan

Social activities and relationships in the Air Force are almost exactly identical to those of polite society in civilian communities. The officer and his spouse will find little need for adjustment from their civilian habits with respect to social activities while on duty in the Air Force. There are a few differences. In general, the income of officers is known to all others: there is no place for bluffing or "keeping up with the Joneses." Another difference is the fact that an air base is in itself a *small* community in which everyone knows almost everyone else. This fact makes for tight-knit social relations and close scrutiny by all of each family's activities. In this sense, the officer and family live in a goldfish bowl. It is well to bear this carefully in mind at all times.

SOCIAL LIFE ON THE BASE

How to be a Good Citizen of the Air Base. The following are a series of tips which if abided by will tend to make the officer and spouse a helpful unit in air base social life. Strive to be on good terms, if not good friends, with all. Choose your close associates with great care. Do not openly express your personal dislikes of individuals. Do not become a member of a clique. Keep your dinner guests, your golf companions, and your bridge parties open to a broad sweep of the families on your air base. Do not attempt to live beyond your means. In repaying your social obligations, do so within your means. If a simple dinner is the best you can do, provide it without excuses. Call on newly arrived families and offer them your assistance. Don't merely sit there for fifteen minutes making conversation. See what you can do to help. When members of officers' families are ill, visit them, offering books, flowers, or magazines if you can afford them. Take an appropriate part in the community activities of the base.

Official Calls. Usually the commander of an air base will hold a reception or open house, perhaps twice a year, for the new arrivals on the base. The old practice of calling on the commander within twenty-four hours after arrival on the air base has been abandoned. Due to the large flow of officers into air bases in these years, it is obviously impractical for the commander to receive each newly arrived officer individually, much as he might like to do so. Practices vary, however, among air bases and with different commanders. It would be wise to inquire on arrival what the desires of the commander of your air base or squadron might be.

Calling Cards. Calling cards are rarely used by officers, but if special circumstances require their use, proceed as follows: Officer's calling cards are of the same size normally used by civilians. The complete name is generally used, as *John Joseph Jones,* although middle initials may be used if desired.

Social Functions. Receptions, dinners, dances, teas, and other such social functions are performed in the Air Force in essentially the same manner as in civilian life.

It is a courtesy to your host or hostess to accept or decline invitations as soon as you possibly can. A formal written invitation should be replied to in the same style as received. If declining an informal invitation, be sure to state clearly a sound reason why you cannot accept it, such as illness in the family, a previously accepted engagement, or military duty.

Often in connection with a reception a receiving line is encountered. The line is based upon the position taken by the senior officer or honored visitor. Other members of the receiving line form on the left of the senior or honored person in order of rank of the officers with their spouses on their left. An aide or an officer acting in that capacity usually greets the guests and introduces them to the honored personage in the receiving line. The officer precedes the spouse in passing through the line. In order to avoid embarrassment and confusion, be certain to announce your name very clearly to the aide so that he may present you accurately to the receiving line. Greet each person in the receiving line before proceeding into the main area of the reception. In greeting the members of the receiving line, repeat their names, as for example, "Colonel Smith, I am very happy to meet you, sir."

At dinner dances, a guest should not fail to invite his dinner partner, his hostess, any lady guest of honor, and house guests to dance, regardless of his dancing ability.

At cocktail parties, circulate as much as possible, conversing with as many people as you can. Do not retire to a corner to tell jokes or discuss flying.

Upon leaving a social function, never fail to bid your host and hostess farewell, expressing thanks for the hospitality you have received. If you must leave prematurely, explain to your host and hostess the reasons therefor.

It is customary to reciprocate an invitation received, regardless of whether you accepted or declined such invitation. In this connection, one extends to house guests of officers invitations to social gatherings to which their host has been invited.

CAPTAIN NOAH MATTHEW JACKSON

AIDE-DE-CAMP TO MAJOR GENERAL SECREST

UNITED STATES AIR FORCE

LIEUTENANT COLONEL AND MRS. NATHAN MENZO NEELY

DEAN STANLEY WILLINGHAM

CAPTAIN

UNITED STATES AIR FORCE

Mary Alice Weible

Captain

United States Air Force

EXAMPLES OF CALLING CARDS.

Seating at a Formal Dinner. The lady of honor at a formal dinner is the wife of the senior officer present, or perhaps a lady whom it is desired to honor especially. The lady of honor is always seated at the right of the host who sits at the head of the table. The lady of next highest rank or honor is at the left of the host. The hostess sits at the foot of the table. The most distinguished (or oldest) gentleman is on the right of the hostess, and the second senior gentleman is on her left. The third senior gentleman sits on the right of the lady of honor. Others are arranged in order of rank, alternating between men and women, and descending in order of rank or honor from the head of the table. The term "rank" in this connection does not necessarily mean military rank, but must be adjudged by the host and hostess in accord with circumstances.

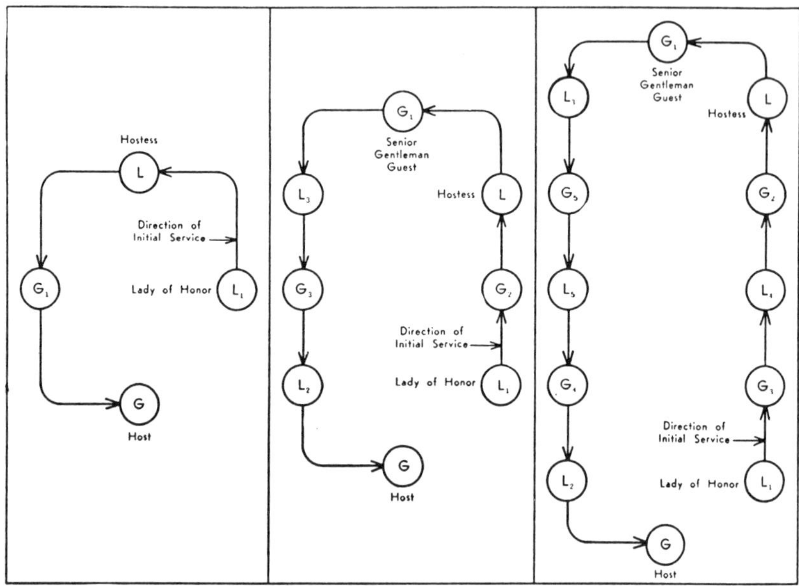

SEATING AND SERVING GUESTS AT DINNER.

The Art of Being a Good House Guest. Much has been written about the art of being a good host. But all too little has been written about being a good house guest, and it needs doing. Thoughtlessness or ignorance sometimes play havoc with cherished friendships. Thoughtfulness will cement and strengthen them. As in all other human relationships, good breeding is shown by a proper regard for the rights of others.

The host and hostess should be informed as nearly as practicable of the exact hour of planned arrival and departure. This will enable them to plan your visit and permit them to plan to resume their normal contacts and activities. Particularly annoying is the invited guest who accepts an invitation and states only that he will arrive during a day. The hostess is left in the dark. Will arrival be morning, afternoon, or evening? Should she remain at home canceling other things she has planned in order to be certain to receive the guest upon arrival? Shall she plan the evening meal for his presence? Equally troublesome is the guest who neglects to announce a definite time as to departure. The hostess will wish to extend all possible consideration to the guest but upon departure will wish to resume the threads of other activities. She may wish to keep or cancel in advance a hairdresser's appointment, attend her bridge club, complete necessary shopping. Uncertainty in these routine arrangements is displeasing. Of course the guest who arrives to stay "for a few days" and fails to inform his hosts even as to the day of planned departure until he descends with packed bags to say his farewells passes entirely beyond the pale. One way to do it is a letter, telegram, or telephone call which might state: "Thanks for your kind invitation. We accept with pleasure. We plan to arrive at your home soon after four Friday afternoon and must start our return before nine on Monday morning." The house guest who accepts a weekend or other invitation of more than short duration without making known these simple things which are usually left to the desires of the guest has taken a firm and certain step which often will deny him from receiving further invitations. Having established these days and dates, bend Heaven and earth to keep them.

A guest must adjust to the conditions of the household. He or she should keep his things picked up and his room tidy. The bathroom should be left in at least as good condition as he found it. A guest should take complete care of his or her room and be similarly thoughtful of the bath. A guest in a servantless home should share in the household work to the extent that is welcome or acceptable. In this way hosts and guests have more uncrowded hours, in which to enjoy one another and the hostess is spared excessive strains.

If plans have been made or suggested, the guest must show his pleasure in sharing them. Adaptability is the password.

Both host and hostess will require more time to be by themselves in order to take care of personal responsibilities. Make it easy for them to do so. Take a walk, write a letter, read a book. This will permit the hostess to order her groceries, or visit the hairdresser, and the host to discharge some small but essential responsibility.

An invitation by a guest to take the family of his host out to dinner is often a welcomed courtesy *but it must not be pressed if once proferred and declined.*

Prior to departure be certain no personal belongings are left behind for the host to package and mail. A remembrance to the hosts in the form of flowers, candy, a book, given either before or after departure, is generally most appreciated. A thank-you letter written a day or two after departure is a necessary courtesy. If there is a maid, leave her a *pourboire*.

Failure to do these things and others which will suggest themselves will subject you to the old saying, "making the hosts twice glad"—glad when the guest arrives, and glad when he departs. Doing them may result in future invitations. The essence is regard for the rights of others.

The Geography of Military Social Life. As soon as the officer and family arrive at an air base they will observe that service personnel are divided into two categories. One of these categories is that group which lives on the base in government furnished quarters. The other category is that group which lives off the base in rental housing obtained in a nearby city. Regardless of which group your family may fall in, you will find that the Officers' Club at the air base is the center of your social activity. At the Officers' Club you will find a seven-day-a-week program of entertainment. There will be nights for bingo, others for dancing, and others for family dinners. Your military social life will largely consist of either visiting the quarters of your friends on the base or off the base or attendance at some entertainment at the Officers' Club.

SOCIAL LIFE IN THE CIVILIAN COMMUNITY

Since some married Air Force officers live in civilian communities rather than on air bases, a substantial amount of their social life pertains to their civilian neighbors. In such circumstances, and if they can afford it, officers are well advised to become active participants in the social activities of their communities. The idea that an officer on duty in a foreign country is an ambassador of the United States is well-advertised. Less emphasized, but by far more important, is the fact that officers living in a civilian community are ambassadors of the Air Force to the American public. Accordingly, Air Force officers should constantly seek to create a good impression on their civilian neighbors. In the main this is accomplished through unimpeachable conduct of the officer and family. It is also helpful, however, if officers become members of the churches, societies, and clubs of the civilian community and participate fully in such social life as the neighborhood offers. It is perhaps more important that Air Force officers discharge their social obligations to their civilian neighbors in a manner reflecting credit on the Air Force than it is that they meet these criteria in respect to other Air Force personnel. In a sense, all social niceties can be summed up in the phrase, "Be a good neighbor." This certainly applies to your civilian friends as well as to your Air Force associates.

PRACTICAL TIPS

The Right Clothes.

	Officer	*Wife*
Official Call	Service Uniform (with coat)	Afternoon dress or suit, hat and gloves
Informal Dinner	Civilian business suit or Service uniform.	Cocktail dress or simple dinner dress
Formal Dinner	Service Dress uniform or Mess Dress uniform or Dinner Jacket	Dinner dress and gloves
Private Cocktail Party	Business suit unless a uniform is suggested on the invitation	Cocktail dress or dressy suit with gloves and, usually, hat
Note: If proceeding from a cocktail party to a more formal function, it is proper to wear to the cocktail party appropriate attire for the more formal affair.		
Official Cocktail Party	Service Dress uniform	Cocktail dress or dressy suit with gloves and usually hat
Barbecue or Other Informal Outdoor Affair	Sport coat, no tie	Blouse and skirt, slacks, but *no shorts*
Official Receptions (not including dinner)	Service Dress uniform or Mess Dress uniform	Dressy afternoon clothes with hat and gloves, or cocktail-type dress with gloves and, usually, hat
Official Reception (including dinner)	Service Dress uniform or Mess Dress uniform	Dinner dress with gloves
At Home or Open House (Before 6 P.M.)	Civilian suit	Afternoon dress or suit with hat and gloves
Parades and Retreats	Service uniform or business suit	Daytime dress or suit with hat and gloves
Private Dinner at Officers' Club	Business suit	Cocktail dress or simple dinner dress without hat

The Right Words. Introducing your wife to any man (except Chiefs of State and very high church dignitaries): "Mary, this is Colonel Brown."

Introducing one lady to another: "Mrs. Jones, may I present Mrs. Green," or "Mary, this is Mrs. Green."

Introducing one officer to another: "Major Smith, this is Captain Brown."

Introducing yourself to an officer senior to you, "Sir, may I introduce myself? I am Captain Jones." (Wait for the other to extend his hand.)

Introducing yourself to an officer of equal or lesser rank: "I'm Captain Jack Jones." (Extend your hand.)

Introducing children or teenagers to adults: "Lieutenant Jones, this is Mary Smith."

Responding to an introduction: "How do you do, Colonel Green."

Thanking host and hostess on departing a social function: "Thank you for a delightful evening."

If you *must* leave a function noticeably early: "Mrs. (Hostess), I'm sorry I must leave early." (Then give reason, and make it good!)

Using titles with names:

Major General Black: "General Black."

Brigadier General White: "General White."

Colonel Smith: "Colonel Smith."

Lieutenant Colonel Jones: "Colonel Jones."

First Lieutenant Brown: "Lieutenant Brown."

Second Lieutenant Green: "Lieutenant Green."

Reaction to Invitations. Respond to an invitation (1) promptly, and (2) in kind. By "promptly" is meant within 24 hours. By "in kind" is meant in the same manner as the invitation is received. If received orally, respond orally. If received in an informal note, respond by informal note (unless a telephone RSVP is specified in the invitational note). If the invitation is received in the form of a formal note or card, respond fully, in writing, on formal note paper, and in the third person: "Lieutenant Jones is happy to accept the kind invitation of Colonel and Mrs. Smith to dinner on 3 January 1972 at seven o'clock."

Always explain a regret to an informal invitation: normally, you do not give reason in declining a formal written invitation. *Do not* decline an invitation issued by your commander unless official duties unavoidably prevent your attendance.

If invited to dinner, and you are a bachelor, inquire of your hostess whether she wishes you to escort anyone to the function. Married persons normally should not attend mixed dinners or cocktail parties when their husbands or wives are unable to attend.

After attending a private social function, it is proper to write a note to your hostess (or make a telephone call) thanking her for the entertainment.

Remember to return the courtesy extended to you. It is not necessary that this be in exactly the same form as the courtesy you received. Inviting your recent host and hostess to be your guests at a college football game is a fully satisfactory reciprocation for a dinner.

THE FORMAL DINING-IN

Certain ceremonies and traditions are gradually becoming a part of the Air Force way of life. One of these is the "dining-in," a formal dinner of a unit or organization.

For the convenience of Air Force officers everywhere, there is included herein material prepared at the Air University at Maxwell Air Force Base, for assistance to the officer who may desire to hold a dining-in for his unit, or whose commander may have assigned him the duty of making preparations for a dining-in.

The dining-in is a formal dinner function for members of an organization or unit.

Background. The custom of dining-in is a very old tradition in England, but is not exclusively military. It is believed that dining-in began as a custom in the monasteries, was adopted by the early universities, and later spread to the military units of the country when the officers' mess was established.

The late General H.H. Arnold probably started the dining-in within the Army Air Corps when he used to hold his famous "wing-dings." The contacts of U.S. Army Air Corps personnel with the British and their dining-ins during World War II gave additional impetus for the growth of this custom in the USAF. It was recognized that those occasions provided situations where ceremony, tradition, and good fellowship could play an important part in the life of military organizations.

Faculty members of the Squadron Officer Course (later designated Squadron Officer School) of the Air University, realizing the value of tradition and ceremony, began having faculty dinings-in. The dining-in was also included in the curriculum for the students. This was also very successful and the custom rapidly spread to Air Force units in other commands.

Purpose. The dining-in provides a situation for officers to see how ceremony and tradition play a part in the life of an Air Force unit. It also provides an occasion for officers to meet socially at a formal military function. It is an excellent means of saying farewell to departing officers and welcoming new officers. Further, the dining-in provides an opportunity to recognize individual and unit achievements. All of these are very useful in building high morale and esprit de corps.

Planning. Preparation for the dining-in should begin well in advance. The date, location, and speaker should be selected, and reservations made. The dinner should be held in a suitable private place dictated by good taste. Details for the various arrangements should be allocated to individuals and their specific duties outlined. An order of events, or agenda, should be prepared.

Guest Speaker and Other Guests. The guest speaker should be military or a civilian who can be expected to address the mess in an interesting manner on an appropriate subject. He should be invited well in advance and advised of what he can expect and what is expected of him. Arrangements should be made for him, and other invited guests, as protocol and custom dictates.

Dress. The dress for the dining-in should be the Air Force Mess Dress. Medals must be worn by all members of the mess and military guests. Civilian guests should wear appropriate civilian attire.

Dining-in Customs. Each member of the mess should arrive within ten minutes after the opening time to meet the guests before dinner is served. When the signal is given for dinner, the members should enter the dining room and stand behind their chairs. There will be no smoking from the time the members enter the dining room until the president of the mess so permits. (Members, unless properly excused, will not leave the mess before the guest of honor and the president have departed.)

The guest of honor and the president of the mess will be the last to join the head table. The president formally opens the mess and continues according to the agenda. The president will remain standing while he is speaking but he will seat the other members of the mess after the toast to the Chief of Staff has been given.

The president's welcoming remarks, after the invocation, will set the tone for the formal part of the agenda. Guests at the head table are introduced by the president of the mess. Other guests of the mess should be introduced by an appropriate member of the mess.

If there is to be an informal portion of the dining-in, such as some form of entertainment, there should be a distinct break between the formal and informal portions.

Each time the mess is adjourned and reassembled, members should stand behind their chairs until the persons at the head table have left the table or are seated.

At any time after the toast to the Chief of Staff, a member may ask to be recognized for an appropriate reason. A typical reason may be that a toast has been forgotten. In such a case the member will stand and ask to be recognized by saying, "Mr. President, I have a point of order." The president will recognize the member by calling his rank and name. The member will, in a polite and forthright manner, tell the president that the toast was not proposed. The president will then ask the member who has the floor, or Mr. Vice, to propose the appropriate toast.

During the dining-in each member should try to pay his respects to the guest of honor. After the mess is adjourned, members should remain at the dining-in until the guest of honor and the president of the mess have left. If there is to be an extensive delay in their leaving, the president may then, at his discretion, allow members to leave. Some unobtrusive signal, such as casing a unit flag, would be an appropriate means of notifying members the dining-in is over. Mr. Vice will be the last member to leave the dining-in.

Presentation of Awards. If individual and unit achievements are to be recognized at the dining-in, an appropriate ceremony should be arranged. This ceremony should take place during the formal portion of the dining-in. A convenient time would be immediately preceding the guest of honor's speech. Under no circumstances should any ceremony follow directly after his speech, because this speech should be the highlight of the dining-in.

Duties of President and Others. The detailed duties of the President of the Mess, the arrangements officer, the host officers, the protocol officer, the mess officer, and Mr. Vice should be specified in advance. For example, the duties of the president of the mess might include the following:

1. Set the date.
2. Determine the location.
3. Secure a suitable speaker. In the invitation to the guest speaker such information as description of the audience, a description of the occasion, and some suggested topic areas will be helpful to the speaker.
4. Arrange for a chaplain to give the invocation. Inform him regarding the invited guests.

5. Appoint any or all of the following:

a. *Arrangements Officer.* (To prepare table arrangements, name cards [head table] and organization identification cards, flags, trophies, public address system, president's gavel and board, dinner chimes at Mr. Vice's place, pencils and pads at head table, guest of honor cup on hand and engraved, awards on hand, photographer available, agenda prepared, guest list available.)

b. *Host Officers.* (To contact designated guest[s] in advance and inform him as to mess customs, dress, place, agenda, and other guests; arrange to meet him on arrival, provide transportation if needed and quarters; introduce guest to president at mess, other guests and members, brief him where he is to sit; when he leaves see that he is escorted to car and given a farewell in behalf of all members of the mess.)

c. *Mess Officer.* (To make dining room reservations, select menu, coordinate timing of food service with head waiter to agree with desires of president of mess; arrange for mess charges to be paid; arrange for bar facilities if president so indicates; if steak is to be served, arrange for use of "steak indicators" except at head table where a waiter should take individual orders.)

d. *Protocol Officer.* (To prepare invitations to all guests for president's signature, brief and assist the host officers as necessary, prepare biographical sketches, prepare follow-up correspondence with guest of honor after the dining-in.)

6. Appoint Mr. Vice. (A junior officer, selected for his wit and ability to speak. Brief him on his duties.)
7. Meet all guests before dinner is served.
8. Open the mess with one rap of the gavel and close it with two raps.
9. When introducing the guests to the mess, leave no room for doubt in the minds of the guests as to when and if they are to acknowledge the introduction. Avoid over-flattering comments.
10. When the president desires that a toast be made, he should call on Mr. Vice. A simple "Mr. Vice" will suffice. (The president will have previously briefed Mr. Vice, who will know which toast is to be proposed.)

Duties of Mr. Vice. The president may assign such duties as the following:
1. Prepare appropriate toasts as directed by the president. Poems and ditties by Mr. Vice are encouraged.
2. Be prepared to propose any other toasts the president may call for during the evening.
3. Be prepared to help the president resolve unforeseen requirements for toast proposals.
4. Sound the dinner chimes at appropriate times.
5. Be the last officer to leave the mess.

Examples of toasts are shown below.

HONORED GUEST SPEAKER:

Most honored Guest,
Your words of wisdom we did hear
And if your footsteps we can follow
This Air Force, this Nation of ours
Will have nothing to fear.

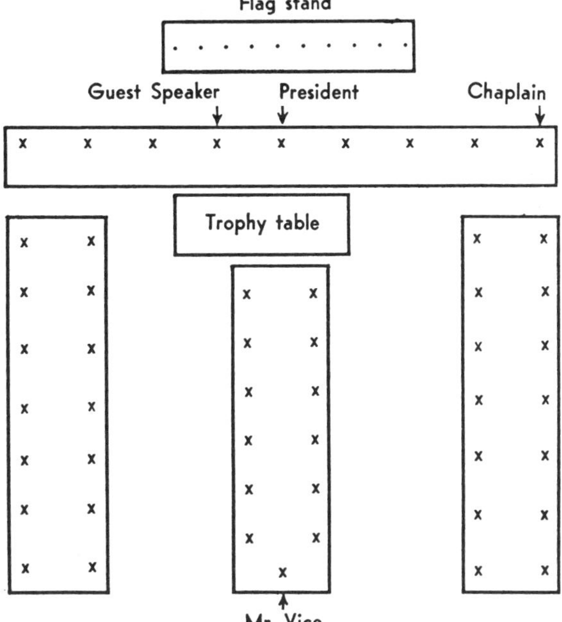

TABLE LAYOUT AND SEATING ARRANGEMENTS.

—"TO OUR HONORED GUEST."
We're glad that you have been our guest
And hope you feel the same.
Your wisdom, be sure, we will embrace,
But for your wit, YOU must take the blame.
—"TO OUR GUEST SPEAKER."

GUESTS:
To our guests from far and wide:
Accept our friendship, joy and pride.
Join our laughter, enjoy our wine
And may your future ever shine.
—"TO OUR GUESTS."

* * *

Honored guests from side to side,
You grace our tables to our pride.
To you honored guests we drink this toast:
We hope you enjoy the evening the most.
—"TO OUR GUESTS."

* * *

TO ALL MEMBERS (UNITS):
To you who join our December crowd
We sing praises long and loud.
Let's take our cups and drink good cheer.

One and all, we're glad you're here.
—"TO OUR MEMBERS."

TO THE OLD AND NEW GUARD:
To those that are leaving
Let all remember
That they, at their beginning,
Were once young and tender,
And to those that now
Hold the future in balance,
Face up to the task
And accept the challenge.
—"TO THE OLD AND NEW"

Agenda. The agenda should be prepared to fit the needs of the specific occasion. A sample is shown below.

WING DINING-IN

DATE

PLACE

1800—Arrive to meet guests.

1827—Informal period ends. Please move into the dining room upon sounding of dinner bell by Mr. Vice. (Do not take glasses to dining room. Wine is served for toasting.) No smoking.

1830—The Mess is formally opened with one rap of the gavel by the President of the Mess. (Remain standing.) (For Air Force units, the commander or senior officer will act as President.)

1833—Invocation by Chaplain . . . .

1835—Toast to the President of the U.S. (Response: "To the President") Toasts to the heads of allied nations, if representatives are present. (Response should be the proper one for that particular country and office.) Toast to the Chief of Staff, USAF, by Mr. Vice. (Response: "To the Chief of Staff.")

1836—President seats the Mess.

1837—Welcoming remarks by the President of the Mess.

1840—Introduction of all guests (if guests are present).

1850—Toast by Mr. Vice to our guests, if applicable (guests remain seated). (Response: "Hear, Hear.")

1851—Cocktail, salad, and steak courses are served. Place a steak indicator in front of your place: pink—rare; blue—medium; yellow—well done. Indicators will be on your table in a glass. Dessert, coffee, and cigars as soon as main course is completed. (The Mess may smoke when the president lights up.)

1935—Individual achievements recognized and awards presented.

1945—Explanation of unit achievements and award of trophies (if appropriate).

2000—Introduction of the speaker by the president.

2005—Message by speaker.

2025—President calls for toast to speaker.

2026—Closing remarks by president.

2030—President adjourns Mess.

2045—Move back to dining room on sounding of bell for entertainment and informal activities. (A senior officer will previously be designated to act as Master of Ceremonies for these activities.)

21

Personal Affairs and Aid for Your Dependents

Man has two supreme loyalties—to country and to family. . . . So long as their families are safe, they will defend their country, believing that by their sacrifice they are safeguarding their families also. —B.H. Liddell Hart

All officers of the United States Air Force may expect sudden changes of station, either temporary or permanent in nature, or changes in their personal status. Personnel affected by such movements may anticipate months or years of separation from their dependents, during which time they may be unable to properly attend to their personal affairs and welfare.

Efficiency of performance varies in direct ratio to an individual's peace of mind and mental stability. No individual desires to expose dependents to distress, want, or insecurity. Every person in the Air Force should arrange his personal affairs so that dependents will be adequately protected and provided for in order to prevent any legal entanglement or embarrassment which could be caused by his absence.

The time to prepare for the sudden separation caused by military transfer is now, while personal matters are moving along in a routine manner, rather than during that turbulent and busy period immediately preceding departure upon a change of station. It is imperative that military personnel keep their personal affairs properly arranged to provide the maximum protection and security for their dependents and that they insure that dependents have ample knowledge of and receive all rights and benefits to which they are entitled. Since individual preparations are determined by the needs of the individuals concerned, it is impossible to prescribe an exact set of preparations applicable to all personnel in the Air Force. The status of some may require only simple adjustments while others may require complicated and minutely detailed arrangements. To assist all personnel in properly arranging their personal affairs, this chapter sets forth in detail the varied subjects that Air Force personnel should consider in providing for the welfare, protection, and security of their dependents at all times.

Air Force Personal Affairs Guides. The Air Force is acutely conscious of the importance to its members of sound guidance in providing assistance in the solution of personal problems. For this reason each base is provided with a Personal Affairs staff. As an aid to individuals there has been issued a series of Air Force Pamphlets. For example, AFP 211-30, *Your Personal Affairs.* In their sum they provide a sound approach to all manner of personal problems.

Members of the Air Force who need advice on personal matters should go first to their Personal Affairs Officer, state their problem candidly and completely, and invite guidance. It will be extended willingly, authoritatively, and without cost. The confidence divulged will be protected.

ALLOTMENTS OF PAY AND DEDUCTIONS

General. Allotments of pay may be made to: (1) individuals or banks for the support of dependents, or savings (including a checking account); (2) life insurance companies for the payment of premiums on life insurance on the life of the allotter; (3) Federal savings and loan associations; (4) a lending institution holding a loan insured by the Federal Housing Administration of the Housing and Home Finance Agency; or the Federal Housing Administration; (5) Air Force Aid Society, and American Red Cross for payment of loans made by these organizations to military personnel. Such allotments may be authorized by all military personnel on active duty wherever serving. (See AFM 177-373.)

Authorization, Discontinuances, and Limitations of Allotments. AF Form 1548 (Allotment Authorization) will be executed in duplicate for the purpose of authorizing, reauthorizing, changing or discontinuing Class S allotments. Class S allotments to a bank for savings (including a checking account), must be made to the credit of one individual only. No allotment for savings (including a checking account) will be made until the allottee has made satisfactory arrangements with the bank for acceptance.

Officers and warrant officers may allot so much of their basic pay, monthly subsistence allowance, and quarters allowance as will leave sufficient balance equal to or greater than the amount of income tax to be withheld in accordance with AFM 177-373 and current Internal Revenue Service regulations.

Emergency Allotments. In exceptional cases, when persons in active service have not made adequate provision for the support of certain dependents, the Secretary of the Air Force may act in their behalf to make allotments for the well-being and protection of their dependents. Such allotments may be terminated at the request of the person whose pay is charged therewith (AFM 177-373).

LEGAL ASSISTANCE FOR MILITARY PERSONNEL (AFR 110-22)

If a member of the Armed Forces has a problem of such a nature that the services of an attorney are desirable or necessary, he or she may go to the nearest Legal Assistance Officer attached to the Judge Advocate's Office on any base.

This officer is in most cases a licensed attorney, or a person trained in the law. The Legal Assistance Officer will counsel with, advise, carry on necessary correspondence, negotiate and draw any legal papers necessary. The LAO is, by regulation, prohibited from entering a formal appearance in a litigated matter.

Wills of Air Force Personnel. *Desirability.* Every person in the Air Force should give consideration to the making of a will, if he or she has not already done so. Whether a will is necessary or desirable and the form it should take depend on the desires and circumstances of the individual, and the laws of the place of execution and of the probable place of probate. State laws govern the execution and probate of wills and the requirements in these respects vary considerably among the various States.

When a person dies without having made a will, he or she is said to have died "intestate." In such event the estate is administered and distributed according to the statute of "Descent and Distribution" of the State of his legal domicile or, in the case of real property, the State or States in which such real property is located. This law thus, in effect, creates a will for the individual, which is administered by an "administrator" appointed by the court, and the estate is distributed to the family members in the manner prescribed by the statute.

If the manner of distribution so prescribed by law does not meet a person's desires or needs, he or she can direct a distribution of his estate according to his or her wishes by making a will.

Review of a Will. An executed will should be reviewed from time to time especially when such events as marriage, birth of children, divorce, or death of a named beneficiary occur. Changes of status such as these often affect the provisions of a will. A will is not effective until death and can be replaced by a new will or changed by a codicil at any time that the testator feels it is necessary. Unless the will is extremely complicated the use of codicils should be avoided. If the will is replaced by a new will, all copies of the previous will should be destroyed. Whenever circumstances change, the officer should review his will with the aid of legal counsel.

Safekeeping. Military personnel, because of their transient status, should not keep the original copy of their will in their possession. After completion, the original copy of the will should be mailed to the named executor or chief beneficiary for safekeeping, or placed in some other secure place where it will be available in the event of death. It is advisable, for reference purposes, to keep a copy of the will, with a notation as to the location of the original will. The principal beneficiary should be advised of the location of the original.

Emergency Will. In the event that an emergency arises which necessitates the immediate execution of a will, the officer should write out his or her desires as to the distribution of possessions and have it attested by three competent witnesses. Such a handwritten will should state the full name, grade, service number, and permanent address of the officer, the exact desires as to the distribution of possessions, the name and address of the person desired to be the executor and the date, place and circumstances of execution. Such a will should definitely be replaced as soon as possible by one prepared with the aid of legal counsel.

Power of Attorney. A power of attorney is a legal instrument whereby one person may designate another person to act in his behalf in legal or personal matters. The person executing the power of attorney is usually referred to as the "principal" and the person to whom the authority is given is usually referred to as the "attorney-in-fact" or simply as the "attorney" for the principal. The authority of a power of attorney may be granted to a member of the principal's family, or to any other person of legal age and capacity; however, when appointing his attorney-in-fact, the principal should select a person in whom he or she has complete trust and confidence. The authority given in a power of attorney, unlike that of a will, becomes invalid upon the death of the principal.

The power of attorney can be made very general and unlimited in scope, or it can be restricted to certain specific functions, depending upon the needs and desires of the principal. A power of attorney may or may not be honored depending upon its acceptance by the individual to whom it may be presented for a transaction in the principal's name. Since the principal is held responsible for the actions of his attorney-in-fact performed within the limits of the power of attorney, it is advisable not to execute power of attorney until a specific need or use for it exists. As a rule, a restricted power of attorney will accomplish the specific needs of the individual.

Preparation and Legal Counsel. The requirements as to preparation of the form and context of a legally effective power of attorney vary considerably under the laws of the various States. For this reason, and in order to properly fulfill the needs and desires of the principal, each power of attorney should be individually prepared under the guidance of legal counsel with due regard to the laws of the State of execution and of the probable place of exercise of the powers granted. Legal counsel should be consulted for the preparation of a power of attorney when the need for such instrument exists.

Estates. *Kinds of Property.* A person accumulates various types of possessions during a lifetime, which become known collectively as an "estate." This estate may consist of "real estate," which is land and any buildings, constructions, fixtures, and improvements erected thereon or attached thereto, "personalty," which includes all items of personal property such as clothes, household furnishings, automobiles, money, stocks, bonds, jewelry, and, in general, any property that is not "real estate," or a combination of both "real estate" and "personalty."

Importance of Joint Tenancy. Air Force personnel, in their arrangements of personal matters for the protection of their dependents, should consider the importance and advantages of arranging title to most of their property by "joint tenancy." In this way, the officer will enable the joint tenant to use and control the property during his lifetime and in the event of his or her death, to obtain full title as survivor without the property being subject to administration through the courts. Property owned in joint tenancy cannot be disposed of by a will if the joint tenant survives him or her. The solution to the officer's problem of providing for the care, welfare, and comfort of dependents in the event of prolonged absence or death, can be more effectively accomplished through the establishment of title to property by "joint tenancy" than it can be done by a power of attorney and a will. The rights of control of property by the joint tenant are full and absolute when title is kept in joint tenancy and upon the death of the officer the property will pass automatically into full possession of the joint tenant without judicial proceedings. However, personal property held in joint tenancy may make it subject to tax because the Soldiers and Sailors Relief Act does not protect dependents.

Sound appraisement on the part of the owner concerning the capabilities and integrity of the contemplated joint owner should be exercised, and, in general, it is *not* advisable to establish joint property title with a person of short acquaintance.

Except as limited by State statutes, titles by joint tenancy with right of survivorship can be granted for various types of property. Deeds establishing ownership to real estate, and bills of sale for personal property such as automobiles, machinery, household goods or livestock can be held in joint tenancy with right of survivorship. Joint bank accounts can be opened, stocks and bonds can be issued to joint owners, with the right of survivorship in each case.

United States Savings Bonds can be issued to joint owners with right of survivorship if the purchaser so requests. A savings bond already issued to a single owner (without a designated beneficiary) is part of the estate but can be reissued during his or her lifetime to joint owners upon proper application to the Treasury Department. Bonds issued to a single owner with a designated beneficiary can be cashed only by the owner during his or her lifetime and on death of the owner do not become part of the estate but go direct to the named beneficiary. Such bonds cannot be reissued without the consent of the beneficiary. A bond held in joint ownership can be cashed by either joint owner with or without the consent of the other joint tenant.

Legal Assistance. Legal advice should be obtained before any individual creates a joint tenancy title on his or her property. Gift, inheritance, or other taxes may often affect the enactment of joint tenancy titles. Competent legal assistance will inform the individual of the advantages and disadvantages of holding the various kinds of property in joint tenancy and will also facilitate the proper establishment of such an estate with due regard to all applicable laws (see AFR 110-22).

Safe-Deposit Boxes. Many banks maintain safe-deposit boxes which are rented to individuals for a yearly fee which varies with the size of the box. When available, a safe-deposit box in a conveniently located reputable bank or trust company is usually the best place to keep valuable papers, such as stocks, bonds, deeds to property, insurance policies, receipts, and the original copy of the individual's will.

Automobile. *Title.* The determination as to the proper owner of an automobile is made, not by possession, but by the certificate of title, or other evidence of ownership, from the State in which the automobile is registered. Most States maintain a bureau of motor vehicles or some other similar agency for this purpose. If the title of the automobile is in the name of one person, such automobile, in the event of the death of the owner, would become part of the estate, subject to his or her will, if executed, or the laws of "Descent and Distribution" of the appropriate State. The actual determination as to disposition of the automobile, pending the result of probate, would rest with the State authorities. Subject to statutory limitations of the State of registration, title to an automobile may be held jointly by a husband and wife. Under such title, in the event of death of one, the automobile would become the property of the survivor. This method is usually the most effective for providing for the use of and title to the automobile for the spouse in case of prolonged absence or death of the officer. It has its disadvantages, however, in the statutory limitations mentioned above, the fact that difficulties are sometimes encountered in effecting transfer of the automobile without the presence and signatures of both joint owners, and in the spouse's being subject to personal property taxes in some states.

Changing of Title. Since State laws govern the transfer of title to an automobile, extreme care should be exercised that all provisions of law relative to such transfer are met. If an individual desires to transfer title to an automobile from himself to his or her spouse and the officer jointly, it is advisable to communicate with the bureau of motor vehicles or other similar agency in the State in which the automobile is registered, requesting the exact procedure necessary to accomplish such action. Upon receipt of the instructions, the individual should utilize the legal assistance facilities in completion of the transfer of title.

Importance of Insurance. In many cases, an automobile is the most valuable single item of property owned by the officer. It is, therefore, advisable to protect such property with adequate insurance. The insurance policy should cover the owner, the spouse and all other persons, who may have occasion to drive the vehicle. Lack of adequate insurance protection may result in extreme financial, legal, and personal difficulties.

Many states have enacted or are enacting so-called "compulsory insurance laws" and strict compliance with these laws is necessary. Officers should not be deceived into a false state of security by possessing only the amount of insurance required by law. In some States, compulsory insurance provides only a minimum of protection rather than adequate insurance protection.

Soldiers' and Sailors' Civil Relief Act. The purpose of the Soldiers' and Sailors' Civil Relief Act is to relieve persons in the military service from worry over the inability to meet their civil obligations by providing adequate representation for the service member during an absence and postponement of certain civil proceedings and transactions until release from such military service. There is nothing in the act which relieves the person from the actual payment of the debts or other obligations, but in the event that the service member is unable to pay premiums on commercial insurance policies, to pay taxes, or to perform other obligations with reference to right and claims to lands of the United States, certain relief may be afforded him or her by this act. In the event of legal action based on a serviceman's breach of obligation, the relief afforded under this act is within the discretion of the court and dependent upon whether the ability of the service member to discharge his or her obligations or to prosecute the action or defense, is materially affected by reason of military service.

Insurance. The payment of premiums on commercial life insurance may be protected under Article IV of this act.

Other Protections. In addition to the protection of commercial life insurance, the Soldiers' and Sailors' Civil Relief Act provides arrangements for adequate legal representation and stays of execution for servicemen on the following legal matters:

(1) Eviction for nonpayment of rent.

(2) Court proceedings arising from mortgages, leases, liens, and other contractual obligations.

(3) Payment of taxes may be deferred in some cases.

(4) An officer is not required to pay local taxes in the State where stationed unless, by some act of the individual, he or she becomes a legal resident, provided the officer maintains legal residence in another State and discharges his or her liability in that State.

Importance of Legal Assistance. This summary on the Soldiers' and Sailors' Civil Relief Act has mentioned only the major points contained in the act. It is very important, therefore, that any officer who is suddenly presented with a tax bill or threatened with legal action of any kind while away from his State of domicile, should immediately report the matter to his Legal Assistance Officer who can then study the case and possibly avert court action through implementation of some of the points of the Soldiers' and Sailors' Civil Relief Act. Too often, individuals delay in notifying the Legal Assistance Officer of their legal difficulties until the case has already reached the litigation stage with the result that the Legal Assistance Officer is unable to assist. The importance of prompt notification and utilization of Legal Assistance facilities in any legal entanglement cannot be overemphasized.

Joint Bank Accounts. *Meaning of Joint Bank Accounts.* A joint bank account is one in which two persons have full authority to perform the functions of depositing and withdrawing funds from the same account. Such an account may be either a savings account or a checking account and is considered to be the joint property of the persons concerned. It is possible in most States to maintain joint accounts and most banks have a specially prepared contract form setting out the legal status of such account in accordance with the State laws which govern the operations of the bank. Such a contract and the accompanying pass book usually contain a phrase somewhat as follows: "John Doe and Mary Doe, jointly or either, with rights of survivorship and not as tenants in common."

Advantages of Joint Bank Accounts. The chief advantage of a joint account is that the funds are readily available to either or both of the parties at any time. This is especially important to Air Force personnel because by maintaining their accounts in such a status, the dependents will be able to obtain the funds therein even though the officer is absent for any reason whatsoever. If the officer carries a bank account in his name only, the dependents may be deprived of the use of the money at a time when it is most urgently needed. Another great advantage is that in the event of death of one of the parties, the survivor automatically becomes sole owner of the money in the account, usually without having the funds frozen while the will of the deceased is being probated. Thus, the spouse or other dependents will have the money available at a time when it is urgently needed.

Allotments to Joint Bank Accounts. Allotments to a bank for deposit in a savings or checking account must be made to the credit of one individual. However, it is permissible to make an allotment to a bank even though the person to whom the allotment is credited is a party in a joint account. Proper arrangements for acceptance and depositing of the allotment must be made with the bank concerned prior to initiation of such an allotment (AFM 177-373).

Credit Unions. The Air Force encourages and assists the establishment of credit union facilities on Air Force installations as cooperative organizations to stimulate systematic savings and create a source of credit for both provident and productive purposes. One is usually to be found on any good-sized base (AFR 170–17).

AGENCIES OF ASSISTANCE TO AIR FORCE PERSONNEL AND THEIR DEPENDENTS

CHAMPUS. For families of active service personnel, this program is extremely important. The CHAMPUS program provides extensive medical care and hospitalization as a right, not as a mere privilege, and at greatly reduced cost to the individual.

Officers are urged to determine upon arrival at any new station, or new residence for family members, the place each would go for medical care or hospitalization—civilian facility or service facility. This includes the situation of a son or daughter away at school or college. Leave nothing to chance and avoid delay if emergency arises.

The CHAMPUS service provided is of a very high standard. The Air Force provides highly capable direction from the top. Still, not every human ailment or frailty is covered, and not every item of expense is paid by the Government. The officer must learn of these exceptions and apply them, if necessary, to his or her own family situation.

Under the CHAMPUS Program from civilian sources, it is accurate to say that the major portion of the costs are paid by the Government for all normal illnesses and needs for surgical care. Partial charges are made, as stated later.

There are definite exclusions for which the Government does not pay the costs. In broad terms, these are chronic situations, cosmetic or voluntary surgery, domiciliary situations, treatment of congenital defects, and conditions which are non-acute.

The Armed Forces CHAMPUS program provides for the inpatient care in civilian hospitals to include outpatient service for spouses and children of active duty personnel. The legislation authorized both a new civilian-hospitalization program and outpatient care from civilian medical facilities for military retirees and their spouses and children, and also for the spouses and children of deceased active or retired military personnel.

Separate assistance is available for mentally retarded and physically handicapped spouses and children of active duty personnel.

Military medical facilities may still be used by dependents and by retirees and their spouses and children on a "space available" basis.

Cost of Civilian Outpatient Care for Spouses and Children of Active Duty and Retired Personnel and for Retirees. The charge to family members of active duty personnel using civilian medical facilities for outpatient care is $50 per year per person (but not more than $100 per family) *and* 20 percent of the remaining outpatient cost.

The charge to retired personnel and to their spouses and children using civilian medical facilities for outpatient care is also $50 per person (but not more than $100 per family unit) *and 25* percent of the remaining outpatient cost.

Cost of Civilian Hospital Care. For spouses and children of active duty military personnel receiving inpatient care in a civilian hospital, there is a basic charge of

$5.50 per day, or the first $25 of the hospital cost, whichever is greater. There are additional charges, as for example, for a private hospital room (if the attending physician certifies a private room is needed, the additional cost is 25 percent of the difference between the cost of the private room and the average cost of a semiprivate room; if the physician does not so certify, the patient pays the additional cost of the private room in full.) For private-duty nursing care, the patient pays the first $100 of cost and 25 percent of charges over $100 when the attending surgeon certifies that this care is needed. Maternity care costs: the first $15 of the physician's charge for a delivery performed in a home or office, if the patient is not hospitalized later incident to the same delivery.

The charge to retired personnel and to their spouses and children (and to the dependents of deceased military personnel) receiving inpatient care in a civilian hospital, is 25 percent of the cost (the Government paying 75 percent of the cost, including physicians' fees).

Note: Retirees and their dependents who become eligible for Social Security Medicare at age 65 will no longer be entitled to the civilian hospitalization and civilian outpatient care provided in CHAMPUS. They will continue to be eligible for "space available" care in military facilities.

Identification. When applying for any kind of medical care—at a service or civilian facility, or to a civilian physician—dependents are required to present their Identification Card (DD Form 1173) as proof of their eligiblity for medical care. If a person uses this card to obtain medical care not entitled, a fine of up to $10,000 and imprisonment for up to five years may be imposed on the offender. A dependent who allows another person to use his or her card unlawfully may be subject to the same penalties.

Caution: Ask the civilian physician if he or she will participate in the CHAMPUS program. If not, another physician should be sought. You may identify those in your community who do participate by inquiring of the American Medical Association, Medical Bureau, or similar offices in your area. Under Service regulations, understood by participating civilian physicians, the Government will pay for authorized care with the understanding that there will be no additional charge above those authorized to the dependent or sponsor for that care. (There have been instances of physicians who have asked and collected from patients sums in addition to the charges they have agreed to accept.)

Dental Care Provided to Service Families. Dental treatment is provided to hospital inpatients who are hospitalized for other authorized care, but only when required as a necessary part of the treatment of the basic medical or surgical condition for which hospitalized. Outpatient treatment of fractures, dislocations, lacerations, and other wounds that are legitimately cared for by dentists may also be paid for. Authorized dental care does not include artificial teeth, bridges, fillings, teeth straightening, or prolonged treatment of the gums.

At Armed Forces and U.S. Public Health Service medical facilities, dental care is provided as follows: In the United States: (1) In an emergency, to relieve pain and undue suffering. Permanent fillings, bridges, and dentures are not authorized. (2) If required for treatment of a medical or surgical condition. (3) Dental care in areas designated "remote" on a facilities-available basis.

Outside the United States dental care may be provided on a facilities-available basis.

FAMILY SERVICES PROGRAM

At each Air Force base an organization exists to aid Air Force personnel and their families in meeting personal problems. The organization consists of volunteers, usually the wives of Air Force officers and airmen stationed at the air base.

The Family Services Center. The focus of operations is known as the Family Services Center, which is an office or building located on the base.

The activities of the Family Services Program include providing assistance to Air Force families in respect to such matters as:

(1) Information concerning the air base and its adjoining community.
(2) Information concerning housing on and off base.
(3) Thrift shops.
(4) Nurseries.
(5) Other air base services and facilities.

The Family Services Center provides help to families in the event of an emergency or casualty by arranging for transportation, baby-sitters, necessary shopping, and other assistance. The Center usually maintains a loan service of household utensils sufficient to meet minimum needs of a family awaiting the arrival of their household goods. At many bases, an FSC Welcoming Committee member calls on newly arrived personnel to offer aid in any degree needed. The Family Services Center operates an indoctrination course to inform spouses of Air Force personnel of essential facts concerning allotments, insurance, social security, retirements, casualty benefits, and the like (AFR 211-24).

AIR FORCE AID SOCIETY (AFR 211-1)

Mission. The Air Force Aid Society is the emergency relief organization for the United States Air Force. Its mission is to improve the morale and welfare of Air Force personnel and their dependents by providing financial assistance in times of emergency. The Aid Society augments the relief available to the Service member through the American Red Cross, and it is not intended that the Aid Society will compete with or replace the work of the American Red Cross. This is because the Congressional charter of the American Red Cross makes it the primary relief organization for the Armed Forces.

Eligibility. The following classes of individuals are eligible for assistance:

(1) USAF personnel on active duty and USAF personnel retired from active duty for length of service or disability, and their dependents.

(2) Dependents of recently deceased USAF personnel who died while on active duty or after they were retired from active duty for length of service or disability.

(3) AF Reserve and Air National Guard personnel on full-time active duty with the USAF for at least 90 days, and their dependents. (Does not include tours of active duty for training, or attendance at service academy or Armed Forces Preparatory School.)

(4) Army, Navy, and Marine personnel and their dependents may receive assistance through the Air Force Aid Society when an AER or NRS office is not readily accessible and it is not feasible to refer the applicant to the Red Cross.

Eligible dependents are the spouse and minor children of Air Force personnel. Parents who are wholly dependent upon the service member may be considered eligible dependents. The existence of a Class D allotment to a relative other than the spouse or child of an airman is acceptable as reasonable evidence of an airman's acceptance of responsibility to support the relative. The existence of a voluntary allotment in an amount constituting substantial support, regular maintenance as a member of an officer's household, or other reliable indication of true dependency will be accepted as evidence that an officer's dependent is eligible for AFAS assistance.

Assistance. Aid Society assistance is rendered in the form of non-interest-bearing loans or cash grants or combinations of both. Assistance may be obtained by those entitled to it by contacting the Aid Society Officer at any Air Force installation or by writing to the Air Force Personnel Center, Randolph AFB, Texas. Persons eligible for assistance who are not connected with, nor living at or near an Air Force Base should make application for assistance from the Aid Society through their local chapter of the American Red Cross.

The Air Force Aid Society is a charitable organization. It receives no Government funds which can be used for relief purposes. The income of the Aid Society comes from gifts and contributions, royalties from books and songs which have been given to the Aid Society, legacies, and interest on the invested capital. The limited resources of the Air Society and the very large number of persons eligible under its charter for assistance, make it mandatory for Aid Society assistance to be limited to cases of emergency only and also make it essential that Aid Society assistance be restricted to those cases which do not rightfully belong to some other welfare organization.

AMERICAN RED CROSS

Consistent with the Congressional charter of the American National Red Cross and under AFR 211-11, the Red Cross has been charged with the primary responsibility of a broad program of volunteer aid to military personnel and their dependents. Army Emergency Relief and the Air Force Aid Society coordinate their functions with the American Red Cross in such a manner as to avoid duplication of effort in providing for the welfare of military personnel and their dependents. The American Red Cross carries out its responsibilities through personnel assigned to military establishments and through the Home Service program of local Red Cross chapters.

AIR FORCE CHAPLAINS

Religious Services. *Marriage.* The chaplain is authorized to perform the marriage rite, provided that all local laws are met and proper legal permission is obtained in each case. Persons contemplating marriage are urged to confer with the chaplain of their faith in order to insure that proper arrangements are made for the marriage.

Funerals. The commander or his representative will assist in making funeral arrangements and the chaplain will conduct appropriate burial services at the interment of members of the military service, active and retired, and for their families, when requested. If the families of personnel who die in the United States request transportation of the body to a home burial ground instead of permitting burial in a post or national cemetery, they will normally be expected to provide a clergyman for the burial services because the chaplain's other duties preclude his absence from the unit for an extended period.

Other Services. Chaplains conduct such other services as are required for the religious guidance of the members of each faith. In this manner, Air Force personnel are afforded every opportunity to attend their individual religious services.

Conferences and Retreats. Chaplains, assisted by civilian religious leaders, are authorized to conduct religious conferences and retreats for Air Force personnel. For the purpose of attending such conferences and retreats, Air Force personnel may be placed on temporary duty. These may be conducted at either military or civilian locations.

For further details, see AFR 265-1.

PERSONAL AFFAIRS OFFICER

Casualty Assistance. The Personal Affairs Officer assists and advises the dependents of deceased service members of the various benefits and privileges to which they are entitled. Such advice includes information regarding six months death gratuity, dependents' compensation, arrears in pay, personal effects, family allowance, burial flag, settlement of Government Life Insurance, and burial allowance. The Personal Affairs Officer is also in a position to refer the dependents to related agencies in regard to civil-service preference, Social Security benefits, States' benefits, transportation of dependents and household goods, and issuance of grave markers.

Other Assistance. Personal Affairs Officers advise military personnel and their dependents as to their rights in securing benefits from the Government and as to sources of information and procedures, but are strictly forbidden by law from acting as an agent or attorney for such persons other than in the discharge of their official duties. In his or her capacity as advisor to military personnel and their dependents, the Personal Affairs Officer maintains liaison with the American Red Cross director, the Air Force Aid Society officer and other agencies, in order that he or she may tactfully bring to the attention of these organizations instances of personnel and their families who are in financial need or in need of other assistance (AFR 211-3).

SCHOLARSHIPS AND CONCESSIONS FOR DEPENDENTS OF AIR FORCE PERSONNEL

Air Force Educational Assistance Program. This program provides through the use of Central Welfare Funds for the granting of merit scholarships and long-term, interest-free loans to dependent children of active duty personnel.

The scholarship aspect provides for up to 30 four-year scholarships annually. Recipients are selected from among those achieving the highest scores on qualifying tests. The loan provision stipulates only that a recipient must have the academic aptitude necessary to assure a "reasonable chance" of completing college. Loans may be continued on a yearly basis throughout the normal four-year academic program, provided academic progress is satisfactory and there is still a need. Loans must be repaid within 10 years of completion of the schooling.

Air Force Aid Society. *The General Henry H. Arnold Student Loan Program.* Those eligible for this program are children (including stepchildren and legally adopted children of an Air Force member in any of the following categories: (1) active member of the Air Force; (2) selected Reserve (Air National Guard,

Category A or B of the USAF Reserve); (3) USAF Reserve and National Guard on continuous active duty during the school term for which assistance is requested spanning the entire term; (4) retired because of length of active duty service, disability, or attainment of age 60 (Reserve Component); and, (4) deceased while on active duty or in retired status. Additionally, the applicant must be a United States citizen enrolled and in good standing at, or accepted for admission to, an approved educational institution on a full-time basis.

Applicants or interested persons who desire additional information should be advised to write to the Director, Air Force Aid Society, National Headquarters, Washington, D. C. 20333.

War Orphans Scholarship, Inc. This organization provides scholarships to dependent orphans (16 years or over) of veterans of World War II. The scholarships are provided in units of $500 and paid through the college or vocational school treasurer as the money is needed to pay for school or living expenses. Dependents desiring to make application or to obtain further information should address such inquiries to War Orphans Scholarship, Inc. Room 219A, Veterans Administration Building, Vermont and H Streets, N.W., Washington, D. C.

The Rockefeller Foundation. Veterans who desire to continue their education may receive assistance through an emergency fellowship granted by the Rockefeller Foundation. Those receiving educational benefits under the GI Bill may receive supplemental aid through the Foundation while those veterans who are ineligible for GI Bill assistance may apply for the full cost of education. All inquiries regarding assistance by the Rockefeller Foundation should be addressed to the organization at 49 West Forty-ninth Street, New York, N.Y.

Other Sources of Scholarships. Many States, universities, colleges, and junior colleges offer scholarships or concessions to children of service personnel. Requests for information regarding such policies should be made to local State authorities or to the institution which the applicant wishes to attend.

APPOINTMENT TO THE FEDERAL MILITARY ACADEMIES

The President is authorized to appoint a limited number of cadets from the United States at large, to be selected from sons and daughters of members of the armed services who were killed or died in service as the result of wounds, injuries, or disease received or aggravated in active service. All such appointees must be otherwise qualified and will be selected in order of merit as established by competitive examination.

Sons and daughters of persons who have been awarded the Medal of Honor, if otherwise qualified, may be appointed by the President as cadets from the United States at large to the Federal Military Academies.

IN-SERVICE HOUSING LOANS

If an officer has been on extended active duty for at least two years, he may finance the purchase of a home by an FHA loan. The law provides a system of mortgage insurance which enables the officer to more easily build or purchase homes. It authorizes the FHA to insure loans up to 97 percent of an appraised value. In addition, the Air Force will pay ½ of 1 percent of the mortgage insurance premium monthly. This is the premium the FHA receives for guaranteeing the mortgage. During the life of the loan, this amounts to a substantial sum.

PAY, ALLOWANCES, AND ALLOTMENTS OF PERSONNEL REPORTED MISSING

Pay and Allowances. Any person who is in active service and who is officially determined to be absent in a status of missing, is entitled, for the period he or she is officially carried or determined to be in such status, to have credited to his or her account the same pay and allowances entitled at the beginning of such period of absence or may become entitled thereafter. Such entitlement shall terminate upon the date of receipt by the Department of the Air Force of evidence that the person is dead or upon the date of death prescribed or determined. Such entitlement, however, shall not terminate upon expiration of term of service during absence and in case of death during absence shall not terminate earlier than the dates stated herein. No pay and allowances accrue to such missing persons for any period during which they may be officially determined absent from their posts of duty without authority, and they shall be indebted to the Government for any payments from amounts credited to their account for such period of absence (AFM 177-373).

Allotments. Allotments, which have been instituted prior to the absence of a missing person who is entitled to accrued pay and allowances, will continue to be paid regularly for a period of 12 months from the date of absence. In the absence of an allotment for support of dependents or payment of insurance premiums, or if an existing allotment is insufficient for its purposes, the Secretary of the Air Force can direct that adequate allotments be instituted. Information listed on DD Form 93 (Record of Emergency Data) will be used as a guide in such determinations. The total of all allotments, however, may not exceed the total pay and allowances to which the missing person is entitled.

When the 12 months' period from date of commencement of absence is about to expire in the case of a missing person, the Secretary of the Air Force shall cause a full review of the case to be made. Following such review and when the 12 months' period has expired, the Secretary of the Air Force is authorized to make a finding of death, or if the person is for any reason assumed to be living, to direct the continuance of the person's missing status. When a finding of death is made it shall include the date upon which death shall be presumed to have occurred for purposes of termination of entitlement to accrual of pay and allowances, settlement of accounts, and payment of death gratuities. Such date shall be the day following the day of expiration of the 12 months' period or in cases where the missing status has been extended, a day to be determined by the Secretary of the Air Force.

Any new allotment or increases in allotments authorized by the Secretary of the Air Force for the well-being of dependents or the payment of insurance may be terminated by the missing person upon return to duty status.

Application by dependents for the initiation of new allotments or changes of existing allotments should be filed with the AF Finance Center, 3800 York Street, Denver, Colorado, on the prescribed form which may be obtained at military installations, or from chapters of the American Red Cross. Applications in any form may be accepted if they satisfactorily establish the identity, relationship, and dependency of the applicant and the need for the increase of allotment requested. The application must indicate allotments and family allowances, if any, being paid to the dependents on whose behalf the application is submitted. It must also include or be accompanied by evidence establishing the need for the allotment or increase requested. The specific amount needed and the date the allotment or increase is desired to be effective must be stated.

BENEFITS AFTER SEPARATION

During the period of military service, each individual accrues certain rights and privileges, under various public laws, which will be available to him or her after discharge or release from active duty. The administration and payment of these benefits is the responsibility of the Veterans Administration. This guide will outline briefly the more important benefits to which the ex-service individual may be entitled. Entitlement to benefits varies with length, time and type of service as well as disabilities incurred in service. It is to be emphasized that in each case where a veteran desires to apply for one or more of these statutory benefits, he should obtain all necessary information from the nearest Veterans Administration installation.

Summary of Principal Benefits. *General Eligibility.* Service personnel presently in the Air Force are eligible for veterans' benefits in one of the following categories:

(1) World War II—Active service between December 7, 1941, and December 31, 1946, both dates inclusive, for pension and compensation purposes, and general provisions of veterans benefits;

—Active service between December 7, 1941, and July 25, 1947, for compensation purposes provided there was service before December 31, 1946, and the disability was incurred or aggravated in service between December 7, 1941, and midnight July 25, 1947;

—Active service between September 16, 1940, through July 25, 1947, for GI bill purposes.

(2) Korea—active service between 27 June 1950 and 1 February 1955. The law grants education expenses, on-the-job training, loan benefits, mustering-out pay, compensation, and unemployment benefits.

(3) Vietnam—active service between 5 August 1964 and 7 May 1975, including all periods of the Vietnam War. The law grants benefits similar to those authorized for Korean War service. See the discussion of the 1966 Benefits Law later in this chapter.

In cases where an individual is eligible in more than one category, the individual dependents may elect to receive benefits under the category which will afford them the greatest award. Besides the requirements as to service, the benefits also require that the period of service be terminated by a discharge under conditions other than dishonorable, except in the case of those benefits discussed later in this chapter.

Table of Principal Benefits. The following table indicates the principal benefits provided by the Federal Government for veterans of World War II and peacetime service. It does *not* show Korean or "Cold War" benefits. Eligibility requirements for each benefit are given later in this chapter.

Civil Readjustment (GI Bill)
- Education
- Loan guaranty
- Vocational rehabilitation

Retention of Government Insurance
Hospitalization and domiciliary care
Disability compensation (service connected)
Disability pension (nonservice connected)
Burial allowance

Burial flag
Burial in National Cemetery
Headstone
Dependents' pension
Employment:
  Civil Service preference
  Reemployment following service
Out-patient treatment and prosthetic appliances

Civil Readjustment. *Eligibility.* The Servicemen's Readjustment Act of 1944 is commonly referred to as "The GI Bill of Rights." It provides certain benefits regarding education, loan guaranty, and readjustment allowances. Public Law 16, Seventy-eighth Congress, provides vocational rehabilitation for disabled veterans of World War II. Eligibility requirements for both of these laws are: (1) Have served in active military service some time between September 16, 1940, and July 25, 1947; (2) have served 90 days or more or be aggravated by, service; and (3) have been discharged under conditions other than dishonorable. *All three conditions must be met.*

The Congress has also authorized similar benefits for Korean service between 27 June 1950 and 1 February 1955, and Vietnam War service between 5 August 1964 and 7 May 1975.

Education and Training. Eligible World War II and Korean War veterans received education or training at Government expense, with tuition, supplies, and in most cases a subsistence allowance provided by the GI Bill. This program is now terminated but see the 1966 Veterans Benefits Law later in this chapter.

Loan Guaranty. Eligible veterans may apply for guaranty of loans, under the provisions of the GI Bill. Three types of loans may be guaranteed. They are (1) to purchase, construct, or improve a home; (2) to buy a farm, stock, feed, farm machinery, seed, and other farm supplies and equipment; or (3) to buy a business or to enable a veteran to undertake a legitimate business venture. The Veterans Administration *does not lend* the money, but administers the loan guaranty benefits by pledging the credit of the United States Government for the repayment of such portions of the loan as may be guaranteed. The actual loan must be obtained through community loan sources, after which the Veterans Administration guarantees the lender against loss.

Investigation. Prior to extending the guarantee, the Veterans Administration is required by law to investigate the loan to determine (1) that the sale price on the property to be purchased is reasonable; (2) that the lending agency meets the approval of the Veterans Administration; (3) that the veteran will properly utilize the property after purchase; (4) that the interest rate on the loan is not excessive; and (5) that the repayment plan is within the financial means of the veteran. Unless all these conditions are found to be satisfactory, the Veterans Administration is not authorized to guarantee the loan.

Vocational Rehabilitation. Eligible disabled veterans may receive education or training at Government expense, with tuition, supplies, and subsistence allowances provided by Public Law 16, Seventy-eighth Congress, provided the Veterans Administration determines that they need vocational training to overcome their service-incurred or aggravated handicap.

Eligible veterans may get training for as long as is necessary to restore their ability to work, up to a total of 4 years. Subsistence allowances conform to those provided in the education benefits under the GI Bill but in addition, the Veterans Administration maintains for disabled veterans a counseling service which enables the veteran to select a course of education or training which will serve best to overcome his disability. Additional allowances may be provided, depending upon the veteran's degree of disability and the number of additional dependents such disabled veteran may have.

Eligibility of Active Duty Personnel. Personnel on active military duty, who meet all eligibility requirements of the GI Bill, may apply for the educational benefits and bonus benefits only. Attendance at the chosen school must not conflict with military duties and subsistence allowances cannot be paid.

Vocational rehabilitation is available only to those veterans who actually incurred or aggravated disabilities during World War II, the Korean War, or the Vietnam War.

Hospitalization and Domiciliary Care. *Hospitalization.* The Veterans Administration operates a network of hospitals for the treatment of ill and disabled veterans. World War II, Korean, and Vietnam War veterans and peacetime veterans, if discharged or separated under conditions other than dishonorable, are entitled to hospitalization by the Veterans Administration. The following priority system is followed:

(1) Emergency cases, which may be taken directly to the nearest Veterans Administration hospital. If possible, the hospital should be contacted by phone or telegraph before the arrival of the patient.

(2) Those suffering from injuries or diseases incurred or aggravated during service.

(3) Other veterans who state under oath that they are unable to pay for hospitalization or treatment of nonservice incurred injury or illness. Such cases, if not emergencies, must wait until a bed is available in the Veterans Administration Hospital.

In implementing this program of medical aid for veterans, the Veterans Administration uses agreements with certain other Federal hospitals and contracts with certain private, State, and municipal hospitals, in order to obtain the use of such installations when the facilities of the Veterans Administration are not conveniently located.

Domiciliary Care. Domiciliary care is made available through "Homes" which are designed to take care of those veterans who have a chronic condition which incapacitates them from earning a living but does not require constant medical attention. Requirements for admission for domiciliary care are essentially the same as for hospitalization.

Application. Applications for either hospitalization or domiciliary care may be completed at any Veterans Administration office.

Transportation. Transportation may be provided at Government expense to the hospital or home, if prior Veterans Administration approval is obtained.

Compensation. Veterans, who are disabled by injury or disease incurred in, or aggravated by, service in either World War II, the Korean War, the Vietnam War, or peacetime in line of duty and not the result of their own willful misconduct, may qualify for disability compensation. The individual need not have been separated from service because of the disability.

Monthly wartime rates range from $58 to $1,130 depending upon the degree of disability, as adjudicated by the Veterans Administration. Compensation includes statutory awards for amputation, blindness or other specified disabilities, which may bring the monthly compensation to a maximum of $3,223.

Out-Patient Treatment and Prosthetic Appliances. *Out-Patient Treatment.* Veterans of World War II, the Korean War, the Vietnam War or peacetime service with service-incurred disease or disability which does not necessitate hospitalization, may be authorized out-patient or "home town" treatment for such ailments. Medical, dental, or surgical attention may be provided under this benefit. Each veteran's eligibility must be determined by the Veterans Administration before such treatments will be authorized. Preliminary examinations may be made in a Veterans Administration clinic, or if such facilities are not conveniently available, by a participating physician or dentist. The treatment, if approved, may then be completed by the Veterans Administration clinic or the veteran may be assigned to an approved private dentist or physician who is participating in the medical program of the Veterans Administration on a fee basis. Out-patient treatment may also include the expense of nursing services and drugs. All applications for such treatment must be approved through the local Veterans Administration office.

Prosthetic Appliances. A veteran's eligibility for prosthetic appliances may be established if: (1) the individual has a service-connected or service-aggravated disability requiring an appliance; or (2) if an appliance is determined necessary as a part of hospital treatment or domiciliary care. Complete information and necessary application forms for procuring or repairing needed prosthetic appliances, are available at any Veterans Administration office.

Disability Pension. Disability pensions are available to World War II, Korean War and Vietnam War veterans of limited income who became permanently and totally disabled by reason not connected with their service in the armed forces. There must have been 90 days service unless the veteran was separated prior thereto for disability incurred in line of duty. The veteran must have been in World War II service prior to 31 December 1946; have been separated under other than dishonorable conditions and the disability must not be the result of his own willful misconduct. In determining the period of active service it is not requisite that the 90 days' service shall have been completed prior to the cessation of hostilities. A period of continuous active service for 90 days which commenced prior to and extended into the period of hostilities or which commenced during the period of hostilities and extended beyond, is sufficient.

In order to qualify for the disability pension, the veteran's income may not exceed $4,960 a year if single, or $14,129 per year if dependents are involved. The monthly rate is based on the percentage of disability.

Employment Rights. *Civil Service Preference.* Preference granted to certain veterans for civil service employment in the Federal Government or the government of the District of Columbia may be either a 5-point preference or a 10-point preference. "Veteran preference" is not a guaranty that the veteran will be appointed to a civil service position but means that eligible veterans will have the number of points of their preference added to the rating they earn on civil service examinations. (See AFP 211-35.)

Eligibility of Peacetime Service Veterans. Peacetime service veterans (with only a few exceptions) are *not* eligible for civil service preference unless they are disabled during such service.

Application for Preference. A veteran who wishes to utilize his eligibility for veteran preference should first apply for a civil service position either through the local State employment office or directly to the agency in which he is interested. Such local employment is for temporary civil service positions only. Periodic civil-service examinations are held in which the veteran may compete for a permanent civil service position. Following the results of such examinations, the veteran's preference will be considered in making appointments.

Reemployment Rights. Veterans must have left other than temporary jobs to enter service in the Armed Forces. After leaving the employment to which they claim restoration, their total service performed must not have exceeded 4 years. Upon honorable separation, the veteran must be qualified to perform the duties of the former job unless a disability sustained during service renders him or her incapable of performing such duties.

Restoration to Former Job in Private Employment or with the Federal Government or to a Job of Like Seniority Status and Pay (unless, in Private Employment, Things Have So Changed as to Make Restoration Impossible or Unreasonable). In general a veteran must apply to his former employer within 90 days after separation from active service (or release from hospitalization of not more than a year); however, Reservists and National Guardsmen who perform an initial period of 3 to 6 months of active duty for training must apply within 31 days of their release from active duty or from hospitalization of not more than a year. An employee who has been given leave of absence for training duty must report for work on the next regularly scheduled workday after being released from duty or from hospitalization of not more than one year, with allowance for necessary travel time.

VETERANS BENEFITS ACT

On 10 July 1974 the President signed into law a bill improving benefits for veterans and aimed primarily at extending benefits of veterans of the Vietnam War. Benefits are described in three categories: (1) Educational benefits; (2) Loan Guaranty benefits; and (3) Other benefits. An important aspect of these benefits lies in the fact that persons presently on active duty with the Air Force are eligible for some of the benefits.

Educational Assistance. *Eligibility.* Any veteran who served on active duty for more than 180 days, any part of which occurred after 31 January 1955 but before 1 January 1977, and who was released under conditions other than dishonorable, or who was released from active duty after 31 January 1955 for a service-connected disability, is eligible. However, those personnel entering military service after 31 December 1976 are not eligible for benefits under this Act. Under the Veterans Educational Assistance Program (VEAP), they may establish an education account for later use on a contributory system whereby monthly contributions by the member are matched on a 2 to 1 basis by the government.

The 181 days required active duty does not include any period when assigned full-time by the Armed Forces to a civilian institution for a course substantially the same as a course offered to civilians; served as a cadet or midshipman at a service academy; or active duty for training in an enlistment in the Army or Air

National Guard or as a Reserve for service in the Army, Air Force, Marine Corps, or Coast Guard Reserve. Six-month enlistees under the reserve program authorized by PL 305, 84th Congress, are also excluded.

Educational Institutions Within the United States. An educational institution approved for training may include any public or private secondary, vocational, correspondence, or business school, junior or teacher's college, normal school, college or university, professional, scientific or technical institution, or any other institution which furnishes education at the secondary school level or above.

Pursuit of an accredited medical residency course would qualify an eligible veteran for educational assistance, provided the course is approved by the State approving agency, and leads to an identified educational or professional objective.

Educational Institutions Outside the United States. A program of education may be pursued only at an approved educational institution of higher learning. The Administrator in his or her discretion may deny or discontinue the educational assistance of any veteran in a foreign educational institution if such enrollment is not for the best interest of the veteran or the Government.

Selection of a Program. Each eligible person may select a program of education at any approved educational institution which will accept and retain him or her as a student or trainee in any field or branch of knowledge which the institution finds the individual qualified to undertake.

Change of Program. A change from one program to another, where the first program is prerequisite to or generally required for entrance into the second program will not be considered a change of program.

Each veteran may make two changes of program. One additional change may be approved if it is found that the program proposed by the veteran is more suitable to his or her aptitudes, interests, and abilities.

Entitlement and Duration of Eligibility. Each eligible person will be entitled to educational assistance for a period of 1½ months or the equivalent in part-time training for each month or fraction thereof of service on active duty after 31 January 1955 but before 1 January 1977, but not to exceed 45 months.

However, if entitlement ends where enrollment is in an educational institution operated on the quarter or semester system, entitlement will be extended to the end of that unexpired quarter or semester. If enrollment is in an educational institution not operated on the quarter or semester system and the period of entitlement ends after a major portion of the course has been completed, entitlement will be extended to the end of the course or for 12 weeks, whichever is the lesser period.

A period of time equivalent to that used under any other law administered by the Veterans Administration will be deducted from the time of entitlement under this law.

Eligibility ceases at the end of 10 years from the date of the veteran's last release from active duty between 31 January 1955 and 31 December 1976.

Allowances. Monthly payments available to veterans are as follows:

Type of Program:	*No Dependents*	*One Dependent*	*Two Dependents*
Institutional:			
Full Time	$342	$407	$464
Three-Quarter Time	257	305	348
Half-Time	171	204	232
Less Than Half-Time	Payments computed at the rate of the established charges for tuition and fees or at the rate of $342 per month for a full-time course, whichever is the lesser.		
While on Active Duty	Same as for less than half-time.		
Correspondence	Cost only.		

Full-time training consists of 12 semester hours or the equivalent, three-quarter time consists of 9 to 11 semester hours or the equivalent, half-time training consists of 6 to 8 semester hours or the equivalent.

A cooperative program is a full-time program of education which consists of institutional courses and alternate phases of training in a business or industrial establishment with the training in such establishment being strictly supplemental to the educational institutional portion.

The educational assistance allowance for a program of education pursued exclusively by correspondence will be computed on the basis of the established charge paid by non-veterans for the same course or courses. The allowance will be paid quarterly on a pro rata basis for the lessons completed by the veteran and serviced and certified by the school.

Application. Application forms will be available at all Veterans Administration offices, active duty stations, and American Embassies in all countries.

Veterans must submit a copy of their separation document, DD Form 214, with the application. Active duty personnel must submit a Statement of Service from their Commanding Officer or designate. To receive the additional allowance for dependents, a certified copy of the public record of marriage, birth, etc., should also accompany the application.

Veterans and service personnel applying for training in the United States should submit their applications to the nearest Veterans Administration Regional office.

For training outside the United States, service personnel should submit their applications and all subsequent communications through their Service Department to the Veterans Benefits Office.

Restrictions. The educational allowance is not payable if the veteran is enrolled in a course paid for by the United States under the provisions of any other law where the allowance would constitute a duplication of benefits from the Federal Treasury. However, if the identity of such Federal funds is lost by being combined with funds from other sources, they are no longer considered a duplication of benefits from the Federal Treasury. A determination will be required on each application for the Veterans Administration educational assistance allowance where a veteran-student receives educational assistance from other sources.

A change of program may not be approved for a veteran where his or her program has been interrupted or discontinued due to his own misconduct, neglect, or lack of application. This restriction may be removed if there exists a reasonable likelihood that there will not be a recurrence of such an interruption or failure to progress.

Any type course which is avocational or recreational in character may not be approved unless the veteran submits justification showing that the course will be of bona fide use in his present or contemplated business or occupation.

Flight training courses meeting Federal Aviation Agency standards may be approved.

Apprentice, or other training on the job, or institutional on-farm training, is authorized. Benefits for veterans with two or more dependents taking OJT are as much as $305 for the first 6 months.

Open circuit television or radio courses may not be approved unless the veteran is enrolled in a program, pursued in residence, leading to a standard college degree, which includes subjects offered through open-circuit television and the major portion of the course requires conventional classroom attendance.

The allowance is not retroactive. That is, no payment may be made for training prior to June 1, 1966.

Wives and Widows Educational Aid. In 1968 a new law extended to wives and widows the same educational assistance previously made available only to the sons and daughters of veterans who died or are permanently and totally disabled as a result of military service. Under the new law $342 a month is provided to spouses taking a full-time educational course at an accredited institution for a maximum of 45 months.

Loan Guaranty Benefits. Under the new law, eligible veterans and servicemen may obtain GI loans made by private lenders for homes and farms and, in certain designated areas, direct loans made by VA for homes and farmhouses.

The following general information summarizes the main parts of the new program. This information, however, is not complete for determining eligibility in every case. Full information may be obtained from Veterans Administration Regional Offices.

Eligibility. Veterans who served on active duty for 181 days or more or were discharged for disability, any part of which occurred after 31 January 1955, and who were discharged or released under conditions other than dishonorable are eligible. Persons whose military service after 31 January 1955 consists of "active duty for training," however, are not eligible.

Members of the U.S. Armed Forces who have served at least two years in active duty status, even though not discharged, are eligible while their service continues without breaks.

Entitlement. For a veteran or serviceman, VA may guarantee a home loan made by a private lender up to $27,500 or 60 percent of the loan, whichever is less. For a farm real estate loan other than for the acquisition of a home, the guaranty may not exceed $4,000 or 50 percent of the loan; and for non-real estate farm loans, the guaranty may not exceed $2,000 or 50 percent of the loan. This means that a lending institution will receive the Government's guaranty which is intended to be in lieu of a down payment or to reduce the down payment which the lender normally requires.

The maximum amounts of entitlement specified above apply to each person who qualifies under the new law. In other words, an individual who is eligible as a service member will not gain additional entitlement when he is released or separated from the service; further, an individual who can qualify as an eligible veteran and who later serves for 181 or more days will not receive additional entitlement for the later period(s) of service.

Purposes of Loans. Loans may be for the purchase of homes; to make alterations, repairs, or improvements in homes already owned and occupied; to purchase farms or farm supplies or equipment; to obtain farm working capital; or to refinance delinquent indebtedness on property to be used or occupied by the veteran as a home or for farming purposes. Direct loans may be made by VA, however, only for the purchase of homes and farmhouses. Business loans are not authorized under the new law.

Amount of Loans. There is no maximum on the amount of a guaranteed loan.

Maturity of Loans. Home loans can be made for a maximum of 30 years, farm real estate loans for a maximum of 40 years, non-real estate loans for a maximum of 10 years.

Down Payment. VA does not require a down payment. Lenders, however, may require down payments.

Liability for Repayment. Loan benefits are not a gift. If the VA loses money because the loan is not repaid as agreed, such loss will be a debt the veteran or service member owes the Government.

Additional Information. Certificates of Eligibility or additional information may be secured from a Veterans Administration Regional Office. VA Form 26-1880, "Request for Determination of Eligibility and Available Loan Guaranty Entitlement," with required supporting documents, should be sent to VA to secure a Certificate of Eligibility. Accurate determinations of eligibility can be made only by VA. No veteran or serviceman should obligate himself for the purchase of a home or farm solely on the basis of this summary. Before undertaking any such obligation, he should ask VA to determine his eligibility. Original discharge certificate or DD Form 214 must accompany his application.

Other Benefits. *Civil Service Preference.* Veterans who served for more than 180 consecutive days after 31 January 1955 are entitled to the provisions of the Veterans Preference Act in Federal employment.

Veterans Administration Medical Care extends to veterans whose sole service was after 31 January 1955 hospital care from the VA on a bed-available basis for treatment of their non-service connected conditions, provided such veterans state under oath their inability to defray the costs of such care.

This law does not extend hospital care to the six months' trainee who volunteered for service after 31 January 1955 and whose entire period of service was in a training status, unless, of course, he was disabled from a disease or injury incurred or aggravated in line of duty. Such disabled veterans are eligible for medical services on the same basis as other veterans with service-connected diseases or disabilities.

Job Counseling and Employment Placement. The job counseling and employment placement service is under the supervision of the U.S. Department of Labor and is administered through the local Veterans Employment Representative in local State Employment Service Offices.

Employment counseling and testing is provided when needed and priority for referral to appropriate training programs and job openings for eligible veterans with first consideration given to the disabled veteran.

Compensation for Disability. Prior laws granted compensation to veterans who are disabled by injury or disease incurred in or aggravated by active service in line of duty and who were released from active duty under conditions other than dishonorable. Payments are based on the degree of disability.

The Veterans Readjustment Benefits Act of 1967 provides the same presumptive provisions that previously were afforded wartime veterans only. These provisions provide that veterans meeting the service requirement of this law will be presumed to have been in sound condition at time of enlistment except for defects noted for purposes of determinative action on claims based on disabilities due to or aggravated by service.

Diseases not detected before release from active duty are presumed to be due to service if they become manifest to a degree of 10 percent after release from active duty as follows:

Chronic disease within one year, tropical diseases and resultant disorders because of therapy administered for the disease or as a preventative—one year; active tuberculosis—three years; multiple sclerosis—seven years; Hansen's disease (leprosy)—three years.

BURIAL BENEFITS (AFR 143-1, AFMs 143-2, 143-3)

What the Air Force Will Do. In the case of officers of the Regular Air Force, Air Force Reserve, or the Air National Guard who die on active duty or directly as a result of active duty, the Air Force will accomplish the following services:

(1) Preparation of remains to include embalming, furnish funeral coach, limousine for relatives, funeral director's services, casket or urn, and cremation, if desired.

(2) Provision of adequate uniforms to clothe the remains.

(3) Provision of an interment flag.

(4) Shipment of the remains to the place designated for burial.

(5) Round trip transportation for military escort to accompany the remains to final destination.

(6) If interment is to be in a national cemetery (a right to which all deceased officers are entitled) the Air Force will reimburse the next of kin to the extent of $75 to defray travel expenses.

(7) A regulation government headstone or grave marker for the grave of an officer interred in a private cemetery may be obtained by applying to The Quartermaster General, Department of the Army, Washington, D.C. This headstone will be shipped prepaid to the railroad station nearest the cemetery. If the officer is buried in a national cemetery the installation of an appropriate headstone will be provided for by the superintendent thereof.

(8) In 1965 the President authorized the military services to provide transportation of the remains of deceased dependents from place of death to the place of burial.

VOTING (AFR 211-19)

Voting Assistance. Air Force personnel are encouraged to exercise the privilege of voting in Federal, State, and primary elections. In order to encourage and assist eligible airmen to vote, the Department of Defense maintains liaison with the various State election authorities to obtain current voting information which, through Air Force channels, is promptly disseminated to all Air Force installations for the information of all personnel. Voting officers are available at squadron level to answer pertinent questions concerning forthcoming elections in the service member's State of domicile, and to provide printed postcard applications for State absentee ballots which are transmitted airmail, postage free, to the appropriate State. Every effort is made to insure that all personnel are protected against coercion of any sort in making their political choices and to maintain the integrity and secrecy of the ballots cast.

Responsibility. Although the Air Force provides information and assistance concerning elections and voting procedures, the actual decision to vote rests with the individual serviceman. Voting must be entirely voluntary and no serviceman will be required or ordered to participate in political elections.

Determination of Eligibility. The determination of eligibility and the specification of requirements for voting are completely governed by the appropriate State. In order to vote, each serviceman must meet such requirements and must be declared eligible to vote by the State in which he desires to exercise his voting privilege.

PERSONAL AFFAIRS RECORD

It is *very important* that you *prepare* and *keep up to date* a record of your personal affairs and property. Both you and your family will probably have more than one occasion to refer to such a record during your military service.

Fill in your record carefully and completely and then, if you wish, make as many copies of your record as you will need for your family, or other persons, *whom you wish to trust with such information. Be sure you consider each item and make the record as complete and accurate as possible.*

Remember, keep your record up to date, as changes or additions occur, by amending the original record or preparing a complete new record, and send copies to whomever you give copies of the original record.

Start now to get the necessary information and to prepare your record, *and don't stop* until it is complete.

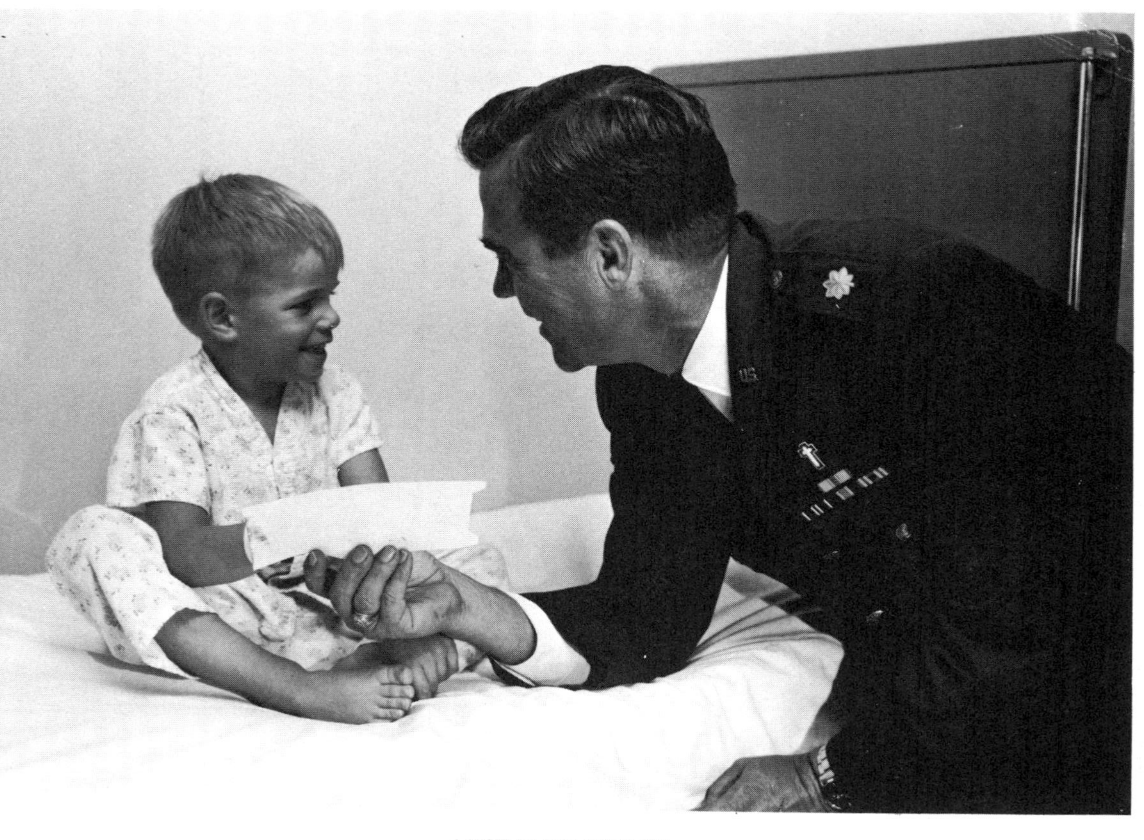

A VISIT BY THE CHAPLAIN.

22

A Security Program for Your Family

He that hath wife and children hath given hostages to fortune. —Francis Bacon

Very few officers entering on extended active duty with the Air Force are interested in the aftermath of their own deaths. Consequently, some do not give serious attention to the question of life insurance and estate planning. Yet every person, regardless of his or her profession, should study these problems with utmost care. Generally speaking, it is only through life insurance programs and a wise investment of savings that most of us can hope to provide adequately for the care of those left behind after death. To regard this subject flippantly, therefore, is to indicate a flippant attitude toward spouses, children, and other dependents—an attitude which is surely not that of a responsible officer of the Air Force.

The Essential Question. The basic essential of a family security program, including life insurance and investments, is to provide an adequate estate for those who are dependent upon the officer. There are literally a hundred different variations of insurance, some inexpensive, some very expensive. Never lose sight of the basic fact that the primary object of life insurance is to provide for the well-being of dependents after your own death. Everything else is incidental. There are insurance policies which promise a rich dividend to you at some selected age. These are called endowment policies. If you can afford them over and above adequate protection for your family against the ever present contingency of your own death, well and good. First, however, first above all, see to it that your insurance program gives your family a reasonable chance for an acceptable standard of living after you have died.

Timing. It will take no belaboring of the point to convince an intelligent officer that the time to take out insurance policies is at the earliest age he or she can afford them. All types of policies involve increased premiums with increased age. Do not become "insurance poor," but *complete* your insurance program as soon as your personal finances permit, certainly while you are a major. Remember that insurance is paid off in dollars, which are subject to inflation; in other words, the purchasing power of insurance dollars may become less and less with passing years. Therefore, along with your insurance program and throughout your active service, you should follow a sound investment program in investment trusts or common stocks.

Educational Policies. Educational policies are designed to produce a given amount of money (say $5,000) for the education of one of your children. Such a policy matures and is payable to you at a stipulated age of your child, or is payable to your child at the time of your death. These are worthwhile policies, but they are relatively expensive.

Ordinary Life Insurance. This form of life insurance obtained from a reputable commercial company is often called "straight life insurance." While it does have a cash surrender value which can be utilized in circumstances of acute financial stress, its main characteristic is that it provides a high level of financial security for your family at a low cost. The proceeds of such a policy may be used to serve the purposes of educational and endowment policies when protection is no longer needed.

Term Insurance. You will hear it said that term insurance is far cheaper than ordinary or straight life insurance. So it is, for a limited period. After a few years, however, term policies must be renewed *at the increased age of the applicant.* Thus, by the time you are 45 years of age, your so-called "cheap" term insurance may have become prohibitive in price. Moreover, upon retirement or later, you may become uninsurable because of physical disability or age. If, however, your need for insurance or for additional insurance is temporary, you may find term insurance best meets this special need.

Government Servicemen's Group Life Insurance. The Congress approved in 1965 a group life insurance program for all members of the Armed Forces. This program authorizes $35,000 life insurance to any person on active duty. The benefits of this insurance will be paid to the beneficiary named by the service member (spouse, children, parents, next-of-kin, or executor) with minimum limitation as to place or manner of death.

The Government Group Life Insurance program costs the service member $5.25 per month, premium payments being deducted from monthly pay and additional costs being met by the Government. The insurance is valid until 120 days after termination of active duty; within the latter period the service member may convert the insurance into a commercial life insurance policy *without physical examination* if so desired. While coverage under this program is optional, its advantages to Air Force members are so obvious and considerable that it should not be necessary to urge it upon them. In no other way can so much life insurance be obtained so inexpensively.

Group Insurance Associations. The officer should also consider the advantages of certain group-plan term policies, especially those designed specifically for the needs of officers, such as those offered by the Armed Force Relief and Benefit Association and the Air Force Association.

Seek Competent Advice. In developing your personal insurance program, seek the advice of a competent estate planner. Insurance agents of reputable agencies are capable of guiding you accurately and expertly to the proper solution of your insurance problem. They are salesmen and are interested in their commissions. For the most part, however, they will give you the facts. It is your duty to your dependents to know these facts. AFP 211-27 is good source material.

DETERMINING A TOTAL SURVIVOR BENEFIT GOAL

Computing or estimating the monthly income which should be provided to survivors of an officer, and the essential duration of that income, is a very individual problem. Each person should reason it out in consideration of his own set of circumstances. For most officers it will be accomplished best by consultation with the spouse. A workable and attainable solution should be reached by logical analysis.

A start may be made after deciding upon answers to some very searching and very personal questions, such as the following:

Protection. At the time of the analysis, is the spouse qualified in some art, trade, or profession, so that he or she could recover promptly to supply all or a major portion of income requirements? Or is he or she untrained for a specific vocation? Consider present age, health, and the spouse's wishes. At what age should it be considered that he or she would be unable to provide a part of the required income, or would not wish to do so?

When there are children to be protected, their age must be considered, with costs of support and education. If there are physical or mental frailties these factors must be evaluated. On the average how much income per month will be required for the support and education of the children, and for how many years?

Clean-up Fund. In the event of the sudden death of the head of the family, how much cash will be needed at once to pay up current bills, installment accounts, and other obligations? Under this heading should be included the expense as it can be estimated of establishing the survivors in a permanent location with the special and heavy costs this task includes.

Present Worth of Estate. Most service families have a tangible net worth which increases through the years preceding retirement. Omit for the present consideration items of personal property which may have material value, but which would not be sold nor of themselves add to income. What is the total upon which survivors could depend for living income from the present family estate? This is the primary base upon which to build the ultimate program.

An annual re-evaluation of this personal estate in an objective manner as would be done by a banker or the trustees of an estate is necessary.

Survivor Benefits. Each individual should study carefully the potential income that would be received by the family in event of death. This study should be repeated every few years, since laws and conditions change. For example: how much would the spouse receive as the death gratuity (six months pay but not in excess of $3000)? How much per month from Veterans Administration

Dependency and Indemnity Compensation or pension? How much from Social Security? How much from insurance policies? How much from the Survivor's Benefit Plan? Consideration of these potential benefits to the family will almost invariably lead to the conclusion that the income provided is insufficient. A sound program of additional insurance and investment in stocks or bonds will be the method chosen by most officers to close the gap. Obviously, the sooner such additional programs are begun, the more time is available to spread their cost and to allow for appreciation of investments. Whatever you decide to do, *begin early.*

Death Gratuity. Upon the death of an active-duty (but NOT a retired) serviceman there is paid a death gratuity of six months' pay, up to a maximum of $3,000. There is also a provision that the gratuity is payable if a member or former member dies of a service-connected cause within 120 days after his discharge or release from active duty or active duty for training. The gratuity includes base and longevity pay, special incentive and hazard pay, but not rental or subsistence allowances. It is not taxable.

Death gratuity will be paid to the surviving spouse, or if there is no surviving spouse, to the surviving children in equal shares. If there is neither spouse nor children, the member may designate one or more parents, brothers, or sisters. If the member has no surviving spouse or children and has not designated a beneficiary, payment will be made to the parents in equal shares, or if there are no surviving parents, to the surviving brothers and sisters in equal shares. No other person will be entitled to the death gratuity.

Arrears of Pay. Upon the death of an Air Force officer, the spouse is entitled to receive whatever arrears of pay would have been due to the officer, provided that the spouse is beneficiary designated on DD Form 93, Record of Emergency Data. The Air Force Finance Center will furnish the beneficiary with the proper forms for application for arrears of pay (DD Form 397).

Accrued Leave Payments. In 1965, authorization was established to pay to survivors of military personnel who die with accrued leave unused and not otherwise compensated for, an amount corresponding to base pay for the number of days of accrued leave on record, up to a maximum of 60 days. This means that the survivors of Air Force officers could receive as much as two months' base pay in addition to other survivor benefits.

SURVIVOR BENEFITS ACT

The Servicemen's and Veterans' Survivor Benefits Act provides benefits for survivors of military personnel who die in service or who die after separation as a result of military service. At the same time, the Act assures protection for service members themselves.

Dependency and Indemnity Compensation. This form of compensation combines benefits which were once paid separately as (1) death compensation, and

---

Additional recommended reading is the book OFFICER'S MANUAL OF PERSONAL FINANCE AND INSURANCE written by Associates in the Social Sciences, USMA, West Point, N.Y. and available from Stackpole Books. Contains valuable information on all personal matters financial—whether to rent or buy a house; all about life insurance, savings programs, stock market, commodity market, mutual fund investments, etc.

(2) as Servicemen's Indemnity payments.

Dependency and Indemnity Compensation, referred to hereafter simply as Indemnity Compensation, will be paid by the Veterans Administration to spouses, orphan children, and dependent parents of those who die as a result of military service.

Eligibility. Spouses, orphan children, and dependent parents of those who die of service-connected causes, either while in service or after separation, may be eligible for Indemnity Compensation.

Application. Application forms for Indemnity Compensation will automatically be sent by VA to the survivors, now on the rolls, of service members and veterans if it appears they are entitled to the benefit.

Applications filed either with the Veterans Administration or the Social Security Administration will constitute applications for benefits from both agencies. Documents such as proof of death, marriage, birth, and dissolution of marriages filed with either agency will be made available to the other agency.

According to marital status and family situation, compensation will be paid in any one, or in any combination of three ways.

Compensation for Spouse. Payable monthly by Veterans Administration to a service member's spouse as long as he or she does not remarry. The amount of compensation is related to the pay of the rank held by the individual at the time of death, under the following formula: $120 plus 12 percent of member's monthly basic pay (rounded to next highest dollar) is the monthly compensation to the widow. The 12 percent is computed based on current active duty pay.

If a spouse is left with dependent children under age 18, additional income will be provided by Social Security benefits.

Compensation for Children. This may be paid by the Veterans Administration for the support of a deceased service member's children in the event their remaining parent dies or is divorced from the service member or if he or she remarries after the partner's death.

The Act provides monthly rates of payment in such cases varying according to the number of children.

All of these payments are in addition to the benefits for which children under age 18 may be eligible under Social Security.

Payment of compensation for support of a child normally stops when the child becomes 18, but may be continued to age 23 if the child is in school. If the orphan becomes incapable of self-support while under age 18, monthly compensation will continue past that age.

Compensation for Parents. The amount of compensation a parent or parents may receive is provided on a sliding scale according to other income.

Since the matter of federal benefits is highly complicated, it is suggested that officers obtain from the Veterans Administration their most up-to-date pamphlet on Federal Benefits.

SOCIAL SECURITY

Members of the Armed Forces and their dependents are eligible for the benefits provided by the Old Age and Survivors Insurance (OASI) program of the Social Security system, on a permanent, contributory basis.

Social Security will provide a monthly income for:

(1) A spouse with children.
(2) Children under age 18.
(3) A spouse at age 60.
(4) Dependent parents.
(5) The service member at age 65, or, at varying reduced rates, at age 62, 63, or 64.

The amount of Social Security benefits paid to survivors depends upon the "average monthly wage" earned by the service member.

Survivors receive a lump sum payment under Social Security equal to three times the monthly retirement benefit for which the deceased was eligible but not exceeding $255.00.

The Social Security benefits for service members and their families are separate from, and in addition to, military retirement pay or any form of compensation benefits paid by the Veterans Administration.

Social Security provides continuing protection, in civilian life as well as during service in the Air Force. Suppose, for example, a man enters service at age 20, after having worked under Social Security as a civilian for two years. Assume he retires at age 50 on a 30-year retirement. He then takes a civilian job covered by Social Security and works until he is 62 years old. The old age benefit he draws can be based on both his years in military service and his years as a civilian worker.

But remember that benefits are based on average monthly wages earned and taxed up to age 62, 63, 64, or 65 or death. Only five years may be eliminated or "dropped out" of the computation for average monthly wages. It is therefore difficult to attain anywhere near the maximum in retirement benefits.

When a person is entitled to more than one benefit, the amount actually payable is limited to the largest of the benefits.

SURVIVOR BENEFIT PLAN

The Survivor Benefit Plan (SBP) replaced a previous program known as the Retired Serviceman's Family Protection Plan. The new Survivor Benefit Plan became effective in 1972. It provides survivor income up to 55% of the member's retired pay. Officers may select the amount of survivor income they wish to provide, sacrificing a portion of their retired pay to pay the premiums. Thus at very low cost an officer may provide a life-long annuity for his beneficiary. Officers retired after 1972 are automatically enrolled in the Survivor Benefit Plan at maximum survivor benefit level if they have a spouse or dependent children, unless they elect not to participate, or to participate at less than the maximum level.

Examples of Survivor Benefit Plan

Base Retired Pay	Payment to Spouse	Monthly Cost to Retiree
$1,000.00	$550.00	$ 72.50
1,500.00	825.00	127.50
1,800.00	990.00	157.50

PENSION FOR DEPENDENTS OF RETIRED PERSONNEL AND VETERANS WHOSE DEATHS WERE NOT SERVICE CONNECTED

A modest survivor pension may be payable to dependents of veterans or retired personnel when retirement was for a reason other than physical disability or death occurred for reason other than a service-connected cause. But the spouse and children of the veteran or retired individual are the only eligibles.

The following payments are not considered "income" as to death pension payments:

Government life insurance proceeds and, in some cases, commercial life insurance proceeds.

The six months' death gratuity.

Donations from public or private relief or welfare organizations.

Payments of pension, compensation, and dependency compensation of the Veterans Administration.

Survivor's Benefit Plan.

Lump sum Social Security death payments.

Payments to an individual under public or private retirement annuity, endowment, or similar plans or programs equal to his contributions thereto.

Proceeds of a fire insurance policy.

Amounts equal to amounts paid by a spouse or child of a deceased veteran for just debts, the expenses of the last illness, and the expenses of burial which are not reimbursed by the Veterans Administration.

For pension and other benefit rates, consult your Personal Affairs Officer.

MUTUAL INSURANCE ASSOCIATIONS FOR SERVICE MEMBERS

There are mutual associations of officers of the military service which have very material importance. Eligible officers may assume with confidence that these organizations are well and carefully administered, and that the services provided are tailored to fit military requirements.

The Armed Forces Relief and Benefit Association. This non-profit service organization offers group life insurance with its attendant low cost to supplement an officer's permanent program. Membership is open to Regular officers and to Reserve officers on extended active duty.

Rates are *very low* because it is group insurance. At death the insurance may be paid in a lump sum; or part in lump sum, part in installments, all in installments, or as an annuity.

While this insurance is limited to active officers, there is provision for conversion to other forms of insurance with the John Hancock Mutual Life Insurance Company, one of the underwriters of the program.

Address: The Armed Forces Relief and Benefit Association
1156 15th Street, N.W.
Washington, D.C. 20005

Air Force Association Life Insurance. This plan provides term insurance at low rates, convertible to commercial insurance without physical examination on separation from the service.

Address: AFA Insurance Division
Mills Building, 1750 Pennsylvania Avenue, N.W.
Washington, D.C. 20006

Air Force Association Flight Pay Protection Plan. This plan insures you against loss of flying pay due to disability resulting from either accident or disease. It provides for payment of monthly benefits up to 24 months for disability resulting from aviation accident and 12 months for other accidents or disease.

Address: AFA Insurance Division
1750 Pennsylvania Avenue, N.W.
Washington, D.C. 20006

United Services Automobile Association. This association was founded to provide automobile insurance at cost for officers and warrant officers of the armed services. In addition to automobile insurance it also makes available a valuable household goods/personal effects policy. Both are arranged to meet the specific conditions of officers.

Eligibility. Policies may be obtained by active and retired officers, Advanced ROTC, OTS/OCS and Academy Cadets, those in certain precommissioning programs, and warrant officers of the several services; officers and warrant officers of National Guard and Reserve components when ordered to extended active Federal duty; the spouses of such deceased officers and warrant officers, so long as their status is not changed by remarriage.

Automobile Coverage. The automobile coverage is broad and similar in scope to policies offered by commercial companies. Its policies cover throughout the United States and in many foreign countries. Shipment coverage is provided upon request.

Household Goods and Personal Effects Policy. The Association also offers insurance on personal effects and household goods. This insures against the hazards of fire, transportation, theft, windstorm, flood, earthquake, and many other perils. There is no adjustment of premium because of relocation of goods. Coverage is worldwide.

Address: United Services Automobile Association
USAA Building, San Antonio, Texas 78288

The Armed Forces Cooperative Insuring Association. The Association offers insurance against fire losses to personal and household effects regardless of station or duty, and at actual cost to members. Coverage is worldwide and applies automatically for damage by named hazards to property wherever located, in government or civilian storage, in use, or in transit. Coverage A provides protection against loss from fire, lightning, smoke, flood, earthquake, explosion, hail, windstorm, or while being transported by common or public carrier including U.S. Government vehicles, vessels, or aircraft. Exceptions include automobiles or other gasoline propelled machines; animals; boats; some types of "collections"; and loss or damage due to hostile or warlike action. It does not protect against loss in parcel post or mail. An additional coverage B against theft, pilferage, sabotage, etc., is available. Premium rates are very low. Settlements are prompt and fair to all concerned.

Eligibility for membership includes any commissioned or warrant officer of the Regular services, or of the Reserve components on active duty except as to the latter a Reserve member must have signified, by contract, an intent to remain on active duty for a period not less than three years. The spouse of a deceased member is permitted to continue the insurance in his or her name until remarriage or death.

Address: The Armed Forces Cooperative Insuring Association
Fort Leavenworth, Kansas 66027

PLANNING A HOSPITAL.

23

Oversea Movement of Your Family and Life Overseas

Travel teaches tolerance. —Benjamin Disraeli

Air Force officers must expect to spend about three years out of every ten at an oversea station. Depending on where the officer is assigned, he or she may go alone to an oversea assignment, leaving the family behind in the United States for some months, the period varying with different oversea stations. This separation of the officer and his family is the result of the limited housing available at oversea bases. On arrival in the oversea area, the officer signs up on a priority list for family housing. This list is governed by the length of the period of separation that has occurred. In some oversea areas officers are permitted to obtain for themselves private rental housing, but this housing must be approved by the oversea command before it can be used as a basis for dependent travel.

ACTION BEFORE DEPARTURE

When alerted for an oversea move, the officer is counseled on travel procedures and tour elections. When dependent travel is authorized, an advance application for concurrent travel should be submitted if the officer intends to move dependents overseas.

Official orders of the Department of the Air Force assigning an officer to an oversea command prescribe the timing, method of transportation, and other essential information about the journey. At the outset the officer should have prepared at least twenty copies of orders, for he or she will need to file them on many occasions incident to movement, pay, travel allowances, shipment of personal property and other matters en route and after arrival.

Upon receipt of orders the officer should report at once to the Base Administrative Officer, Transportation Officer, Finance Officer,

Supply, and Surgeon, receiving from them detailed instructions which he or she must be certain to understand and follow to the letter. The officer must set his or her official and personal house in order so that no dangling, unfinished official or personal business will arise at the last minute before departure or after departure. All personal bills or obligations must be paid, or definite arrangements made to defer payment.

At every step, the officer must bear in mind that his or her spouse, if not authorized to travel concurrently, will have need for certain documents in connection with later movement to the oversea area. The officer must therefore insure that the spouse has all necessary documents such as titles to automobiles, copies of officer's orders, insurance policies, inventories of household goods, children's school records, birth certificates, and the like. The officer should leave with the spouse signed copies of DD Form 1299, "Application for Transportation of Household Goods," together with 20 copies of the officer's travel orders. The officer must also insure that the spouse has sufficient money (an allotment may be indicated) to cover the needs of the family during the period of separation, including the period of movement overseas.

Record of Expenses. The officer or spouse should, from the beginning of all actions connected with the oversea movement, maintain a record of out-of-pocket expenses incurred. Military dependents are reimbursed for out-of-pocket expenses, such as taxi fares, tips, passport fees, visa fees, and other items during transfers. The list of reimbursable expenses is a long and complicated one. Therefore the wise officer or spouse will maintain a record of *all* expenses of any type and after arrival at the oversea station consult with the Finance Officer as to which of these items will result in reimbursement. These reimbursements are in addition to per diem or other direct travel allowances authorized.

Application for Noncurrent Dependent Travel. If concurrent travel was not approved, upon arrival at the assigned station overseas an officer may apply through his or her commander for dependent travel authorization. Approval of this application for dependent travel is given when the oversea commander concerned can insure that adequate family housing, either government or private, will be available for the dependents upon their arrival. The officer must have at least twelve months remaining on his or her tour to move dependents overseas at Government expense.

Receipt of Passport Authorization. After the officer's application for dependent travel is approved by the oversea commander, a letter authorizing the dependents to apply for a no-fee passport is furnished to them.

Application for No-Fee Passport. Upon receipt of passport authorization, dependents should make application for a no-fee passport through the CBPO at the sponsor's base of assignment. When making application, dependents must present:

(1) A birth certificate for each member of the family going overseas;

(2) Two identical photographs not smaller than 2″ × 2″ or taken full face and without hat;

(3) Proof of American citizenship, if not native born.

Photographs. Individuals are required to obtain separate no-fee passports, regardless of age, and accordingly, individual photographs are required.

Processing Passport. The application for passport is then forwarded for processing to the Passport Division, Department of State, Washington, D.C., 20520. Upon issuance, the passport is forwarded to the appropriate aerial port of embarkation, where it will be given to the dependents upon their arrival. Upon receipt of the processed passport, the appropriate port of embarkation will, at the time dependent's priority number comes up, issue port call to the dependent.

Warning. Upon receipt of authorization, dependents should apply for passport with the least possible delay. Dependents who delay applying, following receipt of passport authorization, may deny themselves an earlier port call. In this regard, it is important that dependents have their birth certificates and citizenship evidence on hand so that the passport application is not delayed.

The appropriate aerial port of embarkation will not issue a port call until the passport has been received. If the passport is obtained through nonmilitary channels or if it is inadvertently sent directly to the dependent, it must be dispatched immediately to the appropriate aerial port of embarkation.

Dependent's Travel Orders. At the appropriate time the oversea command will issue to the officer's dependents a letter of instructions similar to the one shown below.

SAMPLE LETTER OF INSTRUCTIONS TO BE FORWARDED TO DEPENDENTS TRAVELING UNACCOMPANIED BY THEIR SPONSOR

(Suggest sponsor's aid in composition so that only pertinent facts are included.)

Dear...: (date)

Inclosed are copies of Special Order No.., this headquarters, dated ...., covering your transportation from the United States to join your sponsor in this area. The Air Force base indicated in these orders has also been advised of your impending travel and is prepared to assist you in any way possible. You should contact the CBPO at the base for assistance. For your information, however, there are certain things we would like to acquaint you with in order to enable you to travel with a minimum of delay and maximum of efficiency. These are:

a. MEDICAL REQUIREMENTS: If you are pregnant, you are requested to furnish the Transportation Officer at the base designated to handle your movement a confirming statement from the attending physician as to expected date of birth of child. Prior to departure from the United States you will likewise be required to be immunized against .... These inoculations may be received at any military installation. A private physician may administer inoculations, but must furnish you with a certificate for presentation to proper authorities. You are required to complete AF Form 1466, Medical and Educational Clearance for Dependent Oversea Travel, before travel is authorized. If this has not been obtained, you are urged to make an appointment with the nearest Air Force medical facility at your earliest convenience. Medical reports and recommendations from civilian physicians may be provided. However, the certificate must be signed and approved by an Air Force medical services evaluator. In some cases, it may be necessary to report in person for further evaluation or examination before clearance can be issued. The purpose is to assure that adequate and necessary care is available in the area of assignment.

b. TRANSPORTATION: (Use applicable subparagraph.) (1) Your sponsor is serving in an "administrative weight restricted" oversea area. Shipment of household goods, not to exceed ________ pounds, is authorized. In addition to household goods, shipment of unaccompanied baggage is authorized. This shipment will not exceed 350 pounds per dependent 12 years of age and older and 175 pounds per dependent under 12 years of age.

(2) Your sponsor is serving in a "full JTR weight" oversea area. Shipment of household goods, not to exceed ________ pounds, is authorized. Part of the shipment may be shipped as unaccompanied baggage. This shipment will not exceed 350 pounds per dependent 12 years of age and older and 175 pounds per dependent under 12 years of age. A "free checkable" baggage allowance, which is normally 66 pounds, is authorized for you and each of your dependents for your travel by air. You should check with the transportation office to determine actual number of pounds authorized in your case. Shipment of your automobile (is) (is not) authorized. Delivery of the automobile to a military water port is your responsibility. This must be accomplished at your own expense. The transportation officer will furnish you with information regarding your household goods, automobile, and personal travel. The transportation officer also makes all necessary arrangements for the shipment of your baggage and household goods. You will be provided with the required transportation requests to enable you to proceed to the proper point of departure from the United States. It is imperative that you travel only during the period indicated in the attached orders. Do not report to the commercial gateway before the time specified in the reservation confirmation furnished by the personnel processing you for oversea movement. (Personally procured travel performed via foreign flag carrier is not reimbursable unless a statement is provided that a US certificated carrier was not available.)

c. (Any additional information deemed necessary for the orderly movement of the individuals should be included in additional paragraphs. Information relative to clothing that should be worn or accompany the individual, what to expect after arrival in the oversea area, instructions on applying for passports and visas if necessary, and related items should be included.)

d. If there are any questions relative to your travel to this area, which have not been covered, it is suggested that the outbound assignments unit in the CBPO of the base designated to handle your movement be queried. He is equipped to provide you with advice and assistance as needed.

e. Transportation has been scheduled for you during (the month indicated in orders). It is important that you travel during (the month indicated in orders). In the event an unforeseen emergency requires you to travel in a month other than indicated, request you advise the outbound assignments unit in the CBPO at (the designated base) who will advise the overseas order writing agency.

Sincerely,

The following paragraph is suggested, for inclusion under c above, where alien spouse and/or alien adopted children are involved.

c. Also inclosed is DD Form 1278, Certificate of Oversea Assignment To Support Application To File Petition for Naturalization, which will serve as your authority to file your petition for naturalization. This form should be filed with your application for petition for naturalization if you have not already filed with the nearest office of the Immigration and Naturalization Service. If you have already filed your application, this form should be forwarded immediately to the appropriate I&NS Office. As soon as citizenship is granted, you should immediately file an application for passport.

(Reference AFR 10-4, volume I, chapter 8, for list of CONUS Air Force activities capable of providing support to unaccompanied dependents.)

Note that the letter of instructions provides that the officer's dependents should make contact with the CBPO at the nearest air base. The CPBO will make application for the family's flight reservations. In due course, the CBPO will inform the officer's dependents of the exact time, date, and flight number of the reservation for aerial travel. The CBPO will further assist the officer's dependents in carrying out all necessary preparations for oversea movement.

Immunizations Required. All dependents transported overseas at government expense are required to receive certain immunizations prior to departure from the United States. At the present time, the immunization required for entry into the oversea commands are:

(1) *Adults.* Smallpox, Typhoid Series, Tetanus Series, and Schick Test.

(2) *Children from Three Months to Six Years.* Smallpox, Typhoid Series, and Series of "DPT" (diphtheria, pertussis, and tetanus).

(3) *Children from Six Years to 15 years.* Smallpox, Typhoid Series, and Series of "DT" (diphtheria and tetanus).

When departure is from a west coast port of embarkation, *all* dependents will be required to complete the Typhus Series of immunization in addition to those listed above.

If immunizations are obtained from a private physician, a complete record should be maintained. Also dependents are advised that immunization requirements may change without prior notice and that current information in this regard may be obtained from the nearest Armed Forces Medical facility. The dependent's record of immunization must be on Public Health Service Form 731 (Yellow). Others will not be accepted.

Shipment of Household Goods. The officer's spouse should consult with the Base Transportation Officer concerning the disposition of household goods. Specific recommendations concerning items of household goods to be shipped to a particular foreign country are included in the numbered AF pamphlets in the 216 series pertaining to that country. This information will doubtless be reinforced by letters from the officer overseas informing the spouse of conditions there. The Transportation Officer will arrange for storage of all or any portion of the family's household goods in commercial facilities, which are located in the general vicinity of the Base. Decisions must therefore be made as to what, if any, items of household goods should be left in storage in the United States. Other household goods will be shipped to the oversea base by any of several surface modes as designated by the Transportation Officer, and will be approximately two months in transit. Some oversea commands limit the weight of household goods that can be shipped to about 2,000 pounds. Check with the Transportation Officer.

ACTIONS ON RECEIPT OF PORT CALL

When the family has received its port call, that is, information as to the exact time for reporting to the APOE, final decisions must be made on the following points:

(1) The exact time, as designated by the officer's spouse, for household goods to be picked up by the commercial carriers for delivery to the oversea base of assignment, or to the commercial storage facility as designated by the Transportation Officer.

(2) What shall be shipped as *unaccompanied baggage.* Unaccompanied baggage is authorized to the extent of 600 pounds per dependent. This baggage may include such items as clothes, pots and pans, light housekeeping items, and collapsible items such as cribs, playpens, and baby carriages when necessary for the care and comfort of an infant. It may be expected that the unaccompanied baggage will arrive in the oversea area approximately one week after the arrival of the dependents. Unaccompanied baggage must be placed in strong containers, such as trunks, which may weigh no more than 300 pounds per container and be of no greater volume than 15 cubic feet.

(3) What shall be carried as *hand baggage.* In this case, 66 pounds per person is authorized. This baggage will go with the traveler on the same aircraft and must be taken to the APOE by the traveler.

(4) What to do about the *automobile.* If the dependent's orders authorized the shipment of a privately owned vehicle (POV), the officer's spouse should drive the car directly to the APOE where arrangements can be made, for a fee of about $30, to have the car delivered to the surface port of shipment for vehicles. It is at this time that the officer's spouse will be required to present the Certificate of Title to the automobile or, if a lien is held on the car, authorization from the holder of the lien permitting the car to be taken out of the United States.

If the privately-owned vehicle is to be delivered directly to the Port of Shipment, the dependent asks the Transportation Officer at the Air Base which Port of Shipment will receive and ship the car. The car is then delivered by the dependent to that Port of Shipment. Depending upon the oversea area to which the dependent is going, one of the following Ports of Shipment for automobiles will be applicable: Brooklyn Army Terminal; Hampton Roads, Va., Army Terminal; Charleston, S.C., Outport; New Orleans Army Terminal; Oakland, Calif., Army Terminal; Seattle, Wash., Army Transportation Terminal Agency; Naval Supply Center, Bayonne, N.J.; and Philadelphia.

Movement to the APOE. The mode of travel of the family to the APOE is determined by whether or not a POV is to be shipped overseas. If travel is to be by automobile, the spouse would be well advised to make reservations ahead of time for necessary overnight stops at hotels or motels.

If travel is to be by commercial means, Transportation Requests should be secured from the Transportation Officer and reservations made for travel to the APOE.

Arrival at the APOE. The officer's spouse should bear in mind that the APOE is a major air base which provides all the facilities common to air bases. He or she should arrange to arrive a few hours prior to the specified time of the flight. It is not necessary, nor is it desirable, that the family arrive the day preceding the specified date of departure for overseas. However, if circumstances make it necessary, over-night facilities can be provided at the APOE. On arriving at the APOE, the dependents should proceed directly to the processing point. The processing-out procedure will require approximately one hour. It is at this time that copies of orders must be shown, and immunization records will be checked. Hand baggage will be tagged, weighed, and loaded on the aircraft. The spouse should keep in hand only those items absolutely necessary during the course of the flight, such as toiletries and infant needs (diapers and formula). The spouse should carry only about $150 in cash; any additional funds should be in the form of travelers checks.

The Flight Overseas. Travel in a MAC aircraft is virtually identical to travel in a commercial airliner. Meals are served; facilities for warming babies' bottles and food are available; smoking is permitted; drinking is forbidden. If the flight makes a stop en route, a snack bar and a small BX facility will be found at hand. If a delay is extended, the hand baggage loaded aboard the airplane will be made available to the passengers. Individuals should wear warm, comfortable clothes.

Arrival at Destination. The MAC flight on which the dependents travel will take them to some central point near their final destination; such as Frankfurt, or Tokyo. At this point, it is necessary to claim hand baggage checked aboard the MAC flight. If it is necessary for the dependents to remain overnight before continuing their journey, transient hotel facilities will be made available to them. Transportation to such transient facilities will be provided by the Air Force. Further travel will be by staff car, bus, or train in most cases. Arrangements for this will have been made by the authorities at the MAC terminal overseas. Upon landing at the MAC terminal overseas the family will be informed of the means of further travel and the schedule of such travel. Necessary intermediate transportation, for example, from the MAC Terminal to the rail station, will be provided by the Air Force. On arrival at the final destination, it is presumed the family will be met by the officer or someone acting for him or her.

Movement of POV. The privately owned vehicle (POV) turned in for shipment at the APOE will move via surface vessels to a seaport in the area to which the family is traveling. The automobile will be approximately one month in transit. It will arrive at such a port as Bremerhaven in Germany, or Yokohama in Japan, where it must be picked up by the officer or his agent. The government does not transport the automobile from the seaport to the duty station of the officer. The officer will be notified by the port authorities of the arrival of the automobile. Usually the officer can obtain a pass to cover absence while accepting delivery of the automobile.

Traveling on Tourist Passports. Authorized dependent travel at the expense of the government is the only travel plan fully endorsed by the Department of the Air Force. Dependents choosing to travel on *tourist passports* for the purpose of visiting their sponsors at oversea bases do so as civilian tourists, traveling commercially and at no expense to the government. It is strongly recommended that dependents electing to travel in the status of civilians fully inform their sponsors of their plans before performing such travel. AFM 36-20 stipulates that the sponsor will serve a longer tour if dependents join the sponsor in the theater at any time in any status, *except* for those dependents traveling in tourist status whose entry has been approved by the Air Force major commander concerned.

Shipment of Pets to Oversea Areas. Certain pets may be shipped at the owner's risk and at no expense to the government through the appropriate port of embarkation on surface carriers belonging to the Military Sea Transportation Service, provided that all requirements are met and space is available.

Exchange Facilities Overseas. Air Force personnel preparing for oversea assignments should *not* assume that they must take with them all things needed for a three-year tour. The exchanges overseas are excellent. They are prepared to meet every conceivable need from clothing (adults, children, and infants) to washing machines, phonographs, and automobiles. There is virtually nothing needed or even desired by an Air Force family that cannot be obtained in the exchanges. Moreover, the local economy of such countries as Germany, Italy, Japan, and England offer many supplementary sources from which family needs can be met, although at increased prices. The exchanges, however, are complete in themselves, since oversea exchanges are allowed to stock and sell to military personnel items denied to exchanges in the United States.

Commissaries Overseas. Like the exchanges, military commissaries overseas meet all food needs. Military families who buy foods at the local markets do so merely because they prefer that course. The commissary can and does meet all normal needs. Food prices in the oversea commissaries are comparable to those in military facilities in the United States.

Medical Care Overseas. The family need not be concerned regarding adequate medical care overseas for adults or children, including pre-natal treatment. Excellent clinics are available in all areas and a general military hospital will be found in the vicinity. Optical and dental services are available for dependents in the military facilities. Do not buy extra sets of glasses for dependents before going overseas. At the oversea clinic the cost is far less.

Schools Overseas. Grade school and high school facilities are provided in all established commands. Standards are closely supervised. Considering the intangible values to be derived by travel and life in an oversea land for a span of a few years, most parents consider that their children have been benefited by the experience. In the unusual case where children's schools are not provided under military control, parents should inform themselves of facilities used by American citizens resident in the area, or ascertain the possibilities for instruction by mail with home tutoring. Unless there are very unusual circumstances indeed, there is no strong reason for Air Force families to be separated merely because of the school situation of their children, because educational facilities for our children have been provided in a manner which has been acceptable to most parents. (See appendix.)

Movement of Dependent Students To and From Oversea Areas. Bona fide student dependents of military personnel stationed outside the continental United States are authorized transportation on a space-available basis to the port of embarkation in the United States to attend school in the United States; and between the United States and oversea port to spend their summer vacations with their families. Only one round trip per year is authorized, and normally will be taken during the summer vacation period. Dependents who pass the age of 21 years while engaged in full-time undergraduate study will be authorized the same transportation.

The following procedures will apply:

(1) The military sponsor overseas will be responsible for requesting a travel authorization from the oversea installation.

(2) The oversea installation will issue the necessary travel authorizations and will forward the authorizations to the student dependent.

(3) If movement is to be by MAC, the students will be directed to report to the APOE serving the particular oversea area for movement on a space available basis. (Student should be prepared for a possible delay at the APOE of several days.)

Postal Service. The Army-Air Force Postal Service provides mail service overseas for Air Force and Army personnel for all commands, bases, and stations. Postal rates are identical with those in the continental United States. Air mail, registry service, and parcel post service are provided.

Merchandise Subject to United States Customs. Service overseas provides an opportunity for the purchase of merchandise made or manufactured in foreign lands which is advantageous as to certain items. For example, cameras manufactured in Germany; pearl necklaces from Japan; objects of art peculiar to the country or station; or items which are sold for prices materially lower than the cost of similar items in the United States, such as the perfumes of France. The officer and spouse who make a real study of these possibilities, purchase wisely, and avoid the purchase of merchandise which will be undesired upon their return to the United States, may acquire unusual possessions which they will cherish throughout their lives.

All such purchases should be made with due regard to customs regulations. The laws and regulations governing the entry of merchandise purchased abroad into the United States are quite liberal and, surprisingly, far more liberal than is the common understanding. The entry or attempt at entry of merchandise purchased abroad without the required customs declaration is an evasion of the law. It is smuggling. It is subject to severe penalties provided in the laws. It is not smart to smuggle. It is much smarter to inform oneself as to what can be brought into the United States openly and legally. For articles of personal use and enjoyment in a household, the quantity authorized for free entry may prove to be entirely sufficient.

Housing in Oversea Areas. The character of housing available to officers overseas varies considerably among the oversea areas. However, government quarters are generally adequate and some are excellent. The standard of government-furnished quarters in any oversea area approximates that of quarters on air bases in the United States. There are exceptions; therefore, it is wise to read the current AF pamphlet on the particular foreign country of interest.

Private housing is an entirely different proposition. The standards may be quite low in such matters as electricity available, effectiveness of the plumbing systems, and heating. The price of private housing is often very high for what the Air Force officer would consider an acceptable house. The cost of utilities in general is two or three times that in the United States. On the other hand, officers who are authorized to find housing on the local economy may receive a station allowance which partly compensates for the increased costs encountered. The amount of the "COLA" (Cost of Living Allowance) varies with oversea stations. As a general matter, it is possible to obtain household help at fairly reasonable rates.

Travel and Recreation Opportunities in Oversea Areas. One of the great attractions of life overseas is the opportunity provided to travel in foreign countries and to take advantage of the recreational facilities available. In Europe, U.S. military personnel may travel about among the interesting countries of this Old World during short periods of leave. By utilizing the facilities of the various U.S. military installations sprinkled over Europe it is possible to visit a vast number of interesting places at a reasonable cost. In the Orient distances become a problem, but nonetheless it is possible for officers to make arrangements to visit countries other than the one to which they are assigned. In Europe and in Japan, facilities for sports of all kinds are not only available, they are close at hand. Where the oversea station is somewhat isolated, such as Korea, it will be found that arrangements have been made to accommodate sportsmen of all types.

LEARNING THE LANGUAGE

Hq USAF announced as a goal that all officers will become proficient in at least one foreign language, and action has been taken to increase language training within the Education Services Program. The officer going overseas is obviously in a much better position to meet this requirement within a foreign environment than is his peer who stays at home. It is an opportunity which should not be passed up, especially since foreign language proficiency has been stated officially to be a professional qualification enhancing the officer's career potential. Off-duty classes, mixing with the host people, and study of the local culture and government are methods which should be used. Even if the language is a difficult one, the officer and family should learn at least the phrases of courtesy and of the social amenities. They will carry one a long way.

LIVING CONDITIONS OVERSEAS

Each foreign country presents the Air Force officer and family with new and sometimes startling living conditions. Conditions vary widely between countries. Therefore each foreign station must be studied separately; generalizations are worthless and misleading. Officers and their spouses who are on orders to a foreign country should read the appropriate Air Force Pocket Guide (AFP 190 series) and the Department's Information pamphlet (AFP 216 series) before making *any* decision regarding the forthcoming foreign service tour. Furniture, housing, cost of living, clothes, pets, currency, operation of appliances, and a thousand other considerations are covered in the two Air Force Pamphlets relative to each country; but they do not usually cover individual bases specifically. Most of the usual "bloopers" will be avoided if these two pamphlets are read carefully.

HAINERBERG HOUSING AREA IN WIESBADEN, GERMANY

They are available through the Personal Affairs Officer or at the Family Services Center of your base.

After consulting the appropriate Air Force pamphlets on life in the particular country to which the family is going, the following points should be considered:

(1) Will your electric equipment work on the current (voltage and cycle) in the oversea area? In the case of voltage, a 220-volt system can be reduced to 110-volts by use of a transformer available at the oversea exchange. A 50-cycle current is more difficult. In general, if the current is 50 cycles (as in France) it would be better to omit shipment of phonographs and electric clocks. However, these items can be adapted to 50-cycle by a competent electrical repairman. Your refrigerators, washing machines, deep freezes, and small electrical appliances will operate satisfactorily on the 50-cycle system.

(2) Should you take your automobile? U.S. types of automobiles may be purchased through exchanges in most areas, and at considerably reduced prices when compared to those in the United States. Think twice about buying a foreign-made automobile. On returning to the United States you will have to pay for the shipment of the foreign-made car. Unless you possess a relatively new automobile, it is a good plan to sell your old car in the United States, and plan to buy a new U.S. made car through the oversea exchange.

(3) Don't "stock up." You will find all necessary items available overseas, and in many cases, available in forms and at prices preferable to the situation in the United States.

As an observation applicable to life in almost all oversea areas, it is possible to say this: You will find the situation not at all difficult, not nearly as "different" as you might have thought. Therefore, relax. You are not going to the moon, but rather, in most cases, to highly civilized and industrialized countries whose cultures were 1000 years old when America was discovered.

Armed Forces Hostess Association. An information service of unique value is provided by the Armed Forces Hostess Association, Room 1A-736, The Pentagon, Washington, D.C. 20310 (OX7-3180/OX7-6857). This is a volunteer association of military wives of all Services, who will provide you on request with considerable information on your overseas base, including appliances, climate, schools, clothing, travel, shipment of pets and other helpful tips. State Department post reports, MAAGs and Missions reports are available also. Include your rank, specific location (APO if possible) and date of departure.

TOURS OF DUTY

The Air Force seeks to insure equitable distribution of oversea assignments to its personnel. See AFR 36-20 for the policies and length of oversea tours. While the policy is on a "first to return, first to go" basis, officers will understand that manning tables and specific requirements must influence actual selections, and that lengths of tours may change with the military situation.

CLAIMS FOR DAMAGED GOODS

When the officer's household goods are delivered at the new station, occasionally items are missing or damaged. In either event the officer should record this fact on the government bill of lading before signing it and returning it to the representative of the mover. As a second step, he or she should go to the Base Claims Officer who will help execute the necessary forms for making the claim against the carrier and against the government.

U-2 AIRCRAFT IS A SINGLE-SEAT, SINGLE-ENGINE AIRCRAFT WITH A SPEED OF 430 MILES PER HOUR, A RANGE OF MORE THAN 3,000 MILES, AND THE CAPABILITY TO OPERATE ABOVE 70,000 FEET.

24

The Department of the Air Force

We must effectively discharge our responsibility to be absolutely certain that the future integrity and combat capability of the Air Force are maintained. How well you fulfill this responsibility will determine to a large extent the future security of our nation. —Curtis E. LeMay, then Chief of Staff, USAF

The Department of the Air Force comprises the totality of all elements of the United States Air Force. It is administered by a civilian Secretary appointed by the President, and is supervised by the Chief of Staff, USAF. To assist the Secretary and the Chief of Staff of the Air Force, the Air Staff functions in the Pentagon at Washington. The Air Staff employs the talents of more than 2,000 Air Force officers and civilian personnel in managing the Air Force.

MISSION

The primary mission of the Air Force is to provide aerospace forces capable of supporting the Nation's objectives in peace and war. The Air Force does this by providing:

- Strategic aircraft and missile forces necessary to prevent or fight a general war;
- Land-based tactical air forces needed to establish air superiority over ground battle areas and air support to ground forces in combat;
- The primary aerospace forces for the defense of the United States against air and missile attack; and
- The primary airlift capability for use by all the Nation's Military Services.

The Air Force also provides the major space research and development support for the Department of Defense and assists the National Aeronautics and Space Administration (NASA) in conducting our Nation's space program. Teamed with the Army, Navy, and Marine Corps, the U.S. Air Force is prepared to fight and win any war if deterrence fails.

PEOPLE

Air Force personnel are positioned on about 134 major Air Force installations—in the United States and 24 friendly countries. There are approximately one million people doing the Air Force job. This includes about:

569,000 active duty military personnel;
244,000 civilian employees; and
246,000 Air National Guard and Air Force Reserve personnel.

COST

The more than $60 billion annual budget of the Air Force is invested roughly as follows:

Aircraft, Missiles and Other Equipment—30 percent.
Personnel (salaries and allowances)—20 percent.
Research, Development, Test and Evaluation—14 percent.
Operation and Maintenance—32 percent.
Military Construction—4 percent.

FUNCTIONS OF THE AIR FORCE

The principal functions of each of the armed services are assigned by agreement between the Joint Chiefs of Staff and approved by the Secretary of Defense. Of the three major services, the Air Force has been assigned primary interest in all operations in the air, and certain missile and space responsibilities. The Air Force is charged with the responsibility of organizing, training, and equipping Air Force units for the conduct of such combat operations. The major combat responsibilities of the Air Force are to execute strategic air warfare, to provide for the air defense of the United States, and to conduct tactical air operations in cooperation with ground forces. In addition to the foregoing, the Air Force must provide air transport for the armed forces and carry out specialized missions, such as aerial reconnaissance. While the Air Force has certain collateral functions assigned to it, such as to interdict enemy sea power through air operations, these collateral functions are not the basis of the organization of the Air Force.

SPACE ACTIVITIES

The Secretary of Defense has designated the Air Force as the military service primarily responsible for the development of space projects in which the Defense Department is interested. Thus the Air Force has developed and provided to NASA all of the rockets on which astronauts have been boosted into space flights. The Air Force operates the Eastern Test Range (Cape Kennedy, Florida) and the Western Test Range (Vandenberg AFB, California) from whence virtually all space flights, manned or unmanned, originate.

USAF SQUADRONS BY TYPE AND NUMBER

MAJOR AIR FORCE SQUADRONS	FY '64	FY '74	FY '79	FY '80	FY '81	FY '82
Bomber	75	28	25	25	25	24
ECM/Reconnaissance	5	1	1	4	4	4
IRBM/ICBM	35	26	26	26	26	26
Tanker	55	38	34	33	33	33
Interceptor	40	7	6	6	6	5
Bomarc	8	—	—	—	—	—
Command, Control & Surveillance	13	8	6	6	6	6

Tactical Bomber	2	—	—	—	—	—
Mace/Matador	8	—	—	—	—	—
Fighter	75	74	79	78	77	79
Reconnaissance	8	13	7	6	6	6
Tanker/Cargo	—	—	—	—	1	1
Tactical Air Control System	1	11	13	9	9	9
Special Operations Force	6	5	5	5	5	5
Tactical Airborne Command Control System	—	—	5	5	5	5
Tactical Electronic Warfare Support	—	—	—	—	1	2
Tactical Airlift	26	17	14	14	14	14
Strategic Airlift	35	17	17	17	17	17
Aeromed Evacuation	5	3	3	3	3	3
Special Mission	2	2	1	1	1	1
Mapping	2	1	—	—	—	—
Weather	6	3	2	2	2	2
Air Rescue & Recovery	12	12	7	7	7	7
Intelligence	—	9	5	6	6	6
Other	20	13	23	23	23	22
TOTAL, USAF	**439**	**288**	**279**	**276**	**277**	**277**
Air National Guard	92	91	91	91	91	91
Air Force Reserve	50	53	53	53	54	54
TOTAL, MAJOR FORCE SQUADRONS	**581**	**432**	**423**	**420**	**422**	**422**

AIR FORCE MAJOR COMMANDS AND SEPARATE OPERATING AGENCIES

Major Commands

Air Force Communications Command
Air Force Logistics Command
Air Force Systems Command
Air Training Command
Alaskan Air Command
Electronic Security Command
Military Airlift Command
Pacific Air Forces
Space Command
Strategic Air Command
Tactical Air Command
U.S. Air Forces in Europe

Separate Operating Agencies

Air Force Audit Agency
Air Force Accounting and Finance Center
Air Force Commissary Service
Air Force Engineering and Services Center
Air Force Inspection and Safety Center
Air Force Intelligence Service
Air Force Legal Services Center
Air Force Manpower and Personnel Center
Air Force Medical Service Center
Air Force Office of Security Police
Air Force Office of Special Investigations
Air Force Service Information and News Center
Air Force Test and Evaluation Center

Direct Reporting Units
Aerospace Defense Center
Air Force Reserve
Air National Guard
Air Force Academy
Air Force Technical Applications Center
Albert F. Simpson Historical Research Center

AIR FORCE ORGANIZATION

The success of our Air Forces in combat will be determined by the perfection of the organization that supports them.

—General of the Air Force Henry H. Arnold.

Before proceeding with the question of Air Force organization, it is necessary to clarify the several Air Force levels of command, to insure a common understanding of subsequent references. These levels of command are as shown below.

Flight. A formation of aircraft or missiles (usually four or more) which is a functional subdivision of a combat squadron; or a functional subdivision or detachment of any squadron; or a parade formation of two or more squads.

Detachment. A flight is the lowest echelon in organization in flying squadrons. Similarly, a detachment is the lowest echelon in non-flying squadrons. The detachment, like the flight, may be administrative, although this is not normally the case.

Squadron. The squadron is composed of a headquarters and two or more flights (detachments). The squadron is the basic Air Force tactical and administrative unit. The squadron is the smallest Air Force unit operated separately.

Group. The group is composed of a headquarters and two or more squadrons and may be tactical (have flying squadrons) or support (have non-flying squadrons) in nature. In either case a group has administrative responsibilities. The tactical group is comparable to an infantry regiment.

Wing. The wing is composed of a headquarters and combat groups and/or squadrons, with necessary support organizations. The wing is capable of completely independent operation.

Air or Missile Division. An Air or Missile Division is normally an operational (tactical) agency. However, an Air Division may be both operational and administrative. An Air Division normally consists of from two to five wings in addition to the Division Headquarters.

Air Force. An Air Force is normally composed of the elements of two or more divisions. The Air Force is usually, but not always, designed to do a particular type job such as bombardment, fighter, or airlift operations.

Major Air Command. The Major Air Command is usually composed of two or more air forces and may or may not be designed for a particular type of air operations. The Air Command may have no air forces within it, as in the case of the Air Force Logistics Command. The Air Command is tailored to the job assigned it.

There really is no standard Air Force organization larger than a wing—actually there are many different types of wings and divisions. The fact is that large units of the Air Force are tailored to accomplish specific missions.

Number of Headquarters Required. Air Force command echelons above the wing level are formed as needed. Circumstances may and often do dictate the omission of certain command levels such as that of the air division, even though an air force or air command may be constituted. The principle is: Get along with the least number of headquarters possible, as they are costly in top-drawer personnel and, if unnecessary, they definitely slow the wheels of progress.

ORGANIZATION OF OPERATING UNITS

Combat-Tactical Units. The basic designated unit of record in the Air Force is the squadron. The combat squadron will be organized and equipped in a manner to achieve the most effective tactical employment of the aircraft or missiles, as determined by the respective commands and Hq, USAF. To insure that both combat and supporting squadrons are capable of coordinated effort, an echelon of command is required for control. The identity and level of the unit established to effect such control will be dependent upon the criteria presented herein for "Combat Wings" and "Combat Groups."

Combat Wing. The organization normally designated to direct the employment of air and space power is the combat wing. The composition of a wing will be governed by the type and number of squadrons required to achieve a predetermined combat potential, consistent with their mission requirements and aircraft or missile characteristics, as approved by Hq, USAF; together with the assembly of field maintenance, major supply, communications, and other extensive facilities and elements required to augment the support normally inherent in the combat squadrons and to effectively operate a typical installation. Combat wings employing small elements from dispersed sites (individual squadrons or flights) may be so organized as to permit the major support activities to be located at the primary installation and operated as a centralized control and service center. Wings will be organized in a manner that will enable them to maintain operational mobility, emergency readiness, and the capability of reassignment between and employment by the respective commands. Therefore, like-type wings and component units will be organized in an identical manner. Variations in manning to meet the requirements of different installations, or those necessitated by geographical location, will be at the discretion of the commands. Command and staff echelons normally will not be established between the wing and combat squadrons.

Combat Group. The unit command established within a wing to control two or more combat squadrons, of identical or dissimilar types, will be identified as a combat group. Such a unit will be established only after it has been determined that peculiarities of the wing's mission, characteristics of employment and location, or the assignment of additional combat squadrons, requires an agency (or agencies) of control apart from the wing's staff. Normally, a combat group will be designed to include integrated direct and base-support elements to permit operations from individual bases. The degree of such integrated support will be dependent upon the command's mission, and its operational and location requirements.

Non-Tactical, Base-Operating Units. A number of Air Force base installations exist for purposes other than the operation of tactical groups and wings. Among others, such purposes include: flying and technical training schools; research,

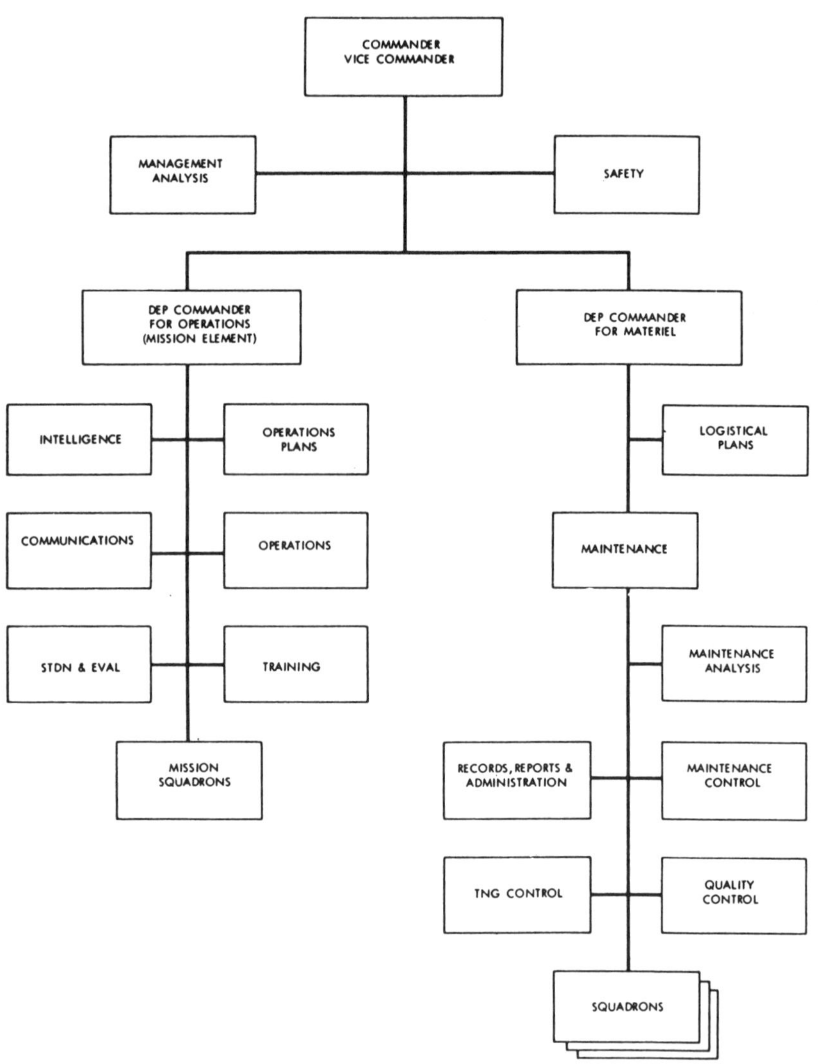

development, and test centers; and airlift operations. The operational peculiarities inherent in the varied missions of these units may require organizational deviations from the more standard tactical wing structure, as well as minor differences in nomenclature. In those instances where the type of unit lends itself to military echelons and nomenclature, the designations of wing, group, and squadron may be used. For activities whose functions, methods of operation, and requirements of manning are not particularly adaptable to military organizations and identities, the following terminology for nondesignated units will be employed, as required, in descending order: deputate, directorate, division, branch, section, and unit.

Technical or Auxiliary Services. The organization of a technical or auxiliary service-type unit (which may be tactical or non-tactical) is contingent upon factors generally differing from those governing other operating units. In the absence of requirements for field maintenance, major supply, and comparable supporting elements at operating locations, the echelon at which a unit of control is established will not be determined by factors of organizational equivalents, unit prestige, and similar considerations. Instead, the status of a wing will be attained only after it has been proven that an operational unit of group stature is not capable of performing the mission, and that the establishment of another group within the wing is mandatory to the accomplishment of the mission. In like manner, a squadron will not be designated as a group unless the unit is of such size and complexity of mission as to exceed that which a squadron-type organization can effectively control.

Depots and Other Logistical Units. The principal organization engaged in the support of USAF units is the Air Force depot. Air Force depots are designed for receipt, storage, and distribution of supplies, together with major maintenance, repair, salvage, procurement, and related logistical functions. Depots are specialized in purpose and therefore restricted to the performance of certain designated

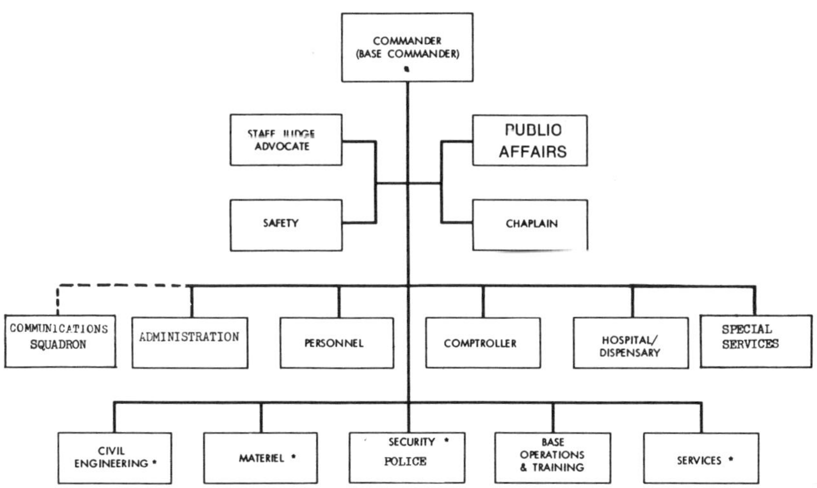

* Functional squadrons as appropriate

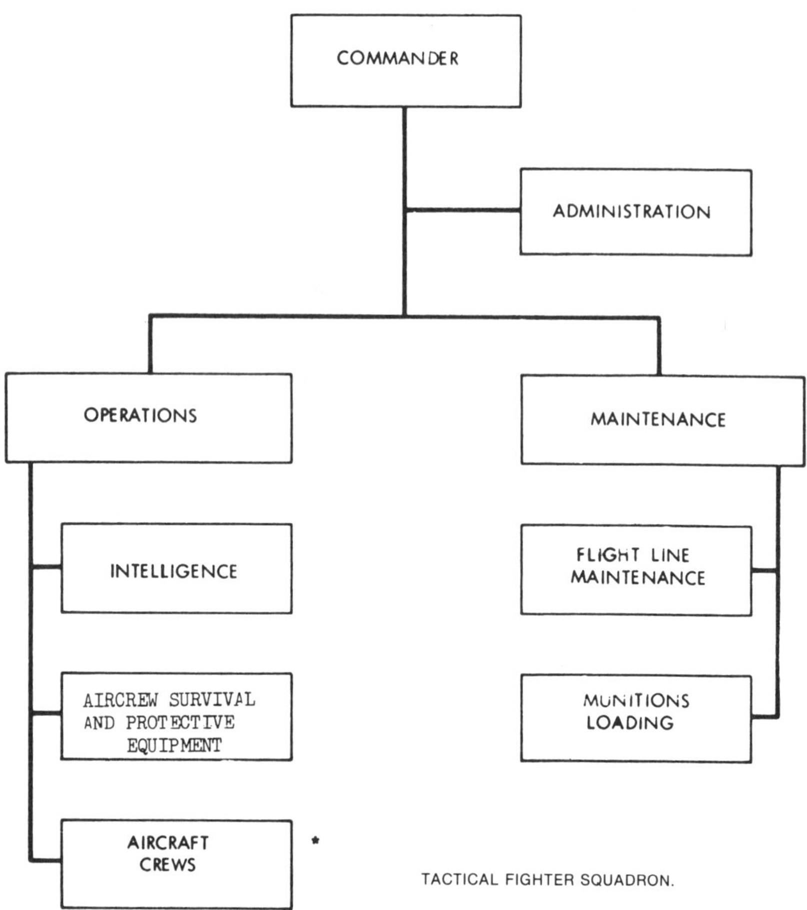

TACTICAL FIGHTER SQUADRON.

functions. Normally, Air Force depots will be fixed-type establishments for zone of interior employment, manned predominantly by civilian personnel, and organized into functional elements identified as directorates, divisions, branches, sections, and units. The number of military personnel required for depot control and administration may be administered by a "Depot Squadron Section." Depot training groups, and other mobile, military-manned logistical units, will be organized as required and attached to an appropriate Air Force depot for training and employment. The official designation of Air Force depots will not contain a suffix "Wing."

Medical Units. *Medical Treatment Units.* An Air Force medical treatment facility, whether established as a "fixed" or "nonfixed" facility, will be operated by a numbered hospital, infirmary, or dispensary unit assigned to the highest echelon exercising command of the respective installation. Fixed medical treatment facilities are established at installations within the continental United States and overseas upon approval of the Chief of Staff, USAF. Nonfixed (mobile air transportable) medical treatment facilities are provided by tactical hospital, infirmary, or dispensary units that are organic or assigned to tactical units, and are established upon approval of the Chief of Staff, USAF, except as otherwise authorized. In accordance with the provisions of AFR 26-2, the official designation of the medical treatment unit will not contain a prefix "headquarters" or a suffix "group," "squadron," or similar nomenclature, but will be identified in a manner consistent with the following examples: 1600th USAF Hospital; 97th Tactical Hospital; 2793d USAF Infirmary.

Other Medical Service Units. Medical treatment units are distinguished from other medical service units in that they are designed and organized to provide personnel and material to operate medical treatment facilities. Other types of medical service units, performing such functions as aeromedical evacuation, medical material, and optical repair, will be designated in a manner consistent with the following examples: 25th Aeromedical Evacuation Group; 43d Medical Materiel Squadron; 4th Optical Repair Flight.

AIR FORCE MISSION STATEMENTS

Aerospace Defense Center. In 1979, the Aerospace Defense Command was reduced from its prior status as a major command of the Air Force, with its resource management of forces transferred to TAC and SAC. The Aerospace Defense Center provides manpower for the U.S. specified command component of NORAD and operates the Space Defense Operations Center in Cheyenne Mountain, Colorado.

HQ: Petersen AFB, Colorado.

Air Force Communications Command (AFCC). AFCC provides three prime services vital to the efficient operation and command and control of United States defense forces: (1) long-haul and short-haul communications; (2) air traffic control and navigational aids, both fixed and mobile; and (3) engineering and installation of ground communications and electronic equipment. Installing, operating and maintaining a global network of communications and a worldwide system of air navigational aids and air traffic control facilities, AFCC helps provide commanders at all levels with the means to operate, control, supply, administer, coordinate and manage the Nation's aerospace forces. AFCC operates onbase and long-haul data and voice networks for Air Force commands. Long-haul circuitry constitutes a major portion of the Defense Communications System (DCS) operated by the military services in support of the Department of Defense.

HQ: Scott AFB, Ill.

Air Force Logistics Command (AFLC). AFLC's prime mission is to keep every weapon system in the Air Force inventory at constant readiness. AFLC is organizationally subdivided into areas of responsibility called Air Logistics Centers (ALCs). In the five industrial-type ALCs located throughout the United States, there is a system manager for each of more than 300 types of missiles, aircraft and support systems, as well as the most complicated of communications-electronics systems. Ogden ALC manages the Minuteman, Oklahoma City the B-52, San Antonio the C-5A, Sacramento the F-111, Warner Robins the C-141, and so on. Through such specialized logistics management, AFLC keeps its procurement, maintenance, supply and transportation functions attuned to the instant needs of customers—the operation commands at home and overseas. AFLC is unique in that more than 90 percent of its people are civilians, doing more than 830 different kinds of highly skilled jobs aimed at keeping the Air Force ready for instantaneous action at any time.

In addition to its five maintenance depots, AFLC operates five specialized facilities: the Military Aircraft Storage and Disposition Center (MASDC), for storage or salvage of all DoD aircraft; the Contract Maintenance Center, where all Air Force maintenance contracts are managed; the Aerospace Guidance and Metrology Center (AGMC), where inertial guidance and navigation equipment is repaired and where the Air Force master standards are maintained and instruments are calibrated; the International Logistics Center; and the Medical Center.

HQ: Wright-Patterson AFB, Ohio.

Electronic Security Command (ESC). The Electronic Security Command is responsible for electronic security in the United States, in Europe, and in the Pacific. The command uses offensive and defensive electronic warfare techniques to defeat opposing electronic systems. It also operates the Air Force Cryptologic Support Center and the Air Force Joint Electronic Warfare Center. The command has about 11,000 military and civilian personnel at about 78 locations in 8 countries.

HQ: San Antonio, Texas.

Air Force Systems Command (AFSC). AFSC is organized to provide the most up-to-date and effective management of Air Force scientific and technical resources and is the single manager of all phases of acquisition of new aerospace systems. Its mission is to accomplish the rapid advancement of aerospace technology and its adaption into qualitatively superior operational aerospace systems. The command helps the Air Force meet the major space requirements of DoD and provides research, development, test, and engineering of boosters, space probes and associated systems necessary to support specific NASA projects and programs arising under basic agreements between DoD and NASA. The commander of AFSC directs the operations of 13 laboratories, 7 divisions and 7 test centers and ranges. AFSC manages 300 locations in the United States and overseas engaged in research, development, test, and engineering.

HQ: Andrews AFB, Md.

Air Training Command (ATC). Recruiting and training comprise the mission of ATC, the free world's largest training system. In the real sense of academic classroom instruction, followed by graduation from a formalized course of study, more than 272,000 students completed technical and medical training courses in fiscal year 1981. More than 2,800 courses of this type are conducted or monitored by ATC on an annual basis. During fiscal year 1981, the Officer Training School, Lackland AFB, Tex., graduated 4,557 officers. In addition, about 2,000 newly commissioned medical officers completed the military indoctrination course; more than 1,500 officers completed undergraduate pilot training, and about 630 completed undergraduate navigator training. Approximately 4,000 personnel from foreign countries were trained under the Security Assistance Program.

HQ: Randolph AFB, Tex.

Air University (AU). Subordinate to ATC, AU is the Air Force's center for professional military education, conducting courses-in-residence for more than 3,000 Air Force officers annually through its facilities at the Squadron Officer School (SOS), the Air Command and Staff College (ACSC), and the Air War College (AWC). (Substantial numbers of U.S. Army and Navy officers, DAF civilians and international officers also attend these schools.) AU's Air Force Institute of Technology (AFIT), located at Wright-Patterson AFB, Ohio, offers courses-in-residence and at civilian colleges and universities in the arts and social sciences, biological and physical sciences, and in logistics, management, medicine and engineering. The Air Force ROTC (AF-ROTC) program, commissioning approximately 2,700 graduates annually from more than 140 U.S. colleges and universities, is a major source of Air Force officers. Air Force Junior ROTC (AFJROTC) conducts an aerospace education course at the high school level. AU is the headquarters for the Civilian Air Patrol, the official voluntary civilian auxiliary of the USAF. AU also conducts courses for commanders, personnel management, comptrollership and for the Air Force Judge Advocate. It operates the USAF Chaplain School, the Academic Instructor School and a comprehensive program for international officers. Additionally, AU conducts the world's largest school, the Extension Course Institute (ECI). Air University also maintains an extensive research capability and the Center for Aerospace Doctrine, Research and Education (CADRE). It also houses one of the country's most complete military libraries.

HQ: Maxwell AFB, Ala.

Alaskan Air Command (AAC). AAC's far-reaching warning system and F-4E Phantoms police the 586,000 square miles of Alaska, parts of which are some 50 miles from Soviet Siberia. In temperatures ranging from -70°F to summer readings of over 100°F, AAC's forces are ready to detect, intercept, and destroy aerial aggressors. AAC provides support for at least 12 Air Force major commands as well as many other DoD services and agencies. It provides the major strength of the Alaskan NORAD Region to protect the northern portion of the North American continent and the polar regions. The command is located at the polar crossroads for aircraft flying between the Orient and Europe; it is on the Great Circle route between the United States and many important locations in the Far East. Several types of transport aircraft are used to supply Alaskan sites.

HQ: Elmendorf AFB, Alaska.

Military Airlift Command (MAC). From its headquarters at Scott AFB, Ill., the Military Airlift Command directs about 1,000 aircraft, 87,000 active-duty people at 300 locations in some 24 countries. MAC is responsible for several major missions which serve the Air Force and DoD: air deployment of combat forces and their support equipment, air logistical resupply of deployed forces, airdrop of troops and battle equipment, aeromedical evacuation; aerial search, rescue and recovery of downed flyers and space hardware; special operations; weather sampling, forecasting, and dissemination; and documentary photography and audiovisual service. Additionally, the command acts as the executive agent to contract commercial airlift for the Department of Defense. The current airlift mission of MAC is to maintain in a constant state of war-readiness the airlift capability necessary to fulfill all tasks assigned by the Joint Chiefs of Staff and the Air Force. As a byproduct of the constant training required for the deployment mission, MAC routinely supports DoD logistic needs on a worldwide basis. The airlift is also used frequently to meet emergencies and relieve suffering caused by natural disasters. MAC's resources are separated into three types of organization: airlift components, technical services and special wings. The three specialized technical services are the Aerospace Audio-Visual Service, Air Weather Service and the Aerospace Rescue and Recovery Service. Special wings are: the 89th Military Airlift Wing, Special Missions, at Andrews AFB, Md.; the 375th Aeromedical Airlift Wing at Scott AFB, Ill., the 1st Special Operations Wing at Hurlburt Field, Fla., and the USAF Airlift Center at Pope AFB, N.C. MAC is also responsible for the training, inspection and safety programs of approximately 51,000 Guard/Reserve forces.

HQ: Scott AFB, Ill.

Pacific Air Forces (PACAF). PACAF, the U.S. Air Force air arm in the Pacific Far East, is responsible, in conjunction with U.S. allies, for maintaining control of the air over the Western Pacific-Far East. It is the air component of the unified Pacific Command (PACOM). PACAF is prepared to react swiftly to any aggression with both defensive and offensive forces. Its forces are composed of supersonic fighters tactical missiles, reconnaissance aircraft, and the light conventional strafing and bombing aircraft most effective in waging special air operations. Responsibility for air defense at the land areas of the Pacific Command is assigned to the Commander in Chief, Pacific Air Forces (CINCPACAF), and carried out through its numbered Air Force commanders. PACAF also assists in the training of friendly air forces in offensive and defensive air tactics and techniques.

HQ: Hickam AFB, Hawaii

Space Command. Formed on September 1, 1982, Space Command is located in Colorado Springs, Colorado. It will become operational in October 1983.

HQ: Peterson AFB, Colorado.

Strategic Air Command (SAC). SAC is the United States' long-range strike force, comprising a mixture of combat aircraft and intercontinental ballistic missiles. During peacetime, SAC's primary mission is to deter nuclear war through its ability to deliver the majority of this Nation's nuclear firepower to any part of the globe, even if subjected to surprise attack. If war should occur, SAC must have the capability to destroy the enemy's war-making powers and will to fight. The command also has the capability of delivering conventional (nonnuclear) weapons with its force of B-52 and FB-111 bombers. A true aerospace force, SAC's blend of intercontinental ballistic missiles and bomber and tanker aircraft is continually maintained at the highest level of alert of any military force in history. Up to 30 percent of SAC bombers and tankers are poised on ground alert and can be launched for survival, under positive control, well within the warning time provided by the Ballistic Missile Early Warning System (BMEWS) and other warning systems. SAC also has the capability to keep a part of its bomber force in the air, within range of enemy targets, on airborne alert 24 hours a day. SAC maintains the world's best strategic aerial reconnaissance force, which includes a long-range high-altitude supersonic reconnaissance aircraft. SAC is not only the largest major command in the Air Force, but also a specified command, operationally responsive to the Joint Chiefs of Staff (JCS).

HQ: Offutt AFB, Nebr.

Tactical Air Command (TAC). TAC is charged with organizing, equipping, training, administrating and operating assigned or attached forces to insure combat ready forces capable of conducting worldwide tactical air operations. As the air component of the United States Readiness Command (USREDCOM), TAC conducts, controls, and coordinates tactical air operations according to tasks assigned by the commander in chief, Readiness Command (CINCRED). In addition, when designated by the JCS, TAC provides air forces for the U.S. Atlantic Command, as was done in the 1965 Dominican Republic operation. TAC is ready on a moment's notice to deploy powerful, tailored packages of airpower nonstop to any spot on the globe for support of the national interests. Support of the U.S. Army is a vital responsibility of TAC. It has a versatile array of aircraft, ranging from supersonic jets to propeller-driven light aircraft, for conducting its missions of air defense, air superiority, close air support, interdiction, and counterair reconnaissance. TAC trains personnel for the tactical air forces overseas and supervises the training of a large percentage of Air National Guard and Air Force Reserve flying squadrons.

HQ: Langley AFB, Va.

United States Air Forces in Europe (USAFE). USAFE, a powerful American overseas air arm, is a primary instrument in the Western line of defense, where it stands constantly alert as a combat-ready force. Its responsibility extends from the United Kingdom to Pakistan. USAFE is the air arm of the unified European Command. Its missions in support of Air Force policy are directed by Headquarters, U.S. Air Force. In time of war, USAFE forces committed to NATO would conduct air operations as directed by SHAPE (Supreme Headquarters Allied Powers Europe). The command's weapon systems are ready for strike, attack, air defense and reconnaissance operations. Strategic and tactical airlift is provided under a joint USAFE-MAC plan of coordinated control. USAFE aircraft earmarked for use in NATO constitute the largest single air contribution of any nation in the Alliance. In addition to theatre-based forces, USAFE dual-based squadrons, now located at stateside bases, are placed under operational control of the U.S. commander in chief, Europe (CINCEUR) upon theater deployment during wartime, contingency or exercise situations. These highly trained units are geared for rapid deployment to European locations set up to accept and service them at a moment's notice. They return to the European theater periodically to keep aircrews familiar with the area and its associated flying problems. The commander in chief, United States Air Forces in Europe (CINCUSAFE), also commands NATO's Allied Air Forces Central Europe.

HQ: Ramstein AB, Germany.

Air Force Accounting and Finance Center (AFAFC). AFAFC performs four major functions: (1) policy guidance on matters pertaining to the accounting and finance program for the Air Force; (2) accounting for all Air Force dollars from initial appropriation by Congress through reports of final expenditure; (3) providing technical supervision, advice and guidance to the worldwide accounting and finance network; and (4) a variety of financial services that can be centralized effectively and economically for Air Force personnel throughout the world. AFAFC maintains all allotment accounts for both active and retired personnel; pays all retired and Air Reserve Forces members; computes monthly pay for all active duty members outside the continental U.S.; and issues all U.S. Savings Bonds purchased by Air Force members on a monthly or quarterly basis. AFAFC is the USAF focal point for providing financial status reports to many other Government activities and agencies.

HQ: Lowry AFB, Colo.

Air Force Audit Agency (AFAA). AFAA is the internal audit organization for the U.S. Air Force. As such, its job is to provide all levels of Air Force management with an independent and objective evaluation of the effectiveness and efficiency with which managerial responsibilities are being carried out. AFAA's range of audit interests is diverse, being concerned not only with financial activities but also with operational and support activities. AFAA auditors examine policies, systems, and procedures relating to the use of resources—men, money, and material.

HQ: Norton AFB, Calif.

Air Force Inspector and Safety Center (AFISC). A separate operating agency, AFISC was one of four Pentagon field extensions redesignated as separate operating agencies of the Air Force on December 31, 1971. It is still a function of the Air Force Inspector General in Washington, D.C., and its commander still carries the added Air Staff position of Deputy Inspector General for Inspection and Safety, HQ USAF. The Center superseded the 1002d Inspector General Group, which had been in existence since January 1950. At that time, Air Force flight safety functions located at Langley AFB, Hampton, Va., and readiness inspection activities headquartered at Kelly AFB, Tex., were brought together under the worldwide mantle of the IG.

HQ: Norton AFB, Calif.

Air Force Manpower and Personnel Center (AFMPC). The Center, the operating arm of the Air Force Deputy Chief of Staff Personnel, is the hub of all personnel actions for airmen and officers below the grade of colonel, and its goal is to "do things for people—not to them." Since moving to Randolph in 1965 after outgrowing its former Pentagon facilities, AFMPC has devised and implemented a continuing series of people-oriented programs. These have been highlighted by improvements in personnel programs, personnel service, and data systems.

HQ: Randolph AFB, Tex.

Air Force Intelligence Service (AFIS). AFIS, as a separate operating agency, provides specialized intelligence service and intelligence to HQ USAF and USAF Commands by directing and conducting intelligence collection activities, by processing and disseminating intelligence information and intelligence, and by conducting programs to insure the adequacy of intelligence equipment and special security systems. A major objective of the AFIS is to play an effective role in insuring the war readiness capability of intelligence resources, human and physical, furnished to combatant commands under USAF and joint war planning documents. The 7602 Air Intelligence Group conducts the worldwide Air Force human source collection effort and handles intelligence aspects of the Air Force in the DoD attache program.

HQ: Washington, D.C.

Air National Guard (ANG). The Air National Guard (ANG) provides the Air Force with some 1,560 aircraft and 2,400 crews with a total of over 97,000 personnel, ready as units, to meet emergency requirements of the active establishment. Some 320 of the then 1,700 mission aircraft and 10,500 personnel were activated in mobilizations of January 26, 1968, to serve in Southeast Asia, Korea, Europe and at various Air Force bases throughout the United States. Air National Guard units, in peacetime, are commanded by the governors of the 50 States in which the units are located, the governor of Puerto Rico, and the commanding general in the District of Columbia. Policy direction and support are received from the Air Force through the National Guard Bureau in the Pentagon. The operations and training supervision is effected by the major Air Force commands gaining each unit in event of mobilization. SAC has 13 Air Guard refueling units. The Air Guard has 75 percent of the total aerospace defense fighter force in the United States today, and the 10 Guard units perform 24-hour alert operations on a continuous basis. There are 19 Guard airlift units. Aircraft of the ANG, under TAC, total some 31 units. These are composed of tactical fighter, airlift, tactical air support, reconnaissance, air refueling, special operations and tactical electronic warfare units. Other units of the Air National Guard include Air Weather Flights and Communications and Electronics Installation Agency units performing daily services for the active establishment on worldwide missions. The Air Guard has a continuous, full-time overseas mission in support of the Air Force, supported by Air Guard crews. Air Guard refueling units have participated in such missions as "Creek Party," with units rotating 2-week assignments at Rhein Main AB, Germany, refueling active force fighter aircraft in USAFE (United States Air Forces in Europe).

HQ: Pentagon, Washington, D.C.

Air Force Reserve (AFRES). Air Force Reserve (AFRES) is a separate operating agency under the USAF Chief of Staff, receiving technical direction and control from the Office of the Air Force Reserve. It is responsible for administering programs affecting some 64,000 nonactive duty members of Ready Reserve programs of the Air Force Reserve. Headquarters AFRES provides training, as well as operational, logistical, budgetary, administrative and personnel support for all AFRES units. It provides the active duty Air Force with Ready Reserve units to meet emergency requirements throughout the world. It exercises command jurisdiction over assigned personnel facilities, property and funds. These include three numbered air forces, 17 wings, 54 flying squadrons, and 138 nonflying units and more than 400 aircraft. The various kinds of nonflying units include aerial port, aeromedical evacuation medical service, mobile maintenance and supply and wartime information security squadrons. Flying units include air refueling, tactical airlift, tactical fighter, special operations, aeromedical airlift, and military airlift (associate), aerospace rescue and recovery, airborne early warning and control and tactical airlift training. These, plus other Reservists with individual assignments, constitute the total Air Force Reserve. Headquarters AFRES has other special responsibilities, such as: acting as liaison with the Selective Service System; formulating and coordinating Air Force plans in conjunction with the U.S. Army; responsibility for providing military assistance in domestic and civil defense emergencies; exercising overall responsibility for Air Force operations in natural disasters including the exercise of operational control over Air Force forces so employed; acting as a single contact in cooperating with U.S. Army headquarters and U.S. Naval districts within the continental United States in carrying out the Air Force responsibilities prescribed in the Basic Plan for Defense (other than aerospace defense) of the United States; assisting other commands in recruiting Reserve personnel to meet augmentation requirements; providing representatives in conjunction with the Army and Navy for the States Reserve Facilities Board; and supervising Air Force cooperation with the Boy Scouts of America program in the United States and Puerto Rico.

HQ: Robins AFB, Ga.

U.S. Air Force Academy (USAFA). The Air Force Academy mission is to provide instruction and experience to each cadet so that he or she graduates with the knowledge and character essential to leadership and with the motivation to become a career officer in the United States Air Force. Toward this end, the Academy strives to attain certain minimum objectives—a basic 4-year education in the sciences, the humanities and other broadening disciplines; an understanding of aerospace power with all its capabilities and the significant role it plays in national defense; high ideals of individual integrity, loyalty and honor; and a sense of responsibility and dedication to honorable service.

Air Force Test and Evaluation Center (AFTEC). AFTEC's mission is to manage the Air Force's Operational Test and Evaluation (OT&E) program. AFTEC recommends Air Force OT&E policy to HQ USAF for approval, as well as plans, directs, controls, evaluates, and reports—independently—on OT&E. AFTEC serves as the principal field command for providing OT&E information to the Secretary of the Air Force (SAF) and the Chief of Staff of the Air Force (CSAF) in preparation for Defense Systems Acquisition Review Council (DSARC) actions and to support those Air Force procurement requests for which OT&E information is statutorily required to be supplied to the Congress. The AFTEC commander commands a separate operating agency which reports directly to the Chief of Staff.
HQ: Kirtland AFB, N. Mex.

Air Force Commissary Service. The Air Force Commissary Service, Kelly Air Force Base, San Antonio, Texas, was activated in 1976 and exercises control of USAF commissaries worldwide. AFCOMS has four elements: a headquarters, a board of directors, 15 stateside complexes and 2 oversea regions, and 137 commissary stores.

Air Force Engineering and Services Center. The Air Force Engineering and Services Center was activated in 1977 at Kelly Air Force Base, Texas. Headquarters are now at Tyndall AFB, Fla. There are four regional offices in San Francisco, Dallas, Atlanta, and Norton AFB, Calif.

Air Force Medical Service Center. The Air Force Medical Service Center is responsible for functions of the Air Force Surgeon General's Office and the Aerospace Medical Division.
HQ: Brooks AFB, Tex.

Air Force Legal Services Center (AFLSC). The Air Force Legal Services Center administers military justice in the Air Force, supervises Air Force claims activities, and directs the Air Force Preventive Law and Legal Assistance Program.
HQ: Washington, D.C.

Air Force Office of Security Police (AFOSP). AFOSP develops the operational policies and practices necessary for the security and defense of Air Force resources and information. Base defense is a major concern.
HQ: Kirtland AFB, New Mexico.

Air Force Service Information and News Center (AFSINC). AFSINC has four directorates: Internal Information, Administration, Hometown News, and Armed Forces Radio and Television.
HQ: Kelly AFB, Texas.

Albert F. Simpson Historical Research Center. The Historical Research Center manages the nation's largest and most valuable collection of documentation of U.S. military aviation history. There are four divisions: reference, research, oral history, and technical services.

HQ: Maxwell AFB, Alabama.

UNIFIED COMMANDS

U.S. Atlantic Command (LANTCOM). The commander in chief, Atlantic Command (CINCLANT) is responsible for all joint U.S. military action in the Atlantic Ocean area. With headquarters in Norfolk, Va., the Atlantic Command staff is composed of senior military officers from all U.S. services.

Established December 1, 1947, the Atlantic Command was formed to facilitate greater administrative and operational control of U.S. military forces in the Atlantic.

The commander in chief, Atlantic Command (CINCLANT), is aided by component commands when designated by the Joint Chiefs of Staff. The Army Forces Atlantic (ARLANT) and Air Forces Atlantic (AFLANT) consist of Army and Air Force units which would be integrated into the Atlantic Command when deemed necessary by the Joint Chiefs of Staff. Both of these component commands have their own commanders, and while the forces are not permanently assigned, they could be melded into the Atlantic Command on a moment's notice. The April 1965 Dominican Republic operation is a prime example of how the Atlantic Command functions when fully activated.

The Atlantic commander in chief (CINCLANT), in his capacity as commander in chief, U.S. Atlantic Fleet, controls all U.S. Navy and Fleet Marine Forces in the Atlantic area on a permanent basis. Combining this with designated Army (ARLANT) and Air Force (AFLANT) units, the Atlantic commander in chief has the capacity to summon and deploy to a contingency area any of the three—or a combination of them—to counteract any emergency.

Additionally, the commander in chief, Atlantic Command and U.S. Atlantic Fleet, is also Supreme Allied Commander, Atlantic (SACLANT), under the North Atlantic Treaty Organization and as such is commander of designated allied forces from the six NATO nations which contribute deep sea forces to the defense of the Atlantic community in time of war.

Aerospace Defense Command (ADCOM) and North American Aerospace Defense Command (NORAD). ADCOM, headquartered in Colorado Springs, Colo., is a specified command consisting of U.S. air defense forces assigned to it. ADCOM is commanded by the same U.S. Air Force general officer who heads the Aerospace Defense Center (ADC) and NORAD, an international command consisting of both United States and Canadian air defense forces. NORAD is the single agency responsible for air defense of the United States and Canada. NORAD has operational control of forces provided by its component commands: U.S. Air Force Aerospace Defense Command, and the Canadian Forces Air Defense Command.

U.S. European Command (USEUCOM). USEUCOM, located in Stuttgart, Germany, is the senior U.S. military headquarters in Europe. The U.S. commander in chief Europe (USCINCEUR), who is also NATO's Supreme Allied Commander Europe (SACEUR), exercises operational control through his deputy commander in chief (DCINC) over the following service components: the U.S. Army Europe (USAREUR), U.S. Naval Forces Europe (USNAVEUR) and the U.S. Air Forces in

Europe (USAFE). USCINCEUR also commands assigned Military Assistance Advisory Groups (MAAG) and Missions in Europe and North Africa. The principal mission of the U.S. European Command is to support USCINCEUR in his responsibilities as Supreme Allied Commander Europe, and to honor the U.S. commitment to NATO.

The USEUCOM area of responsibility extends through 17 NATO and non-NATO countries in Europe and North Africa.

Because of the magnitude of his duties as SACEUR, USCINCEUR has delegated broad powers for the direction of USEUCOM to his DCINC. Assisted by a purely U.S. Joint Staff, the DCINC exercises operational command of U.S. forces through component commanders who are responsible for planning, deploying, directing and coordinating U.S. forces to ensure they are constantly ready to conduct war in either a nuclear or conventional environment.

Major forces assigned to USEUCOM include: the U.S. Army Europe (USAREUR), with combat-ready divisions dual-based in Germany and the U.S.; the U.S. 6th Fleet, with carrier, amphibious and service task forces deployed in the Mediterranean; and the U.S. Air Forces in Europe, with air squadrons assigned to three numbered air forces in the United Kingdom, Spain and Germany—the 3d, 16th, and 17th.

Since it is envisioned that any war in Europe would be a NATO war, strong staff and command relations between the U.S. commands and NATO commands are most essential. This has been provided for in both plans and organization. Only the air defense units of Army and Air Force are under the operational control of SACEUR in peacetime. All other forces remain under national control until released to SACEUR for wartime operations.

U.S. Pacific Command (PACOM). PACOM has the broad responsibility of defending the United States against attack throughout the Pacific. PACOM supports and advances United States interests throughout the Pacific, part of the Arctic Ocean, the Bering Sea, Japan, Republic of Korea, non-Communist countries of Southeast and South Asia, and most of the Indian Ocean.

PACOM extends from California to Pakistan (more than 10,000 miles) and from part of the Arctic Ocean to Antarctica. More than half the people on earth live under 36 flags in the PACOM area. PACOM's present size was attained January 1, 1972, under the worldwide unified command reorganization. Its basic organizational structure stems from July 1957 when DoD consolidated all U.S. forces in the Pacific and Far East under the commander in chief, Pacific (CINCPAC), who reports directly to the Chairman of the Joint Chiefs of Staff. Headquarters are at Camp H. M. Smith near Honolulu, Hawaii. CINCPAC exercises operational command of PACOM forces through component and unified commanders. Principal component commanders are: U. S. Army, Pacific (USARPAC), U. S. Pacific Fleet (PACFLT), and Pacific Air Forces (PACAF). Subordinate unified command headquarters are in Japan and Korea. Also reporting to CINCPAC are chiefs of U.S. Military Assistance and Advisory Groups throughout the area, as well as designated representatives in the Republic of the Philippines, Guam and Trust Territory of the Pacific Islands, and Australia. Some 400,000 men and women representing all military branches serve with PACOM. The Air Force component command, PACAF, has two numbered air forces: the 5th in Japan and the 13th in the Philippines.

U.S. Readiness Command (USREDCOM). USREDCOM, established on January 1, 1972, replaced the U.S. Strike Command. Under operational control of the

Commander in Chief, U.S. Readiness Command (USCINCRED), with Headquarters at MacDill AFB, Fla., are approximately six divisions of the Army Forces Command and almost 50 tactical air squadrons—fighter, reconnaissance, airlift, and special operations—of the Tactical Air Command. These forces contribute greatly to our ability to deal efficiently and rapidly with any limited war in a manner and on a scale best calculated to bring the limited war to a conclusion while minimizing the risk of hostilities broadening into general war.

The USREDCOM mission includes the provision of a general reserve of combat-ready ground and air forces to reinforce other unified commands. To assure that these forces are combat-ready, USREDCOM conducts frequent joint training exercises in the U.S. and makes available joint force elements for joint and combined exercises overseas.

Aside from their vital training function, USREDCOM exercises provide a testing ground for examining current joint operating concepts and developing recommendations on joint doctrine for consideration by the Joint Chiefs of Staff (JCS).

USREDCOM forces are also on call to respond to humanitarian and mercy missions anytime and anywhere, as in the Nicaraguan earthquake in December 1972, just as its predecessor, the U.S. Strike Command, performed numerous mercy missions in Africa, Southern Asia, and in Mexico, during the 10 years of its existence.

U.S. Southern Command (USSOUTHCOM). The hub of all United States military activities in Central and South America is Quarry Heights in the Canal Zone, the site of USSOUTHCOM headquarters.

Within its 7½ million square mile area of responsibility, USSOUTHCOM directs U.S. military efforts and advances a broad range of programs in Latin America that are vital to the security of the Western World.

RANDOLPH AIR FORCE BASE, TEXAS.

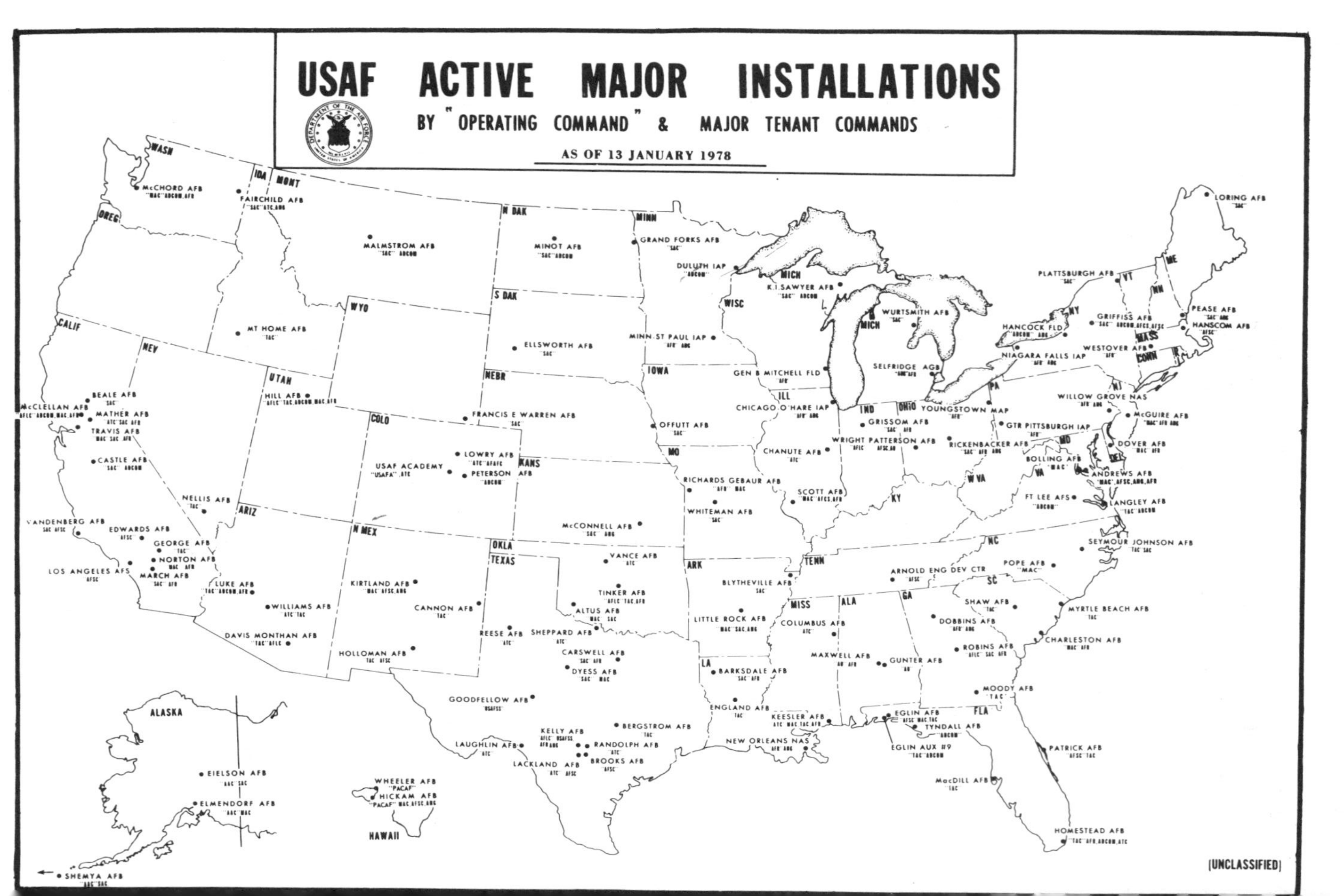

USAF ACTIVE MAJOR INSTALLATIONS
BY "OPERATING COMMAND" & MAJOR TENANT COMMANDS
AS OF 13 JANUARY 1978
McCHORD AFB
FAIRCHILD AFB
MALMSTROM AFB
MT HOME AFB
MINOT AFB
GRAND FORKS AFB
DULUTH IAP
K.I. SAWYER AFB
WURTSMITH AFB
SELFRIDGE AGB
ELLSWORTH AFB
MINN ST PAUL IAP
GEN B MITCHELL FLD
CHICAGO O'HARE IAP
GRISSOM AFB
WRIGHT PATTERSON AFB
YOUNGSTOWN MAP
GTR PITTSBURGH IAP
RICKENBACKER AFB
PLATTSBURGH AFB
GRIFFISS AFB
HANCOCK FLD
NIAGARA FALLS IAP
WESTOVER AFB
PEASE AFB
HANSCOM AFB
LORING AFB
WILLOW GROVE NAS
McGUIRE AFB
DOVER AFB
BOLLING AFB
ANDREWS AFB
LANGLEY AFB
FT LEE AFS
SEYMOUR JOHNSON AFB
BEALE AFB
McCLELLAN AFB
MATHER AFB
TRAVIS AFB
CASTLE AFB
HILL AFB
FRANCIS E WARREN AFB
LOWRY AFB
USAF ACADEMY
PETERSON AFB
OFFUTT AFB
CHANUTE AFB
SCOTT AFB
RICHARDS GEBAUR AFB
WHITEMAN AFB
NELLIS AFB
VANDENBERG AFB
EDWARDS AFB
GEORGE AFB
NORTON AFB
MARCH AFB
LOS ANGELES AFS
LUKE AFB
WILLIAMS AFB
DAVIS MONTHAN AFB
KIRTLAND AFB
CANNON AFB
HOLLOMAN AFB
McCONNELL AFB
VANCE AFB
TINKER AFB
ALTUS AFB
REESE AFB
SHEPPARD AFB
CARSWELL AFB
DYESS AFB
GOODFELLOW AFB
KELLY AFB
BERGSTROM AFB
RANDOLPH AFB
BROOKS AFB
LAUGHLIN AFB
LACKLAND AFB
BLYTHEVILLE AFB
LITTLE ROCK AFB
BARKSDALE AFB
ENGLAND AFB
NEW ORLEANS NAS
KEESLER AFB
COLUMBUS AFB
MAXWELL AFB
GUNTER AFB
ARNOLD ENG DEV CTR
POPE AFB
SHAW AFB
MYRTLE BEACH AFB
DOBBINS AFB
ROBINS AFB
CHARLESTON AFB
MOODY AFB
EGLIN AFB
TYNDALL AFB
EGLIN AUX #9
PATRICK AFB
MacDILL AFB
HOMESTEAD AFB
ALASKA
EIELSON AFB
ELMENDORF AFB
SHEMYA AFB
HAWAII
WHEELER AFB
HICKAM AFB
WASH
OREG
CALIF
IDA
MONT
NEV
UTAH
WYO
ARIZ
COLO
N MEX
N DAK
S DAK
NEBR
KANS
OKLA
TEXAS
MINN
IOWA
MO
ARK
LA
WISC
ILL
MICH
IND
OHIO
KY
TENN
MISS
ALA
GA
FLA
SC
NC
VA
W VA
MD
DEL
PA
NJ
NY
CONN
MASS
VT
NH
ME
(UNCLASSIFIED)

AIR BASES OUTSIDE CONUS: NORTH; CENTRAL AMERICA; ATLANTIC.

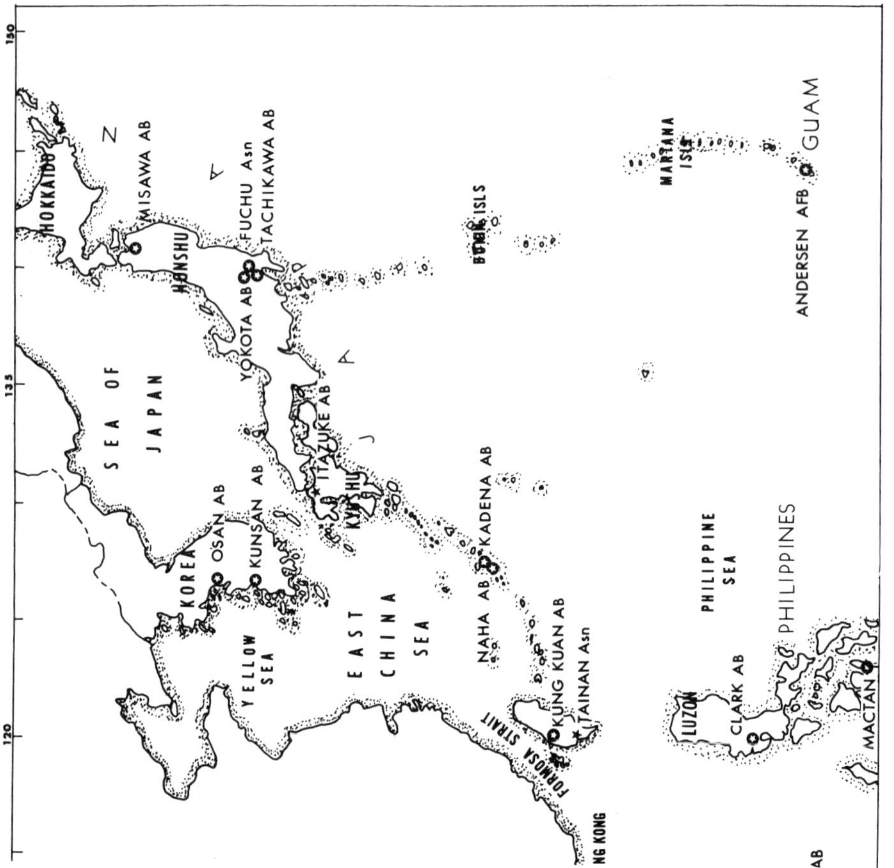

AIR BASES OUTSIDE CONUS: PACIFIC.

USSOUTHCOM's primary mission is to provide defense for the Panama Canal. Other major missions of the command include: administering the military assistance program for Latin American countries; directing United States participation in hemisphere defense exercises; conducting mapping and charting activities; and directing disaster relief and search and rescue operations.

The commander in chief, U.S. Southern Command (USCINCSO), with his joint staff of Army, Navy, Air Force and Marine personnel, exercises operational control over the Army, Naval and Air Forces assigned for the defense of the Panama Canal. Because these minimal forces are not required at all times to man positions in defense of the Canal, they may be used in accomplishing the collateral tasks that make up the balance of the USSOUTHCOM mission.

The commander in chief (USCINCSO) is represented in Latin American countries by a senior officer of one of the armed services. He is the principal military advisor to the U.S. ambassador and directs the military assistance and advisory efforts of the Army, Navy and Air Force section chiefs as they work with their counterparts of the host country military.

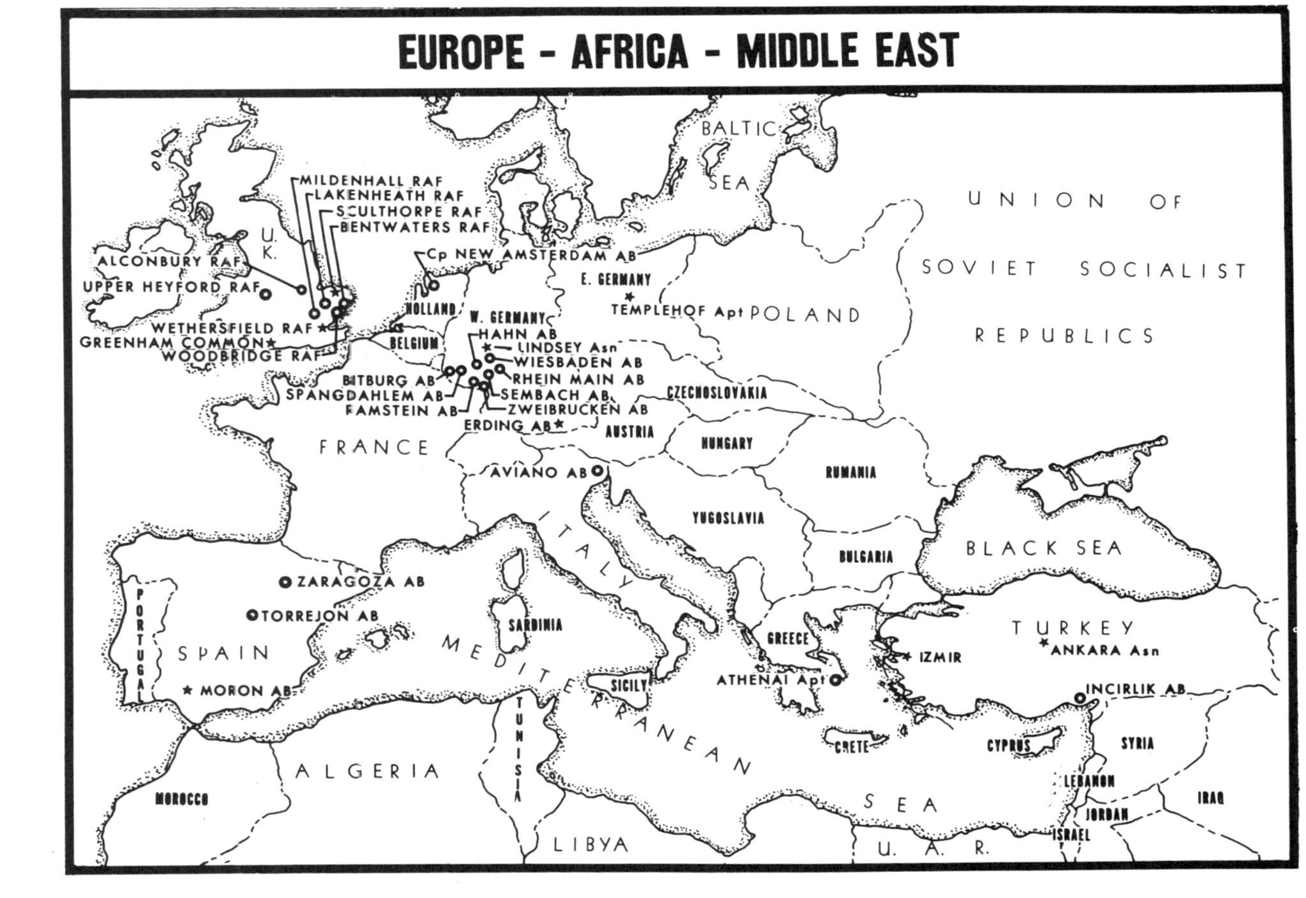
EUROPE - AFRICA - MIDDLE EAST
MILDENHALL RAF
LAKENHEATH RAF
SCULTHORPE RAF
BENTWATERS RAF
U. K.
ALCONBURY RAF
UPPER HEYFORD RAF
WETHERSFIELD RAF
GREENHAM COMMON
WOODBRIDGE RAF
Cp NEW AMSTERDAM AB
HOLLAND
BELGIUM
W. GERMANY
HAHN AB
LINDSEY Asn
WIESBADEN AB
RHEIN MAIN AB
SEMBACH AB
ZWEIBRUCKEN AB
BITBURG AB
SPANGDAHLEM AB
RAMSTEIN AB
ERDING AB
E. GERMANY
TEMPLEHOF Apt
POLAND
CZECHOSLOVAKIA
AUSTRIA
HUNGARY
RUMANIA
YUGOSLAVIA
BULGARIA
BALTIC SEA
UNION OF SOVIET SOCIALIST REPUBLICS
FRANCE
AVIANO AB
ITALY
BLACK SEA
TURKEY
ANKARA Asn
IZMIR
INCIRLIK AB
GREECE
ATHENAI Apt
ZARAGOZA AB
TORREJON AB
PORTUGAL
SPAIN
MORON AB
SARDINIA
SICILY
MEDITERRANEAN SEA
CRETE
CYPRUS
SYRIA
LEBANON
JORDAN
ISRAEL
IRAQ
MOROCCO
ALGERIA
TUNISIA
LIBYA
U. A. R.

COMMAND INSIGNIA

STRATEGIC AIR COMMAND

TACTICAL AIR COMMAND

AIR TRAINING COMMAND

AIR FORCE
LOGISTICS COMMAND

AIR FORCE
SYSTEMS COMMAND

MILITARY
AIRLIFT COMMAND

AIR FORCE
COMMUNICATIONS COMMAND

ELECTRONICS
SECURITY COMMAND

US AIR FORCES
IN EUROPE

PACIFIC AIR FORCES

ALASKAN AIR COMMAND

THE PENTAGON.

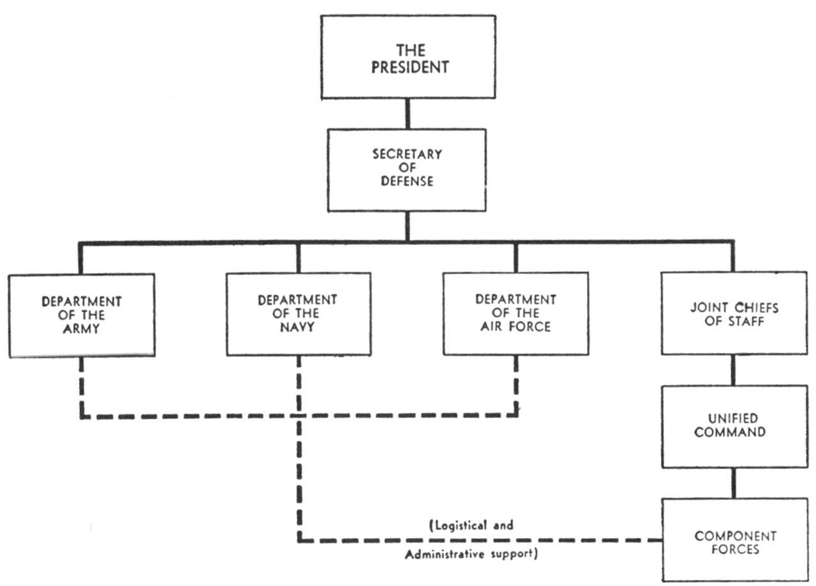

ARMED FORCES COMMAND LINES.

25

The Air Reserve Forces

The reservist is twice the citizen. —Winston Churchill

The Air Reserve Forces are the *Air National Guard* and the *Air Force Reserve.* All our Reserve Forces are placed in one of three categories: Ready Reserve, Standby Reserve, and Retired Reserve. This is a force of close to 580,000 men and women using an annual budget of well over a billion dollars.

THE READY RESERVE

By "Ready Reserve" is meant that portion of our reserve forces which can be ordered to active duty under conditions short of a Congressional declaration of war or national emergency. The total Ready Reserve, including the Guard, is just over 202,000 people. The *Air National Guard* is a Ready Reserve force, composed entirely of units organized and trained to be used as units. The Ready Reserve portion of the *Air Force Reserve* includes this type of unit; but also includes individual Reservists assigned to specific mobilization positions with Regular Air Force organizations, and other individuals earmarked as attrition replacements.

THE STANDBY RESERVE

In addition to the Ready Reserve, there is within the Air Force Reserve a Standby Reserve which can be ordered to active duty *only* when Congress declares war, or national emergency. This totals about 33,000 personnel. Some of these Standby Reservists would not be available because of critical needs of the government or the civilian community as a whole.

Although a number of Standby Reservists participate in correspondence courses or train with Ready Reserve units, the majority of them would need refresher training in order to function effectively in their Air Force specialties.

WHY RESERVES?

First, the question of why we have Reserve units. One reason is that the Constitution and the Congress say that we will. But another very good reason is that reserve units provide an economical way to maintain a surge expansion capability for the Air Force during peacetime.

The Reserves—even the Standby and Retired Reserves—provide the Air Force a strong base of grass roots community support. They are an effective influence in favor of Air Force recruiting as well. By talking about the Air Force, the Reservists help to interest young people in Air Force careers.

The primary reason for Reserves is wartime support. Until a few years ago the Air Force, like the other U.S. Military Services, traditionally had relied on its Reserve components to supply the extra manpower and capability needed for wartime expansion. Today our Air Reserve Forces—the Air National Guard and the Air Force Reserve—are much more than a wartime mobilization force; they are, in fact, a vital part of the aerospace force in being.

SMOKE RISING FROM RIM DIRM AIRFIELD IN NORTH VIETNAM AFTER STRIKE BY USAF FIGHTER BOMBERS.

Under current concepts for their use, the "Ready Now" Reserves are very much a part of our deterrent strength. Their readiness and usefulness is vital to the Air Force mission and, therefore, to the security of the nation. When the Air Force says its Reserves are "Ready Now" it means that they can fight today. They can be and have been called up in hours.

The use of Reserve forces to help keep the peace is sometimes thought to be a "new" concept. But it really isn't. The idea of the citizen-soldier was born as a "ready now" concept. As far back as 1622, the citizens of the Plymouth Colony formed a local militia which they called the "Train Band." As rapidly as their populations could support them, the "train bands" were formed in other nearby colonial settlements. The members of these organizations were not professional soldiers. They were farmers, weavers, silversmiths, and merchants. Their military obligation was a 24-hour-a-day affair; and there was hardly a period from 1622 until after the American Revolution when some of the members of the "train band" were not serving as soldiers on a full-time basis. There was no active establishment, the citizen-soldier was the sole defense of his community.

Our country and economy grew, however, as our population grew, and a standing military force became a necessity if only to prevent constant disruption of the more and more complex national structure. With the oceans as effective barriers to sudden attack, our citizen-soldiers had time on their side and could become, in fact, Reserves—to be called upon only when the size and strength of the regular establishment had to be increased for war.

This, then, became the pattern for our military forces: an active duty establishment strong enough to fend off the first attack, and a reserve which could be mobilized and trained to expand the regular force to war-winning size.

But military weapons and concepts of war are constantly changing. Today our Air National Guard and Air Force Reserve are much more than just a wartime mobilization force. They are, truly, an integral and vital part of our total aerospace power in being.

COMPOSITION OF RESERVES

To give you an idea of how large a part they play in the Air Force plan, we have 145 flying squadrons—91 in the Air National Guard and 54 in the Air Force Reserve.

Additionally, we have a wide variety of non-flying units—communications, medical, aerial port squadrons, and many other types. Altogether there are some 163,000 Reservists in the Ready-To-Go flying and non-flying organizations. All have specific tasks to perform in the war and contingency plans of the major commands. Each major command of the Regular Air Force gets a sizeable augmentation force from the Reserve.

Tactical Air Command, for example, has a reserve force which consists of 38 tactical fighter units, 8 tactical reconnaissance units, 2 special operations' squadrons, and two tactical control groups. TAC also receives one fighter-interceptor squadron and one aircraft control and warning squadron in Puerto Rico for use by the Southern Command.

Military Aircraft Command gets 35 tactical airlift squadrons, 17 military airlift associate squadrons, six air rescue squadrons, 51 aerial port squadrons, and 39 weather flights.

The Air Force Communications Command is represented by more than 190 communications electronics units which provide critical air-traffic control capabilities, as well as installation repairing, and restoring communications, navigational aids, and air traffic control equipment.

Pacific Air Forces have one fighter-interceptor squadron and three aircraft control and warning squadrons. Alaskan Air Command has one tactical airlift squadron.

Reserve medical units of various kinds are also distributed among the commands.

Air Force missions and responsibilities are so broad in scope and of such variety that they can be accomplished only by a completely responsive, instantly reacting, and highly flexible force. They can be accomplished only through the full use of all resources available to the total Air Force, Reserve as well as Regular.

RESERVE TRAINING

Realistic training for wartime readiness is now par for the course for Reserve units and personnel. Airlift units carry high-priority cargo while training their personnel in airlift skills. Fighter and reconnaissance units take the place of regular units in support of joint Army-Air Force exercises and maneuvers. Air Defense units fly air alert intercept missions under active Air Force control.

Management has been improved to give Air Force combat commands responsibility for supervising and training and operational inspection of those reserve units which would be assigned to them in wartime. Standards of performance are identical for regular and reserve units. Each aircrew, regardless of component, is required to meet the same rigid requirements as regular Air Force crews before it can be designated "combat ready."

The operations room of a good reserve outfit is almost indistinguishable from that of a good regular outfit—except perhaps on week days. Our Reserve combat crews receive the normal 48 inactive duty training periods and 15 days of active duty each year plus 36 additional flying training periods. Even this is not enough for all flying requirements, and Reserve personnel engage in training whenever they can get time off from their jobs—because these part-time airmen have to meet exactly the same training requirements that the full-time Air Force crews must meet. And meet them they do. They pass their Operational Readiness Inspections exactly like active force units. Some of the Reserve units pass with exceptionally high scores.

How is this possible with a part-time operation? A great deal of the Reserve success is due to the stability of personnel. It takes a while longer to train Reservists, but the Reserves do not have the problem of annual turnover that plagues regular units.

Another asset, and an extremely important one, is the technician program. The Air Technicians of the Guard and the Air Reserve Technicians of the Reserve have a dual status: they are full-time employees in a civilian capacity, as well as being reservists who train with the units on weekends and during annual active duty. In other words, they wear two hats. They make up about 14 percent of the total Reserve personnel strength, and provide a permanent, hard-core base of skill and experience, not only for the day-to-day maintenance and operating functions, but also for training younger personnel. This valuable resource of experience is augmented by a handful of active Air Force advisors who assist in training and assure standardization with the requirements of the gaining commands of the Regular Air Force.

COLD WAR USES OF AIR RESERVE FORCES

The rather revolutionary Air Force concepts for reserve forces first proved their validity in the Fall of 1961. During the Berlin crisis the President ordered to active duty 36 Air National Guard and Air Force Reserve flying squadrons and an Air National Guard tactical control group—a total of some 27,000 Reservists. Very soon after they reported to active duty, ten fighter squadrons, a reconnaissance squadron, and a complete radar tactical control group were on duty in Europe.

Hardly had the units recalled for Berlin returned to their homes when a new trouble spot appeared. On October 22, 1962, President Kennedy alerted the world to the existence of a Soviet build-up of offensive missiles in Cuba. Late on the night of October 27, 24 Air Force Reserve troop carrier squadrons and supporting units were ordered to active duty, 400 aircraft and some 15,000 personnel in all.

This was the swiftest reserve mobilization since the American Revolution and the battles of Lexington and Concord. Ninety-three percent of the men in these units reported for duty the next morning. Their aircraft were operationally ready, and they could have flown a combat mission that day. Secretary of Defense McNamara called it "a fantastic performance." "This," he said, "is the standard of performance that has been built into the Air Force's Reserve and Guard programs."

But the story of our part in the Cuban crisis was more than the story of the mobilized units. Many units and thousands of individuals who were not ordered to active duty practically came on duty anyway—voluntarily. Air National Guard fighter squadrons which might have been needed were at a peak of readiness. Air Force Reserve recovery units at civilian airfields and unused airstrips supported aircraft dispersal operations of SAC, TAC, and ADC on a purely voluntary basis. The Commander of the Strategic Air Command expressed his personal thanks to the Reservists for the support they gave his forces.

Air National Guard C-97 squadrons and Air Force Reserve C-124 squadrons fitted their over-water training flights into the worldwide system of Military Airlift Command, flying essential cargo and filling gaps left by active forces concentration on direct support of Cuban crisis actions.

Even before the mobilization for Cuba, Air Force Reserve troop carrier units had airlifted close to a million pounds of essential defense materiel to Florida to support the build-up.

Twice in little more than a year, the Air Reserve Forces proved their effectiveness as a part of our nation's deterrent strength. Twice their quick response helped to prevent open hostilities. Twice, they demonstrated the validity of the Air Force's "total force" policy of using "Ready Now" Reserves.

In 1965 Reservists responded within hours to the emergency airlift requirement dubbed "Power Pack" to evacuate U.S. citizens and foreign nationals from the Dominican Republic.

By the end of June Reservists had flown 1,747 missions, carried 5,115 passengers and airlifted more than 4,000 tons of cargo, logging almost 16,000 flying hours.

In the summer of 1966 the ANG was flying 200 cargo trips a month in support of the Military Airlift Command—half of them in direct support of Southeast Asia operations. 18,000 Guardsmen were involved.

The Reserve Forces contributed heavily to our fighting posture in the Vietnam War. All units were brought to top readiness. Additionally, Reserve Forces transport aircraft were used on training missions to supplement Regular airlift to Vietnam.

In 1973 the AFR flew 300 missions in support of Israel during the Israeli-Egyptian war.

OTHER CONTRIBUTIONS OF THE RESERVES

There is also a day-to-day payoff that comes from our Reserve forces. The in-being capability of today's citizen-airmen pays real dividends in productive effort for the Air Force in peacetime as well as in crisis. The Reservists are a valuable asset to their gaining commands. In recent years, the Reserve Forces provided more than half of all the Air Force support for Army airborne training and air-dropped more than 100,000 paratroopers.

The Reserve airlift squadrons need oversea training flights to develop proper standards of operational readiness. The extra airlift of cargo which they provide is clear profit for the nation. To give an idea of the size of this profit, 16 Air National Guard airlift squadrons airlifted more than 12 million pounds of MAC cargo overseas during one year. In fact, in this and other programs, the Air Reserve Forces provided nearly half of the total Air Force requirement for strategic airlift crews in recent years, and 60 percent of the tactical airlift.

Another contribution of the Reserves is to air defense. Each of the 10 Air National Guard fighter-interceptor units keeps two aircraft and four aircrews on runway alert around the clock. Their mission—Air Defense identification and intercept missions under the North American Aerospace Defense Command. In Hawaii, the Air National Guard fighter squadron and two radar squadrons are the principal air defense of the island State.

Some non-pay Reservists are assigned to flights in which they all have the same profession. These units also help the Air Force in its day-by-day mission. Judge Advocate General area representatives receive training credit for performing legal work at Air Force installations too small to have their own JAG section. Research and Development Flights provide special project assistance which would require costly consultant services otherwise.

These are only a few of the peacetime benefits that come as a by-product of Reserve training. This kind of realistic training helps to maintain the high degree of responsiveness and quick reaction capability which added so effectively to our strength in Berlin, Cuba, and, more recently, in Southeast Asia.

BASIC FUNCTION OF THE RESERVES

As General LeMay, then Air Force Chief of Staff, stated during the Berlin crisis, *It is pretty obvious that the regular establishment forces can't do all the job—today or in the future. I see a continuing need for Ready Reserve Forces that can pitch in and help the active units.*

In peacetime, the Air National Guard is commanded by the governors of the various States; it is supported by the Air Force through the National Guard Bureau, under Air Force policies and regulations. It is trained under the supervision of the gaining commands, SAC, MAC, TAC, AFSC, and AFCC, who also have the inspection responsibility.

The Air Force Reserve is a direct reporting unit of the Air Force and has the same gaining command relationship as the Guard.

Both components are governed by Air Force policies, regulations, and training directives and both are responsive to training standards and performance requirements prescribed by the gaining command.

The Air Force program for Reserve forces is unique. It puts into practice the concept of reliance on a "ready now" augmentation force. As General LeMay told the Congress in 1964: *The Air Force thinks of the Air National Guard and the Air Force Reserve in the same manner as it does of its regular units. As nearly as possible, we expect the same rapid response from them. They are subjected to identical readiness tests. We need a 'Ready Now' combat capability in the Air Reserve Forces because we depend on them to augment the active force in times of crisis.*

EC-135 AIRBORNE COMMAND POST.

26

Photos of Air Force Weapons Systems

F-15S OVER ALASKA.

HOUND DOG MISSILE IN FLIGHT JUST AFTER LAUNCH FROM UNDER WING OF B-52.

F-4C PHANTOM FIGHTER DROPPING TWO 750-POUND GENERAL PURPOSE BOMBS IN NORTH KOREA.

F-5 FREEDOM FIGHTER EN ROUTE TO NORTH VIETNAM TO STRIKE TARGET IN "SKOSHI TIGER" TEST OF SMALL VS. LARGE FIGHTER TYPES.

DEFENSE COMMUNICATIONS SATELLITE.

THE A-7-D

SUPERSWIFT F-104A "STARFIGHTER" PACKS A "SIDEWINDER" GUIDED MISSILE ON EACH WING.

THE VERSATILE F-111. WITH FULL WINGS SPAN IT HAS SHORT TAKEOFF CAPABILITY, YET CAN FLY SUPERSONIC AT ANY ALTITUDE AND WITH WINGS NEAR FUSELAGE. DESIGN SPEED IS MORE THAN MACH 2.

BOEING B-52 BOMBER.

GENIE AIR-TO-AIR MISSLE.

THE C-5-A.

C-123 TROOP CARRIER AT TAN SON NHUT AIR BASE, SOUTH VIETNAM.

C-130 "HERCULES" TRANSPORT AT TAN SON NHUT AIR BASE, SOUTH VIETNAM.

C-141 "STARLIFTER."

T-37 JET TRAINER.

T-38 "TALON" JET TRAINER.

X-15 ROCKET-POWERED RESEARCH AIRCRAFT, LANDING ON ROGERS DRY LAKE, EDWARDS AFB.

U-2 RECONNAISSANCE PLANE.

THE AWACS AIRCRAFT

JET FIGHTERS IN C-5.

F-4E WITH AUXILIARY FUEL TANKS, IN FLIGHT OVER MT. McKINLEY, ALASKA.

TWO UH-1N HELICOPTERS.

KC-10 REFUELING A C-5.

COBRA SKIN . . . TITAN MISSILE LAUNCH FROM VANDENBERG AFB, CA.

KC-135 REFUELING AN F-16.

LAUNCH OF THE SPACE SHUTTLE COLUMBIA, APRIL, 1981.

Appendix A
USAF Bases

Air power is a thunderbolt launched from an egg-shell invisibly tethered to a base.
—Herman Nickerson, Arms and Policy, 1945

With the permission of *Air Force Magazine*, the following guide to USAF bases in the United States is reproduced.

Altus AFB, Okla. 73521; within Altus city limits. Phone (405) 482-8100; AUTOVON 866-1110. MAC base. 443d Military Airlift Wing; training for C-141 and C-5 crews; basic flight engineer course; 340th Air Refueling Gp. (SAC); 2002d Communications Sqdn. (AFCC). Base activated Jan. 1942, inactivated May 1945, reactivated Jan. 1953. Area 4,113 acres. Altitude 1,376 ft. Military 3,508; civilians 876. Payroll $71.3 million. Housing: 163 officer; 637 NCO; 12 transient (4 VOQ, 4 VAQ, 4 transient). 30-bed hospital.

Andersen AFB, Guam 96334; 16.8 mi. N of Agana. Phone (671) 366-1110; AUTOVON 343-1110. SAC base. Hq. 3d Air Div., 43d Strategic Wing. Base activated as North Field, 1945; renamed Oct. 7, 1949, in memory of Brig. Gen. James Roy Andersen, reported missing on a flight from Guam to Hawaii, Feb. 26, 1945. Area 20,500 acres, including off-base facilities. Altitude 525 ft. Military 3,801; civilians 645. Payroll $73.3 million. Housing: 243 officer and 1,508 NCO; transient 206. Clinic, outpatient care only. 63-bed hospital at Naval Regional Medical Center, Agana, Guam.

Andrews AFB, Md. 20331; 11 mi. SE of Washington, D.C. Phone (301) 981-9111; AUTOVON 858-1110. MAC base. 1776th Air Base Wing; Hq. Air Force Systems Command: 76th Airlift Div.; 89th Military Airlift Wing; 113th Tactical Fighter Wing (ANG); 459th Tactical Airlift Wing (AFRES); 2045th Communications Gp. (AFCC); Det. 11, 1361st Audiovisual Sqdn. Base activated June 1943; named for Lt. Gen. Frank M. Andrews, military air pioneer, WW II commander, European theater, killed in aircraft accident May 3, 1943, in Iceland. Area 4,216 acres. Altitude 279 ft. Military 5,360; civilians 3,236. Payroll $195.2 million. Housing: 392 officer; 1,696 NCO: 273 transient (incl. 82 temp. living quarters for incoming personnel, 141 VOQ, 50 TAQ). 250-bed hospital.

Arnold AFS, Tenn. 37389; approx. 7 mi. SE of Manchester. Phone (615) 340-5011; AUTOVON 882-1520. AFSC station; site of Arnold Engineering Develop-

At the end of each entry in this Guide to Bases are data on base population and facilities, designated by the following symbols: M and C—assigned military and civilian personnel, including, where applicable, contractor, BX, and nonappropriated fund employees; TP—total military and civilian annual payroll; O, N, T/G—on-base Officer, NCO, and Transient/Guest housing units; H ( ), D—hospital, dispensary medical facilities with number of hospital beds in parentheses. In some instances, information was not available.

ment Center, free world's largest complex of wind tunnels, jet and rocket engine test cells, space simulation chambers, and hyperballistic ranges, which support the acquisition of new aerospace systems by conducting research, development, and evaluation testing for USAF, other services, and government agencies. Base activated Jan. 1, 1950; named for Gen. H. H. "Hap" Arnold, wartime Chief of the AAF. Area 40,118 acres. Altitude 950 to 1,150 ft. Military 145; civil service 216; contractor employees 3,600. Payroll $100 million. Housing: 24 officer; 16 NCO; 48 transient. Dispensary.

Barksdale AFB, La. 71110; in Bossier City. Phone (318) 456-2252; AUTOVON 781-1110. SAC base. Hq. 8th Air Force; 2d Bomb Wing. Base is also site of 917th Tactical Fighter Gp. (AFRES), flying A-10s. The 917th TFG is the only AFRES A-10 replacement training unit. In spring 1981 it became first USAF installation to receive the KC-10 Extender tanker aircraft. Base named for Lt. Eugene H. Barksdale, WW I airman killed in Aug. 1926, in crash near Wright Field, Ohio. Area 22,000 acres (20,000 acres reserved for recreation). Altitude 167 ft. Military 5,532; civilians 1,526. Payroll $121.4 million. Housing: 205 officer; 828 NCO; 29 transient. 65-bed hospital.

Beale AFB, Calif. 95903; 13 mi. E of Marysville. Phone (916) 634-3000; AUTOVON 368-1110. SAC base. 14th Air Div.; 9th Strategic Recon Wing; 7th Missile Warning Sqdn.; 100th Air Refueling Wing; 1883d Communications Sqdn. (AFCC). Beale is the only USAF base having SR-71, U-2, and TR-1 reconnaissance aircraft. Originally US Army's Camp Beale, became AF installation in Nov. 1948; became AFB in Dec. 1951. Named for Brig. Gen. E. F. Beale, Indian agent in California prior to Civil War. Area 23,204 acres. Altitude 113 ft. Military 4,170; civilians 560. Payroll $81.8 million. Housing: 395 officer; 1,342 NCO; 45 transient. 30-bed hospital.

Bergstrom AFB, Tex. 78743; 7 mi. SE of downtown Austin. Phone (512) 479-4100; AUTOVON 685-4100. TAC base. Hq. 12th Air Force; Hq. 10th Air Force (AFRES); 67th Tactical Recon Wing (host) with RF-4C recon operations; 602d Tactical Air Control Wing; 924th Tactical Airlift Gp. (AFRES); also F-4D fighter operations; TAC NCO Academy West. Base activated Sept. 22, 1942; named for Capt. John A. E. Bergstrom, first Austin serviceman killed in WW II, died Dec. 8, 1941, at Clark Field, Philippines. Area 3,998 acres. Altitude 541 ft. Military 3,900; civilians 650. Payroll $75.7 million. Housing: 92 officer; 612 NCO; 190 transient. 30-bed hospital.

Blytheville AFB, Ark. 72315; 4 mi. NW of Blytheville. Phone (501) 762-7000; AUTOVON 637-1110. SAC base. 42d Air Div.; 97th Bomb Wing. Base activated June 1942; inactivated Feb. 1947; reactivated Aug. 1955. Area 3,092 acres. Altitude 254 ft. Military 2,735; civilians 335. Payroll $36.7 million. Housing: 200 officer; 825 NCO; 79 transient. 25-bed hospital.

Bolling AFB, D.C. 20332; 3 mi. S of US Capitol. Phone (202) 545-6700; AUTOVON 227-0101. MAC base. 1100th Air Base Wing; Air Force Office of Scientific Research (AFSC); Air Reserve Personnel Center Operating Location; Air Force Chief of Chaplains; US Air Force Office of History. Activated Oct. 1917; named for Col. Raynal C. Bolling, assistant chief of air service, killed in France during WW I. Area 604 acres. Altitude 16 ft. Military 1,562; civilians 1,157. Payroll $38 million. Housing: 296 officer; 1,100 NCO; 168 transient (including 69 VAQ, 84 VOQ, 15 guest quarters).

Brooks AFB, Tex. 78235; 7 mi. SE of San Antonio. Phone (512) 536-1110; AUTOVON 240-1110. AFSC base. Home of Aerospace Medical Div., USAF School of Aerospace Medicine; USAF Occupational and Environmental Lab, USAF Human Resources Lab; tenant units include the USAF Medical Service Center, a security squadron, and a communications group. Base activated Dec. 8, 1917; named for Cadet Sidney J. Brooks, Jr., killed Nov. 13, 1917, on his final solo flight before commissioning. Area 1,330 acres. Altitude 600 ft. Military 1,450; civilians 960. Payroll $46 million. Housing: 70 officer; 100 NCO; 8 transient. Dispensary.

Cannon AFB, N.M. 88101; 7 mi. W of Clovis. Phone (505) 784-3311; AUTOVON 681-1110. TAC base. 27th Tactical Fighter Wing, F-111D fighter operations. Activated Aug. 1942 under the Army Air Corps. Deactivated May 1947. Reactivated July 1951 under the Air Force; named for Gen. John K. Cannon, WW II commander of all Allied Air Forces in Mediterranean theater. Area 3,780 acres. Altitude 4,295 ft. Military 3,866; civilians 409. Payroll $62.5 million. Housing: 149 officer; 863 NCO; 104 transient. 25-bed hospital.

Carswell AFB, Tex. 76127; 7 mi. WNW of downtown Fort Worth. Phone (817) 735-5000; AUTOVON 739-1110. SAC base. 19th Air Div.; 7th Bomb Wing (SAC); 301st Tactical Fighter Wing (AFRES). Activated Aug. 1942; named Jan. 30, 1948, for Maj. Horace S. Carswell, Jr., native of Fort Worth, WW II B-24 pilot and posthumous Medal of Honor recipient. Area 2,750 acres. Altitude 650 ft. Military 4,921; civilians 1,233. Payroll $76 million. Housing: 128 officer; 679 NCO. 125-bed hospital.

Castle AFB, Calif. 95342; 8 mi. NW of Merced. Phone (209) 726-2011; AUTOVON 347-1110. SAC base. 93d Bomb Wing. Conducts training of all SAC B-52G and H and KC-135 aircrews. Also houses 84th Fighter Interceptor Sqdn. (TAC), and is site of Castle Air Museum. Activated Sept. 1941; named for Brig. Gen. Frederick W. Castle, WW II B-17 pilot and Medal of Honor recipient. Area 2,777 acres. Altitude 188 ft. Military 5,052; civilians 400. Payroll $100.9 million. Housing: 92 officer; 842 NCO; 388 transient (incl. 108 VAQ, 276 VOQ, and 4 transient quarters). 25-bed hospital.

Chanute AFB, Ill. 61868; 14 mi. N of Champaign at Rantoul, Ill. Phone (217) 495-1110; AUTOVON 862-1110. ATC base. Chanute Technical Training Center provides training in missile and aircraft mechanics, aerospace ground equipment, life support, metallurgy and nondestructive inspection, weather forecasting, weather equipment, and fire protection and rescue. Chanute Technical Training Display Center is base museum. Base activated May 1, 1917; named for Octave Chanute, aeronautical engineer and glider pioneer who died in 1910. Area 2,125 acres. Altitude 735 ft. Military 7,500; civilians 1,700. Payroll $111 million. Housing: 142 officer; 1,518 NCO; 38 transient. 60-bed hospital.

Charleston AFB, S.C. 29404; in North Charleston. Phone (803) 554-0230; AUTOVON 583-0111. MAC base. Joint-use airfield. 437th Military Airlift Wing and 315th MAW (AFRES Assoc.). Also 1968th Communications Sqdn.; Det. 1, 48th Fighter

Interceptor Sqdn. (TAC); and Det. 7, 1361st Audiovisual Sqdn. Base activated June 1942; inactivated Feb. 1946, reactivated Aug. 1953. Area 3,772 acres. Altitude 45 ft. Military 7,081 (incl. AFRES); civilians 1,667. Payroll $81 million. Housing: 142 officer; 813 NCO; 75 trailer spaces; 472 transient (150 VOQ, 322 VAQ). Dispensary.

Columbus AFB, Miss. 39701; 10 mi. NNW of Columbus. Phone (601) 434-7322; AUTOVON 742-1110. ATC base. 14th Flying Training Wing, undergraduate pilot training. Base activated in 1941 for pilot training. Area 4,606 acres. Altitude 214 ft. Military 2,972; civilians 906. Payroll $57.3 million. Housing: 262 officer; 558 NCO. 15-bed hospital.

Davis-Monthan AFB, Ariz. 85707; within city limits of Tucson. Phone (602) 748-3900; AUTOVON 361-1110. TAC base. Hq. 836th Air Div.; 355th Tactical Training Wing, A-10 combat crew training; 390th Strategic Missile Wing (Titan II) (SAC). Also site of AFLC's Military Aircraft Storage and Disposition Center; 23d Tactical Air Support Sqdn.; 41st Electronic Combat Sqdn.; and 868th Tactical Missile Training Sqdn. Base activated in 1927; named for two local aviation accident victims—1st Lt. Samuel H. Davis, killed Dec. 28, 1921; and 2d Lt. Oscar Monthan, killed Mar. 27, 1924. Area 11,000 acres. Altitude 2,705 ft. Military 5,762; civilians 1,290. Payroll $95.7 million. Housing: 215 officer; 1,040 enlisted. 70-bed hospital.

Dover AFB, Del. 19901; 4 mi. SE of Dover. Phone (302) 678-7011; AUTOVON 455-1110. MAC base. 436th Military Airlift Wing and 512th MAW (AFRES Assoc.). Dover is largest air cargo port on East Coast. Base activated Dec. 1941; inactivated 1946; reactivated in Feb. 1951. Area 3,600 acres. Altitude 28 ft. Military 4,900; civilians 1,300. Payroll $96.9 million. Housing: 229 officer; 1,327 NCO; 297 transient. 30-bed hospital.

Duluth International Airport, Minn. 53814; 5 mi. NW of Duluth. Phone (218) 727-8211; AUTOVON 825-0011. TAC base. 23d NORAD Region; Hq. 23d Air Div. (TAC); SAGE Control Center (NORAD); 4787th Air Base Gp. (TAC); 148th Tactical Recon Gp. (ANG). Activated Mar. 1951. Area 1,139 acres. Altitude 1,429 ft. Military 1,036; civilians 287. Payroll $19 million. Housing: 70 officer; 361 military; 24 transient. Dispensary.

Dyess AFB, Tex. 79607; WSW border of Abilene. Phone (915) 696-0212; AUTOVON 461-1110. SAC base. 12th Air Div. and 96th Bomb Wing (SAC); 463d Tactical Airlift Wing (MAC); 1993d Communications Sqdn. (AFCC); 417th Field Training Det. (ATC). Base activated Apr. 1942; deactivated Dec. 1945; reactivated Abilene Air Base, Sept. 1955. In Mar. 1956 renamed for Lt. Col. William E. Dyess, WW II fighter pilot known best for his escape from a Japanese prison camp, killed in P-38 crash at Burbank, Calif., Dec. 1943. Area 6,058 acres. Altitude 1,789 ft. Military 4,978; civilians 462. Payroll $119.8 million. Housing: 150 officer; 849 NCO; 128 transient. 40-bed hospital.

Edwards AFB, Calif. 93523; 20 mi. E of Rosamond. Phone (805) 277-1110; AUTOVON 350-1110. AFSC base. Site of Air Force Flight Test Center (AFFTC), which conducts new and follow-on testing of aircraft and related avionics and weapon systems. AFFTC also operates the USAF Test Pilot School, which trains pilots and flight test engineers. Also the site of the Air Force Rocket Propulsion Laboratory, US Army Aviation Engineering Flight Activity, and the NASA Dryden Flight Research Facility. Edwards is the primary landing site for all Space Shuttle test and evaluation flights. Base activated Sept. 1933; named for Capt. Glen W. Edwards, killed June 5, 1948, in crash of YB-49 "Flying Wing" experimental bomber. Area 301,000 acres. Altitude 2,302 ft. Military 3,657; civilians 4,737. Payroll $178 million. Housing: 558 officer; 2,997 enlisted; 92 transient. 15-bed hospital.

Eglin AFB, Fla. 32542; 2 mi. SE of Valparaiso; 7 mi. NE of Fort Walton Beach. Phone (904) 881-6668; AUTOVON 872-1110. AFSC base. AF Armament Div.; AF Armament Lab; 3246th Test Wing; 39th Aerospace Rescue and Recovery Wing; 33d Tactical Fighter Wing; Tac Air Warfare Center; 919th Special Operations Gp. (AFRES); Air Force Armament Museum. Base activated in 1935; named for Lt. Col. Frederick I. Eglin, WW I flyer killed in aircraft accident, Jan. 1, 1937. Area 464,980 acres. Altitude 85 ft. Military 8,865; civilians 4,400. Payroll $227.1 million (includes AFRES). Housing: 322 officer; 2,014 NCO; 84 transient. 160-bed hospital.

Eielson AFB, Alaska 99702; 26 mi. SE of Fairbanks. Phone (907) 372-1181; AUTOVON (317) 377-1291. AAC base. 343d Composite Wing; 18th Tactical Fighter Sqdn.; 343d Combat Support Group is host unit. Air defense, search and rescue for AAC, and close air support for ground forces; 6th Strategic Wing (SAC) tanker operations; communications for AFCC; Arctic Survival School (ATC). Activated Oct. 1944; named for Carl B. Eielson, Arctic aviation pioneer, died Nov. 1929. Area 35,000 acres (approx.). Altitude 534 ft. Military 2,670; civilians 300. Payroll $67.7 million. Housing: 148 officer; 1,015 NCO; 20 transient. Dispensary.

Ellsworth AFB, S.D. 57706; 11 mi. ENE of Rapid City. Phone (605) 342-2400; AUTOVON 747-1110. SAC base. 44th Strategic Missile Wing; 28th Bomb Wing, including SAC postattack command and control system sqdn. Activated July 1954; named for Brig. Gen. Richard E. Ellsworth, killed Mar. 18, 1953, in crash of RB-36. Area 5,675 acres. Altitude 3,600 ft. Military 6,053; civilians 774. Payroll $105.7 million. Housing: 419 officer; 1,474 NCO; 142 transient. 30-bed hospital.

Elmendorf AFB, Alaska 99506; bordering Anchorage. Phone (907) 552-1110; AUTOVON (317) 552-1110. AAC base. Hq. Alaskan Air Command; 21st Tactical Fighter Wing; NORAD Region Control Center; Rescue Coordination Center; 531st Aircraft Control and Warning Gp.; 18th Tactical Fighter Sqdn.; 43d Tactical Fighter Sqdn.; 1931st Communications Gp. (AFCC); 6981st Electronic Security Sqdn. (ESC); 616th Military Airlift Gp. (MAC); 17th Tactical Airlift Sqdn. (MAC); 71st Aerospace Rescue and Recovery Sqdn. (MAC); 11th Weather Sqdn. (MAC); plus varied US Army and Navy activities. 21st Combat Support Gp. (AAC) is host unit. Base activated July 1940; named for Capt. Hugh M. Elmendorf, killed Jan. 13, 1933, at Wright Field, Ohio, while flight-testing a new type of pursuit plane. Area 13,400 acres. Altitude 118 ft. Military 6,209; civilians 1,464. Payroll $128.1 million. Housing: 356 officer; 1,839 NCO; 140 transient. 140-bed hospital.

England AFB, La. 71301; 5 mi. W of Alexandria. Phone (318) 448-2100; AUTOVON 683-1110. TAC base. 23d Tactical Fighter Wing, A-10 fighter operations. Base activated Oct. 1942; named for Lt. Col. John B. England, WW II P-51 pilot and ace credited with 17.5 victories, killed Nov. 17, 1954, in France in F-86 crash. Area 2,282 acres. Altitude 89 ft. Military 3,173; civilians 540. Payroll $53.8 million.

Housing: 109 officer; 491 NCO; 44 transient. 40-bed hospital.

Fairchild AFB, Wash. 99011; 12 mi. WSW of Spokane. Phone (509) 247-1212; AUTOVON 352-1110. SAC base. 47th Air Div.; 92d Bomb Wing (SAC); 3636th Combat Crew Training Wing (ATC); 141st Air Refueling Wing (ANG); Det. 24, 40th Rescue and Weather Reconnaissance Wing (MAC); Det. 1, 4000th Aerospace Applications Gp. (SAC); and 2039th Communications Sqdn. (AFCC). Base activated Jan. 1942. Named for Gen. Muir S. Fairchild, USAF Vice Chief of Staff at his death in 1950. Area 6,127 acres. Altitude 2,462 ft. Military 4,000; civilians 1,700. Payroll $72.3 million for civilian and active-duty military and $12.5 million for ANG. Housing: 502 officer; 1,079 NCO; transient incl. 60 VOQ and 62 VAQ, no family quarters. 45-bed hospital.

Francis E. Warren AFB, Wyo. 82001; adjacent to Cheyenne. Phone (307) 775-1110; AUTOVON 481-1110. SAC base. 4th Air Div.; 90th Strategic Missile Wing. Base activated July 4, 1867; under Army jurisdiction until 1947 when reassigned to USAF. Home of the first Atlas-D ICBM missile wing (1960-65); named for Francis Emory Warren, Wyoming senator and early governor. Base has 5,872 acres, plus 200 Minuteman III missile sites distributed over more than 12,600 sq. mi. Altitude 6,142 ft. Military 3,664; civilians 578. Payroll $47.5 million. Housing: 211 officer; 620 NCO; 36 transient. 32-bed hospital.

George AFB, Calif. 92392; 6 mi. NW of Victorville. Phone (714) 269-1110; AUTOVON 353-1110. TAC base. Hq. 831st Air Div.; 37th Tac Fighter Wing, home of TAC's "Wild Weasel" F-4G squadrons; 35th TAC Fighter Wing, "Pave Spike" F-4E sqdn.; F-4 transitional and upgrade training; German Air Force training in F-4. TAC F-106 detachment. Base activated in 1941; named for Brig. Gen. Harold H. George, WW I fighter ace killed Apr. 29, 1942, in Australia in aircraft accident. Area 5,347 acres. Altitude 2,875. Military 5,169; civilians 454. Payroll $85.3 million. Housing: 229 officer; 1,212 NCO; 198 Senior NCO; transient 40 TLQs. 46-bed hospital.

Goodfellow AFB, Tex. 76908; 2 mi. SE of San Angelo. Phone (915) 653-3231; AUTOVON 477-2011. ATC base. 3480th Technical Training Wing; USAF Technical Training School. Base activated Jan. 1941; named for Lt. John J. Goodfellow, Jr., WW I fighter pilot killed in combat Sept. 17, 1918. Area 1,127 acres. Altitude 1,877 ft. Military 1,394; civilians 576. Payroll $20 million. Housing; 3 officer; 96 NCO; 86 transient (23 VAQs, 63 VOQs).

Grand Forks AFB, N.D. 58205; 16 mi. W of Grand Forks. Phone (701) 594-6011; AUTOVON 362-1110. SAC base. 319th Bomb Wing; 321st Strategic Missile Wing (Minuteman III). Base activated in 1956. Area 6,912 acres. Altitude 911 ft. Military 5,116; civilians 469. Payroll $89.4 million. Housing: 418 officer; 1,695 NCO; 243 transient. 30-bed hospital.

Griffiss AFB, N.Y. 13441; 1 mi. NE of Rome. Phone (315) 390-1110; AUTOVON 587-1110. SAC base. 416th Bomb Wing. Major tenant is Rome Air Development Center (RADC), part of AFSC. Base also houses headquarters of AFCC's Northern Communications Area; 485th Communications and Installations Gp. (AFCC); and 49th Fighter Interceptor Sqdn. (TAC). Base activated Feb. 1, 1942; named for Lt. Col. Townsend E. Griffiss, killed in aircraft accident, Feb. 15, 1942 (the first US airman to lose his life in Europe while in the line of duty during WW II). Area 3,896 acres. Altitude 504 ft. Military 3,871; civilians 2,870. Payroll $109.8 million. Housing: 175 officer; 558 NCO; 140 transient. 70-bed hospital.

Grissom AFB, Ind. 46971; 7 mi. S. of Peru. Phone (317) 689-5211; AUTOVON 928-1110. SAC base. 305th Air Refueling Wing; 434th Tactical Fighter Wing (AFRES); 931st Air Refueling Gp. (AFRES). Activated Jan. 1943 for Navy flight training; reactivated June 1954 as Bunker Hill AFB; renamed May 1968 for Lt. Col. Virgil I. "Gus" Grissom, killed Jan. 27, 1967, at Cape Kennedy, Fla., with other Astronauts Edward White and Roger Chaffee, in Apollo capsule fire. Area 2,810 acres. Altitude 800 ft. Military 2,552; civilians 968. Payroll $47.6 million (SAC only). Housing: 276 officer; 1,852 NCO; 138 transient. Dispensary.

Gunter AFS, Ala. 36114; 4 mi. NE of Montgomery. Phone (205) 279-1110; AUTOVON 921-1110. ATC station. Hq. Air Force Data Automation Agency and site of Air Force Data Systems Design Center; Air Force Logistics Management Center; USAF Extension Course Institute; USAF Senior NCO Academy. Base activated Aug. 27, 1940; named for William A. Gunter, longtime mayor of Montgomery and airpower exponent, died 1940. Area 348 acres. Altitude 220 ft. Military 1,430; civilians 1,048. Payroll included in Maxwell entry. Housing: 118 officer; 206 NCO; 107 transient.

Hancock Field, N.Y. 13225; 10 mi. NNE of Syracuse. Phone (315) 458-5500; AUTOVON 587-9110. TAC base. 4789th Air Base Gp., host unit, supports 21st NORAD Region; Hq. 21st Air Div. (TAC); 113th Tactical Control Flight (ANG); 174th Tactical Fighter Wing (ANG); 3513th USAF Recruiting Sqdn. Base activated Sept. 1942 as Syracuse Army Air Base, renamed Mar. 1952 for Clarence E. Hancock (1885–1949), prominent local citizen and member of US House of Representatives. Area 765 acres. Altitude 421 ft. Military 884; civilians 315. Payroll $20 million. Housing: 61 officer; 167 NCO; 17 transient; two temporary lodging facilities for families. Clinic.

Hanscom AFB, Mass. 01731; 17 mi. NW of Boston. Phone (617) 861-4441; AUTOVON 478-4441. AFSC base. Hq. Electronic Systems Div. (AFSC), manages development and acquisition of command control and communications systems. Also site of AF Geophysics Lab, center for research and exploratory development in the terrestrial, atmospheric, and space environments. Base has no flying mission; transient USAF aircraft use runways of Laurence G. Hanscom Field, state-operated airfield adjoining the base. Named for a pre-WW II advocate of private aviation, killed in a lightplane accident in 1941. Area 887 acres. Altitude 133 ft. Military 1,848; civilians 3,025. Payroll $118 million. Housing: 289 officer; 406 NCO; 16 transient. Dispensary.

Hickam AFB, Hawaii 96853; 6 mi. W of Honolulu. Phone (808) 422-0531 (Oahu military operator); AUTOVON 430-0111. PACAF base. Hq. Pacific Air Forces; 15th Air Base Wing, support organization for Air Force units in Hawaii and throughout the Pacific; 154th Tactical Fighter Gp. (ANG); Hq. Pacific Communications Div. (AFCC); 1st Weather Wing (MAC); 834th Airlift Div. (MAC). Base activated Sept. 1937. Named for Lt. Col. Horace M. Hickam, air pioneer killed in crash Nov. 5, 1934, at Fort Crockett, Tex. Area 2,731 acres. Altitude sea level. Military 5,100; civilians 2,000. Payroll $128 million (includes Hickam and Wheeler AFBs and

Bellows AFS). Housing: 535 officer; 1,940 NCO. Clinic.

Hill AFB, Utah 84056; 7 mi. S of Ogden. Phone (801) 777-7221; AUTOVON 458-1110. AFLC base. Hq. Ogden Air Logistics Center. Furnishing logistics support for Minuteman and Titan II missiles; BOMARC drone and Maverick missiles; Walleye; laser and electro-optical guided bombs; emergency rocket communications systems; MX missile; F-4, F-16, and F-101 systems manager; air munitions, aircraft landing gears; wheels, brakes, tires, and tubes; photographic and aerospace training equipment; and COM-10. Also home of 388th Tactical Fighter Wing; 508th Tactical Fighter Gp. (AFRES); 6545th Test Gp. (AFSC), which includes management of Utah Test and Training Range and RPV test programs. Base activated Nov. 1940. Named for Maj. Ployer P. Hill, killed Oct. 30, 1935, testflying the first B-17. Area 7,000 acres; manages 961,896 acres. Altitude 4,788 ft. Military 5,280; civilians 14,319. Payroll $421 million. Housing: 263 officer; 882 NCO; 8 transient. 35-bed hospital.

Holloman AFB, N.M. 88330; 6 mi. SW of Alamogordo. Phone (505) 479-6511; AUTOVON 867-1110. TAC base. Hq. 833d Air Div.; 49th Tactical Fighter Wing (F-15 operations); 479th Tactical Training Wing (T-38 fighter lead-in training); 4449th Mobility Support Sqdn. (Harvest Bare); and 82d Tactical Control Flight. 6585th Test Group (AFSC) conducts test and evaluation of aircraft and missile systems and operates the Central Inertial Guidance Test Facility, the High Speed Test Track Facility, and the Radar Target Scatter (RATSCAT) Site. Base activated in 1942; named for Col. George V. Holloman, guided-missile pioneer, killed in B-17 crash in Formosa, Mar. 19, 1946. Area 57,530 acres. Altitude 4,092 ft. Military 5,737; civilians 1,371. Payroll $110 million. Housing: 192 officer; 1,360 NCO; 250 transient. 25-bed hospital.

Homestead AFB, Fla. 33039; 5 mi. NNE of Homestead. Phone (305) 257-8011; AUTOVON 791-0111. TAC base. 31st Tactical Fighter Wing; F-4D fighter operations and training; site of ATC sea-survival school; 726th Tactical Control Sqdn. (TAC); Naval Security Group Activity; 482d Tactical Fighter Wing (AFRES); and 301st Aerospace Rescue and Recovery Sqdn. (AFRES). Base activated Apr. 1955. Area 3,491 acres. Altitude 7 ft. Military 6,508; civilians 1,172. Payroll $96 million. Housing: 321 officer; 1,294 NCO; 299 transient (214 VAQ, 83 VOQ). 80-bed hospital.

Hurlburt Field, Fla. 32544; 8 mi. W of Fort Walton Beach. Phone (904) 881-6668; AUTOVON 872-1110. TAC base, though part of the Eglin AFB (AFSC) reservation. Home of 1st Special Operations Wing, focal point of all USAF special operations; USAF Special Operations School; MC-130E (Combat Talon), AC-130H (Spectre Gunship), HH-53H (Super Jolly), and UH-1N (Iroquois) helicopter sqdns.; TAC's only special operations combat control team and special operations weather team; 4442d Tactical Control Gp., including USAF Air Ground Operations School, 823d Civil Engineering Sqdn. (Red Horse). Base activated in 1943; named for Lt. Donald W. Hurlburt, WW II pilot killed Oct. 2, 1943, in a crash on Eglin reservation. Altitude 35 ft. Military 3,534; civilians 390. Payroll $60 million. Housing: 74 officer; 306 NCO; 341 transient. Clinic only at Hurlburt, but 200-bed hospital at Eglin main base.

Indian Springs AF Auxiliary Field, Nev. 89018; 45 mi. NW of Las Vegas. Phone

(702) 897-6201; AUTOVON 682-6201. TAC base. 554th Combat Support Sqdn.; Det. 1, 57th Fighter Weapons Wing; provides bombing and gunnery range support for tactical operations from Nellis AFB; manages construction of realistic target complexes; supports US Department of Energy research activities. Base activated in 1942. Area 1,652 acres. Altitude 3,124 ft. Military 343; civilians 13. (Payroll included in Nellis AFB entry.) Housing: 78 officer and NCO quarters; 40 trailer spaces. Dispensary.

Keesler AFB, Miss. 39534; located in Biloxi. Phone (601) 377-1110; AUTOVON 868-1110. ATC base. Hq. Keesler Technical Training Center (communications, electronics, personnel, and administrative courses); Keesler USAF Medical Center. Hosts MAC and AFRES weather recon units. TAC airborne command and control sqdn., AFCC installation gp., and AFCC NCO Academy/Leadership School. Base activated June 12, 1941; named for 2d Lt. Samuel R. Keesler, Jr., WW I aerial observer, killed in action Oct. 9, 1918, near Verdun, France. Area 3,600 acres. Altitude 26 ft. Military 13,726; civilians 3,650. Payroll $209 million. Housing: 430 officer; 1,527 NCO; 68 transient. (350 VOQ units on space availability, tech training students occupy many units.) 337-bed hospital.

Kelly AFB, Tex. 78241; 5 mi. SW of San Antonio. Phone (512) 925-1110; AUTOVON 945-1110. AFLC base. Hq. San Antonio Air Logistics Center; Hq. Electronic Security Command; AF Electronic Warfare Center; AF Cryptologic Support Center; Joint Electronic Warfare Center; USAF Service Information and News Center; AF Commissary Service; 433d Tactical Airlift Wing (AFRES); 149th Tactical Fighter Gp. (ANG). Base activated May 7, 1917; named for Lt. George E. M. Kelly, first Army pilot to lose his life in a military aircraft, killed May 10, 1911. Area 3,925 acres. Altitude 689 ft. Military 4,210; civilians 18,560. Payroll $452.8 million. Housing: 46 officer; 368 NCO. 3-bed dispensary.

Kirtland AFB, N.M. 87117; S of Albuquerque. Phone (505) 844-0011; AUTOVON 244-0011. MAC base. 1606th Air Base Wing. Major agencies and units include AF Contract Management Div. (AFSC); AF Test and Evaluation Center; AF Weapons Laboratory (AFSC); Office of the Chief of Security Police; New Mexico ANG; 1550th Aircrew Training and Test Wing (MAC); Defense Nuclear Agency Field Command; Naval Weapons Evaluation Facility, Sandia Laboratories; Lovelace Biomedical and Environmental Research Institute; Department of Energy's Albuquerque Operations Office; AFSC NCO Academy; AF Directorate of Nuclear Surety; 150th Tactical Fighter Gp. (ANG); 1960th Communications Sqdn.; 3098th Aviation Depot Sqdn. and Det. 1, 1369th Audiovisual Sqdn. These agencies furnish contract management; nuclear and laser research, development, and testing; operational test and evaluation services; advanced helicopter training; and HC-130 search and rescue training. Base activated Jan. 1941; named for Col. Roy S. Kirtland, air pioneer and commandant of Langley Field in the 1930s, died May 2, 1941. Area 51,330 acres. Altitude 5,352 ft. Military 4,876; civilians 12,090. Payroll $479.2 million. Housing: 124 officer; 2,010 NCO; 380 transient (211 VOQ, 169 VAQ). Dispensary and 40-bed hospital.

K. I. Sawyer AFB, Mich. 49843; 20 mi. S of Marquette. Phone (906) 346-6511; AUTOVON 472-1110. SAC base. 410th Bomb Wing; 46th Air Refueling Sqdn.; 87th Fighter Interceptor Sqdn. (TAC); 2001st Communications Sqdn. (AFCC). Base activated in 1959; named for Kenneth I. Sawyer, who proposed site for county airport, died in 1944. Area 5,224 acres. Altitude 1,220 ft. Military 3,678; civilians 417. Payroll $60 million. Housing: 337 officer; 1,356 NCO; 40 BOQ units; 224 transient (incl. 20 fully furnished efficiency apartments and 200 trailer spaces in housing section). 50-bed hospital.

Lackland AFB, Tex. 78236; 8 mi. WSW of San Antonio. Phone (512) 671-1110; AUTOVON 473-1110. ATC base. Provides basic military training for airmen; technical training of basic, advanced security policy/law enforcement personnel; patrol dog-handler courses; training of instructors, recruiters, and social actions/drug abuse counselors; USAF marksmanship training; Officer Training School; Defense Language Institute-English Language Center; Wilford Hall USAF Medical Center. Base activated in 1941; named for Brig. Gen. Frank D. Lackland, early commandant of Kelly Field flying school, died 1943. Area 6,828 acres, incl. 4,017 acres at Lackland Training Annex. Altitude 787 ft. Military 19,860; civilians 4,891. Payroll $202.3 million. Housing: 106 officer; 619 NCO; 1,257 transient. 1,000-bed hospital.

Langley AFB, Va. 23665; 3 mi. N of Hampton. Phone (804) 764-9990; AUTOVON 432-1110. TAC base. Host unit 1st Tactical Fighter Wing, F-15 fighter operations; Hq. Tactical Air Command; 5th Weather Wing (MAC); 2d Aircraft Delivery Gp. (TAC); 460th Reconnaissance Technical Sqdn. (TAC); 6th Airborne Command and Control Sqdn. (TAC); US Army TRADOC Flight Det.; 48th Fighter Interceptor Sqdn. (TAC). Base activated Dec. 30, 1916; is the oldest continuously active AFB in the US; named for aviation pioneer and scientist Samuel Pierpont Langley, who died in 1906. NASA Langley Research Center is located across base. Area 3,500 acres. Altitude 10 ft. Military 8,681; civilians 2,319. Payroll $173.3 million. Housing: 384 officer; 1,259 NCO; 202 transient. 65-bed hospital.

Laughlin AFB, Tex. 78840; 6 mi. E of Del Rio. Phone (512) 298-3511; AUTOVON 732-1110. ATC base. 47th Flying Training Wing, undergraduate pilot training. Base activated Oct. 1942; named for 1st Lt. Jack T. Laughlin, B-17 pilot killed over Java, Jan. 29, 1942. Area 4,008 acres. Altitude 1,080 ft. Military 2,600; civilians 500. Payroll $46 million. Housing: 255 officer; 348 NCO; 39 transient. 15-bed hospital.

Laurence G. Hanscom AFB (*see Hanscom AFB*).

Little Rock AFB, Ark. 72099; 12 mi. NE of Little Rock. Phone (501) 988-3131; AUTOVON 731-1110. MAC base, 314th Tactical Airlift Wing; 308th Strategic Missile Wing; SAC Titan II ICBM support base; 189th Air Refueling Gp. (ANG); Det. 9, 1365th Audiovisual Sqdn. Base activated in 1955. Area 6,919 acres. Altitude 310 ft. Military 6,195; civilians 800. Payroll $111 million. Housing: 313 officer; 1,222 NCO; 380 transient (160 VAQ, 220 VOQ). 30-bed hospital.

Loring AFB, Me. 04751; 4 mi. W of Limestone. Phone (207) 999-1110; AUTOVON 920-1110. SAC base. 42d Bomb Wing. Base activated Feb. 25, 1953, as Limestone AFB; renamed for Maj. Charles J. Loring, Jr., F-80 pilot killed Nov. 22, 1952, in North Korea; posthumously awarded Medal of Honor. Area more than 9,000 acres. Altitude 746 ft. Military 3,276; civilians 762. Payroll $54.5 million. Housing: 654 officer; 1,364 NCO; 12 transient; 4 VIP. 20-bed hospital.

Los Angeles AFS, Calif. 90009; in metropolitan Los Angeles area, city of El Segundo, one mi. S of Los Angeles IAP. Phone (213) 643-1000; AUTOVON 833-1110. Headquarters of AFSC's Space Division, which manages the development, launch, and on-orbit control of DoD's space program. Support unit is 6592d ABGp. Station activated Dec. 14, 1960, 23 tenant units on station; also provides support to 41 off-station units/activities. Military 2,010; civilians 1,300. Payroll $67 million. Housing units under construction nearby in San Pedro; the first of 300 scheduled to be occupied in late 1982. Station is site for one of the five prototype AF Family Support Centers.

Lowry AFB, Colo. 80230; 6 mi. E of Denver. Phone (303) 370-1110; AUTOVON 926-1110. ATC base. Technical Training Center; Air Force Accounting and Finance Center; Air Reserve Personnel Center, and the 3320th Correction and Rehabilitation Sqdn. Lowry Technical Training Center conducts training in avionics, aerospace munitions, air intelligence, logistics, and audiovisual fields. Base activated Feb. 26, 1938; named for 1st Lt. Francis B. Lowry, killed in action Sept. 26, 1918, near Crepion, France, while on a photo mission. Area 1,863 acres on base and 3,833-acre training annex 25 mi. E of Lowry. Altitude 5,400 ft. Military 8,841; civilians 5,011. Payroll $154 million. Housing: 94 officer; 772 NCO; 40 transient. Dispensary.

Luke AFB, Ariz. 85309; 20 mi. WNW of Phoenix. Phone (602) 856-7411; AUTOVON 853-1110. TAC base. 832d Air Div., 405th Tactical Training Wing; 58th Tactical Training Wing; Hq. 26th NORAD Region; Hq. 26th Air Div. (TAC); 302d Special Operations Sqdn. (AFRES). Luke, the largest fighter training base in the free world, conducts training of USAF aircrews in the F-4C and F-15, German students in the F-104G, and foreign training in the F-5 (at nearby Williams AFB). Base activated in 1941; named for 2d Lt. Frank Luke, Jr., observation balloon-busting ace of WW I and first flyer to receive the Medal of Honor, killed in action Sept. 29, 1918, near Murvaux, France. Area 4,197 acres plus 2,700,000-acre range. Altitude 1,101 ft. Military 6,000; civilians 2,000. Payroll $140 million. Housing: 80 officer; 786 NCO; 40 transient. 105-bed hospital.

MacDill AFB, Fla. 33608: adjacent to Tampa city limits. Phone (813) 830-1110; AUTOVON 968-1110. TAC base. Hq. US Readiness Command, Rapid Deployment Joint Task Force; 56th Tactical Fighter Wing conducts replacement training in the F-4D and the F-16. The wing has completed 75% of its conversion to the F-16. Base activated Apr. 15, 1941; named for Col. Leslie MacDill, killed in an aircraft accident Nov. 8, 1938, near Washington, D.C. Area 5,631 acres. Altitude 6 ft. Military 6,242; civilians 1,350. Payroll $141 million. Housing: 58 officer; 746 enlisted; 350 transient. 75-bed USAF regional hospital.

Malmstrom AFB, Mont. 59402; 4 mi. E of Great Falls. Phone (406) 731-9990; AUTOVON 632-1110. SAC base. 341st Strategic Missile Wing; Hq. 24th Air Div. (TAC); SAGE Region Control Center (NORAD). Base activated Dec. 15, 1942; named for Col. Einar A. Malmstrom, WW II fighter commander killed in air accident Aug. 24, 1954. Site of SAC's first Minuteman wing. Area 3,573 acres, plus about 23,000 sq. mi. of missile complex. Altitude 3,525 ft. Military 4,334; civilians 496. Payroll $72.3 million. Housing: 294 officer; 1,112 NCO; 107 transient. 29-bed hospital.

March AFB, Calif. 92518; 9 mi. SE of Riverside. Phone (714) 655-1110; AUTO-

VON 947-1110. SAC base. Hq. 15th AF; 22d Bomb Wing; 452d Air Refueling Wing (AFRES); 303d Aerospace Rescue and Recovery Sqdn. (AFRES). Base activated Mar. 1, 1918; named for 2d Lt. Peyton C. March, Jr., who died in Texas of crash injuries Feb. 18, 1918. Area 7,117 acres. Altitude 1,530 ft. Military 4,149; civilians 1,414. Payroll $177.1 million. Housing: 103 officer; 608 NCO; 177 transient. 145-bed hospital.

Mather, AFB, Calif. 95655; 12 mi. ESE of Sacramento. Phone (916) 364-1110; AUTOVON 828-1110. ATC base. DoD executive manager for navigator training (USAF, Navy, Marine Corps basic navigation training). Only navigator training base; also trains USAF electronic warfare officers and navigator-bombardiers. 320th Bomb Wing (SAC); 940th Air Refueling Gp. (AFRES); 3506th Recruiting Gp. Base activated 1918; named for 2d Lt. Carl S. Mather, killed in midair collision, Jan. 30, 1918, in Texas. Area 5,800 acres. Altitude 96 ft. Military 4,900; civilians 2,100. Payroll $121 million. Housing: 447 officer; 864 NCO; 40 transient. 75-bed hospital.

Maxwell AFB, Ala. 36112; 1 mi. WNW of Montgomery. Phone (205) 293-1110; AUTOVON 875-1110. ATC base. Hq. Air University, professional education center for USAF; site of Air War College, Air Command and Staff College, Squadron Officer School, Leadership and Management Development Center, Academic Instructor and Foreign Officer School; Hq. Air Force ROTC; Hq. Civil Air Patrol-USAF; Community College of the Air Force; 908th Tac Airlift Gp. (AFRES). (The Senior NCO Academy and Extension Course Institute are at Gunter AFS.) Base activated 1918; named for 2d Lt. William C. Maxwell, killed in an air accident Aug. 12, 1920, in the Philippines. Area 2,523 acres. Altitude 168 ft. Military 4,079; civilians 2,178. Payroll $153 million. Housing: 275 officer; 388 NCO; 1,029 transient (971 VOQ and 58 VAQ). 90-bed hospital.

McChord AFB, Wash. 98438; 8 mi. S of Tacoma. Phone (206) 984-1910; AUTOVON 976-1110. MAC base. 62d Military Airlift Wing; Hq. 25th Air Div. (TAC); 318th Fighter Interceptor Sqdn. (TAC); SAGE Region Control Center (NORAD); 446th Military Airlift Wing (AFRES Assoc.) Base activated May 5, 1938; named for Col. William C. McChord, killed Aug. 18, 1937, while attempting a forced landing at Maidens, Va. Area 4,609 acres. Altitude 322 ft. Military 5,703; civilians 1,703. Payroll $128.2 million. Housing: 111 officer; 882 NCO; 284 transient. Dispensary.

McClellan AFB, Calif. 95652; 9 mi. NE of Sacramento. Phone (916) 643-2111; AUTOVON 633-1110. AFLC base. Hq. Sacramento Air Logistics Center, logistics management, procurement, maintenance and distribution support for such USAF weapon systems as F-111, FB-111, A-10, T-39; surveillance and warning systems, Space Transportation System, communication-electronics equipment, radar sites, and generators; maintenance support for F-4 and F-106 aircraft. Associate units include 41st Rescue and Weather Recon Wing (MAC); 2049th Communications Gp., and 1849th Electronics Installations Sqdn. (AFCC); 1155th Technical Operations Sqdn. (AFSC); 431st Fighter Weapons Sqdn. (TAC); Hq. 4th Air Force (AFRES); Defense Logistics Agency; and US Coast Guard Air Station, Sacramento (DOT). Named for Maj. Hezekiah McClellan, pioneer in Arctic aeronautical experiments, killed in crash May 25, 1936. Area 2,625 acres. Altitude 76 ft. Military 3,600; civilians 13,900. Payroll $367.3 million. Housing: 168 officer; 507 NCO; 21 transient. Dispensary.

McConnell AFB, Kan. 67221; 5 mi. SE of Wichita. Phone (316) 681-6100; AUTOVON 962-1110. SAC base. 381st Strategic Missile Wing; 384th Air Refueling Wing; 184th Tactical Fighter Gp. (ANG). Base activated June 5, 1951; named for Capt. Fred J. McConnell, WW II B-24 pilot who died in a crash of private plane Oct. 25, 1945; and for his brother, 2d Lt. Thomas L. McConnell, also a WW II B-24 pilot, killed July 10, 1943, during attack on Bougainville in the Pacific. Area 2,608 acres. Altitude 1,371 ft. Military 4,058; civilians 753. Payroll $63 million. Housing: 149 officer; 445 NCO; 133 transient. 20-bed hospital.

McGuire AFB, N.J. 08641; 18 mi. SE of Trenton. Phone (609) 724-1110; AUTOVON 440-0111. MAC base. 438th Military Airlift Wing; Hq. 21st Air Force; N.J. ANG; N.J. Civil Air Patrol; 170th Air Refueling Gp. (ANG); 108th Tactical Fighter Wing (ANG); 514th Military Airlift Wing (AFRES Assoc.); the MAC NCO Academy East; and Air Force Band of the East. Base adjoins Army's Fort Dix; formerly Fort Dix Army Air Base. Activated as AFB in 1949; named for Maj. Thomas B. McGuire, Jr., P-38 pilot, second leading US ace of WW II, holder of Medal of Honor, killed in action Jan. 7, 1945, in the Philippines. Area 3,552 acres. Altitude 133 ft. Military 4,886; civilians 1,714. Payroll $115.9 million. Housing: 442 officer; 1,312 NCO; 620 transient (186 VOQ, 244 VAQ, 160 transient family units, 30 transient). Dispensary and 163-bed hospital.

Minot AFB, N.D. 58705; 13 mi. N of Minot. Phone (701) 727-4761; AUTOVON 344-1110. SAC base. 57th Air Div.; 91st Strategic Missile Wing; 5th Bomb Wing; 5th Fighter interceptor Sqdn. (TAC). Base activated Feb. 1957. Area 5,050 acres, plus additional 19,324 acres for missile sites. Altitude 1,650 ft. Military 5,648; civilians 588. Payroll $85.9 million. Housing: 543 officers; 1,927 NCO; 104 transient. Dispensary, also 40-bed military hospital in city of Minot.

Moody AFB, Ga. 31699; 10 mi. NNE of Valdosta. Phone (912) 333-4211; AUTOVON 460-1110. TAC base. 347th Tactical Fighter Wing, F-4E fighter operations. Base activated June 1941; named for Maj. George P. Moody, killed May 5, 1941, while test-flying Beech AT-10. Area 6,050 acres. Altitude 233 ft. Military 2,610; civilians 486. Payroll $52.9 million. Housing: 61 officers; 245 NCO; 41 transient. 25-bed hospital.

Mountain Home AFB, Idaho 83648; 56 mi. SE of Boise. Phone (208) 828-2111; AUTOVON 857-1110. TAC base. 366th Tactical Fighter Wing, F-111A fighter and EF-111A electronic counter-measures operations. Base activated April 1942. Area 6,639 acres. Altitude 3,000 ft. Military 4,205; civilians 683. Payroll $65 million. Housing: 242 officer; 1,296 NCO; 105 transient. 20-bed hospital.

Myrtle Beach AFB, S.C. 29577; S of Myrtle Beach. Phone (803) 238-7211; AUTOVON 748-1110. TAC base; shares runway with Myrtle Beach Jetport. 354th Tactical Fighter Wing. A-10 fighter operations. Served as Army air base, 1941–47; USAF base since 1956. Area 3,793 acres. Altitude 25 ft. Military 3,138; civilians 463. Payroll $55.5 million. Housing: 130 officer; 670 NCO; 65 trailer lots; 116 transient. 32-bed hospital.

Nellis AFB, Nev. 89191; 8 mi. NE of Las Vegas. Phone (702) 643-1800; AUTO-

VON 682-1800. TAC base. Tactical Fighter Weapons Center, host unit. F-4E, F-5E, F-15, F-16, F-111, A-10, T-38, UH-1N operations; 57th Fighter Weapons Wing; USAF Thunderbirds Air Demonstration Sqdn.; 4440th Tactical Fighter Training Gp. (Red Flag); 554th Operations Support Wing; 554th Range Group; conducts advanced tactical fighter training and realistic combat training for DoD; provides test and evaluation of air tactics and new equipment. Tenant units: 474th TFW; EWCAS/ Joint Task Force; 4450th Tactical Training Gp.; 3096th Aviation Depot Sqdn.; 2069th Communications Sqdn. Base activated July 1941; named for 1st Lt. William H. Nellis, WW II P-47 fighter pilot, killed Dec. 27, 1944, in Europe. Area 11,274 acres, with ranges totaling 3,012,770 acres. Altitude 2,171 ft. Military 9,256; civilians 1,026. Payroll $162 million. Housing: 168 officer; 1,329 NCO; 100 trailer spaces; 1,075 transient (incl. 846 VAQ, 204 VOQ, 25 TLQ). 40-bed hospital.

Norton AFB, Calif. 92409; 59 mi. E of Los Angeles, within San Bernardino corporate limits. Phone (714) 382-1110; AUTOVON 876-1110. MAC base. 63d Military Airlift Wing; Hq. AF Inspection and Safety Center; Hq. Defense Audiovisual Agency; Hq. AF Audit Agency; Hq. Aerospace Audiovisual Service (MAC). Also Ballistic Missile Office (AFSC); 445th Military Airlift Wing (AFRES Assoc.); MAC NCO Academy West and 22d Air Force NCO Leadership School. Base activated March 2, 1942; named for Capt. Leland F. Norton, native of San Bernardino, WW II A-20 attack bomber pilot, killed in action May 27, 1944, near Amiens, France. Area 2,407 acres. Altitude 1,156 ft. Military 5,396; civilians 2,798. Payroll $161 million. Housing: 56 officer; 208 NCO; 339 transient. Clinic.

Offutt AFB, Neb. 68113; 8 mi. S of Omaha. Phone (402) 294-1110; AUTOVON 271-1110. SAC base. Hq. Strategic Air Command; 55th Strategic Reconnaissance Wing; 544th Strategic Intelligence Wing; AF Global Weather Center (MAC); 3d Weather Wing (MAC); and 3902d Air Base Wing. Base activated 1888 as Army's Fort Crook; landing field named in 1924 for 1st Lt. Jarvis J. Offutt, WW I pilot, died Aug. 13, 1918, from injuries received at Valheureux, France. Area 1,914 acres. Altitude 1,048 ft. Military 12,464; civilians 2,110 (incl. 468 contractor personnel). Payroll $220.1 million. Housing: 882 officer; 1,798 NCO; 60 transient. 65-bed hospital.

Patrick AFB, Fla. 32925; 2 mi. S of Cocoa Beach. Phone (305) 494-1110; AUTOVON 854-1110. AFSC base. Operated by the Eastern Space and Missile Center in support of DoD, NASA, and other agency missile and space programs. Major tenants are Equal Opportunity Management Institute; AF Technical Applications Center; 549th Tactical Air Support Gp.; and 2d Combat Communications Gp. (AFCC). Activated in 1940, base is airhead for Cape Canaveral AFS. Named for Maj. Gen. Mason M. Patrick, chief of AEF's Air Service in WW I and chief of the Air Service/Air Corps, 1921–27. Area 2,341 acres. Altitude 9 ft. Military 3,281; civilians 6,697. Payroll $69.5 million. Housing: 247 officer; 1,401 NCO. 25-bed hospital.

Pease AFB, N.H. 03801; 3 mi. W of Portsmouth. Phone (603) 436-0100; AUTOVON 852-1110. SAC base. 45th Air Div.; 509th Bomb Wing; 157th Air Refueling Gp. (ANG). Base activated 1956; named for Capt. Harl Pease, Jr., WW II B-17 pilot and Medal of Honor recipient, killed Aug. 7, 1942, during attack on Rabaul, New Britain Island. Area 4,374 acres. Altitude 101 ft. Military 3,580; civilians 545. Payroll $68.7 million. Housing: 138 officer; 1,073 NCO; 50 trailer spaces; 128 transient. 70-bed hospital.

Peterson AFB, Colo. 80914; 7 mi. E of Colorado Springs. Phone (303) 591-7321; AUTOVON 692-7011. SAC base. Home of 46th Aerospace Defense Wing (SAC), which supports Hq. North American Air Defense Command/Aerospace Defense Command (NORAD/ADCOM) Combat Operations Center in Cheyenne Mountain; Aerospace Defense Center; the Air Force Academy; and Fort Carson, Colo. Base activated in 1942; named for 1st Lt. Edward J. Peterson, killed Aug. 8, 1942, in aircraft crash at the base. Area 1,176 acres. Altitude 6,200 ft. Military 3,357; civilians 1,086. Payroll $92 million. Housing: 106 officer; 384 NCO; 40 transient. Dispensary.

Plattsburgh AFB, N.Y. 12903; adjacent to Plattsburgh, N.Y. Phone (518) 563-4500; AUTOVON 689-1110. SAC base. 380th Bomb Wing, medium bomber and tanker operations with FB-111 and KC-135. 4007th Combat Crew Training Sqdn. trains all FB-111 combat crews for SAC. Second oldest active military installation in the U.S., established 1814; AFB since 1955. Area 3,305 acres. Altitude 235 ft. Military 3,700; civilians 660. Payroll $62.5 million. Housing: 230 officer; 1,412 NCO. 20-bed hospital.

Pope AFB, N.C. 28308; 12 mi. NNW of Fayetteville. Phone (919) 394-0001; AUTOVON 486-1110. MAC base. USAF Airlift Center; 317th Tactical Airlift Wing; 1st Aeromedical Evacuation Sqdn.; 1943d Communications Sqdn.; 53d Mobile Aerial Port Sqdn. (AFRES). Base adjoins Army's Fort Bragg and provides intratheater airlift support for airborne forces and other personnel, equipment, and supplies. Base activated 1919; named for 1st Lt. Harley H. Pope, WW I flyer, killed Jan. 6, 1919, when his JN-4 "Jenny" ran out of fuel near Fayetteville and crashed. Area 1,750 acres. Altitude 218 ft. Military 3,834; civilians 617. Payroll $71 million. Housing: 89 officer; 370 NCO; 216 transient. Dispensary.

Randolph AFB, Tex. 78148; 20 mi. ENE of San Antonio. Phone (512) 652-1110; AUTOVON 487-1110. ATC base. 12th Flying Training Wing, T-37 and T-38 pilot instructor training. Major tenants are Hq. Air Training Command; Air Force Manpower and Personnel Center; Occupational Measurement Center; Office of Civilian Personnel Operations; and Hq. USAF Recruiting Service. Base activated June 1930; named for Capt. William M. Randolph, killed Feb. 17, 1928, when his AT-4 crashed on takeoff at Gorman, Tex. Area 2,901 acres. Altitude 761 ft. Military 5,532; civilians 2,354. Payroll $179 million. Housing: 203 officer; 816 NCO; 13 transient. Dispensary.

Reese AFB, Tex. 79489; 6 mi. W of Lubbock. Phone (806) 885-4511; AUTOVON 838-1110. ATC base. 64th Flying Training Wing, undergraduate pilot training. Base activated in 1942; named for 1st Lt. Augustus F. Reese, Jr., P-38 fighter pilot killed in Sardinia, May 14, 1943. Area 3,597 acres. Altitude 3,338 ft. Military 2,537; civilians 568. Payroll $61 million. Housing: 108 officer; 289 NCO. 28 transient. 10-bed hospital.

Robins AFB, Ga. 31098; at Warner Robins; 18 mi. SSE of Macon. Phone (912) 926-1110; AUTOVON 468-1110. AFLC base. Hq. Warner Robins Air Logistics Center; Hq. Air Force Reserve (AFRES); 2853d Air Base Gp.; 19th Bomb Wing (SAC);

5th Combat Communications Gp. (AFLC); 3503d Recruiting Gp.; 1926th Communications Sqdn. (AFCC). Base activated March 1942; named for Brig. Gen. Augustine Warner Robins, an early Chief of the Materiel Division of the Air Corps, died June 16, 1940. Area 8,856 acres. Altitude 294 ft. Military 3,843; civilians 15,262. Payroll $428.8 million. Housing: 249 officer; 1,147 NCO; 40 transient. 40-bed hospital.

Sawyer AFB *(see K. I. Sawyer AFB).*

Scott AFB, Ill. 62225; 6 mi. ENE of Belleville. Phone (618) 256-1110; AUTOVON 638-1110. MAC base. 375th Aeromedical Airlift Wing; Hq. Military Airlift Command; Hq. Air Force Communications Command; Hq. Aerospace Rescue and Recovery Service; Hq. Air Weather Service. Also, Defense Commercial Communications Office; Environmental Technical Applications Center; USAF Medical Center, Scott; 7th Weather Wing; 932d Aeromedical Airlift Gp. (AFRES Assoc.); Airlift Communications Division; and 375th Air Base Gp. Base activated June 14, 1917; named for Cpl. Frank S. Scott, first enlisted man to die in an air accident, killed Sept. 28, 1912, at College Park, Md. Area 3,000 acres. Altitude 453 ft. Military 6,714; civilians 3,122. Payroll $196.7 million. Housing: 404 officer; 1,469 NCO, plus 120 spaces for privately owned trailers; 283 transient. 175-bed hospital; 100-bed aeromedical staging facility.

Seymour Johnson AFB, N.C. 27531; adjacent to Goldsboro. Phone (919) 736-0000; AUTOVON 488-1110. TAC base. 4th Tactical Fighter Wing, F-4E fighter operations with dual-based commitment to NATO; 68th Bomb Wing (SAC); 2012th Communications Sqdn. (AFCC). Base activated June 12, 1941; named for Navy Lt. Seymour A. Johnson, native of Goldsboro, killed March 4, 1941, in crash in Maryland. Area 4,281 acres. Altitude 109 ft. Military 5,155; civilians 837. Payroll $86 million. Housing: 524 officer; 1,221 NCO; 138 transient (includes 46 VOQ, 92 VAQ). 30-bed hospital.

Shaw AFB, S.C. 29152; 10 mi. WNW of Sumter. Phone (803) 668-8110; AUTOVON 965-1110. TAC base. Hq. 9th Air force (TAC); 363d Tactical Fighter Wing, F-16 fighter and RF-4C recon operations; 507th Tactical Air Control Wing, manages 407L/485L tactical air control systems. Base activated Aug. 30, 1941; named for 2d Lt. Ervin D. Shaw, one of the first Americans to see air action in WW I, killed in action in France July 9, 1918, when his Bristol fighter was shot down during a reconnaissance mission. Area 3,269 acres; supports another 8,038 acres. Altitude 244 ft. Military 5,132; civilians 576. Payroll $88 million. Housing: 389 officer; 1,315 NCO; 16 transient. 40-bed hospital.

Shemya AFB, Alaska (APO Seattle 98736); located at western tip of the Aleutian Islands chain, midway between Anchorage, Alaska, and Tokyo, Japan. Phone (907) 572-3000; AUTOVON (317) 572-3000. AAC base. Activated in 1943. Shemya was used as a bomber base in WW II. The International Date Line has been bent around Shemya so the local date is the same as elsewhere in the U.S. Area about 4.5 mi. long by 2.5 mi. wide. Altitude 270 ft. Military 547. Payroll $7.5 million. Housing: 70 transient. Dispensary.

Sheppard AFB, Tex. 76311; 4 mi. N of Wichita Falls. Phone (817) 851-2511; AUTO-

VON 736-1001. ATC base. Sheppard Technical Training Center provides resident courses in aircraft maintenance, civil engineering, communications, missile, comptroller, transportation, and instructor training. The 3785th Field Training Gp. provides specialized and advanced training at 70 field training detachments and 19 operating locations worldwide. The School of Health Care Sciences provides training in medicine, dentistry, nursing, biomedical sciences, and health services administration. 80th Flying Training Wing conducts undergraduate pilot training and instructor training for the Euro-NATO Joint Jet Pilot Training Program. The wing trains allied fighter pilots for 12 NATO countries. Base activated June 14, 1941; named for Morris E. Sheppard, U.S. Senator from Texas, died in 1941. Area 5,000 acres. Altitude 1,015 ft. Military 7,729; civilians 3,476. Payroll $147.5 million. Housing: 200 officer; 1,087 NCO. 378-bed hospital.

Tinker AFB, Okla. 73145; 8 mi. SE of Oklahoma City. Phone (405) 734-7321; AUTOVON 735-1110. AFLC base. Hq. Oklahoma City Air Logistics Center, furnishes logistic support for bombers, jet engines, instruments, and electronics; Electronics Installation Center; 3d Combat Communications Gp.; 552d Airborne Warning and Control Wing (TAC); 507th Tactical Fighter Gp. (AFRES). Base activated March 1941; named for Maj. Gen. Clarence L. Tinker. On June 7, 1942, at the end of the Battle of Midway, General Tinker's LB-30 (an early model B-24) apparently went down at sea after attacking retreating enemy ships. Area 4,277 acres. Altitude 1,291 ft. Military 5,700; civilians 16,500. Payroll $440 million. Housing: 110 officer; 422 NCO. 30-bed hospital.

Travis AFB, Calif. 94535; at Fairfield, 50 mi. NE of San Francisco. Phone (707) 438-4011; AUTOVON 837-1110. MAC base. Hq. 22d Air Force; 60th Military Airlift Wing; 349th Military Airlift Wing (AFRES Assoc.); 307th Air Refueling Gp. (SAC); David Grant Medical Center. Base activated May 25, 1943; named for Brig. Gen. Robert F. Travis, killed Aug. 5, 1950, in a B-29 accident. Area 6,170 acres. Altitude 62 ft. Military 9,016; civilians 2,347. Payroll $193.2 million. Housing: 241 officer; 1,926 NCO; 584 transient (incl. 40 transient living quarters, 204 VOQ, 188 VAQ, 83 Aerial Port quarters with cooking facilities, 69 Aerial Port quarters without). 290-bed hospital.

Tyndall AFB, Fla. 32403; 13 mi. E of Panama City. Phone (904) 283-1113; AUTOVON 970-1110. TAC base. Home of the USAF Air Defense Weapons Center and the 325th Combat Support Group. The base is a single DoD unit for centralization of operational and technical expertise on air defense. Conducts weapons-firing programs and evaluation for fighter-interceptor pilots; tests new air defense-related equipment and tactics. Home of the biennial Project William Tell fighter interceptor weapons meet, which tests the mission skills of the best air defense fighter units. Tenants include Air Force Engineering and Services Center; 3625th Technical Training Sqdn. (ATC); 678th Air Defense Gp. (TAC); 2021st Communications Sqdn. (AFCC); and TAC NCO Academy East. Location of the first Region Operations Control Center scheduled for activation in late 1982. Base activated Dec. 7, 1941; named for 1st Lt. Frank B. Tyndall, WW I fighter pilot, killed July 15, 1930, in crash of P-1 near Mooresville, N.C. Area 28,000 acres. Altitude 18 ft. Military 4,250; civilians 1,198. Payroll $90 million. Housing: 142 officer; 929 NCO. 80-bed hospital.

U.S. Air Force Academy, Colo. 80840; 10 mi. N of Colorado Springs. Phone (303) 472-1818; AUTOVON 259-3110. Direct reporting unit; activated April 1, 1954, at Lowry AFB, Colo. Moved to permanent location Aug. 1958. Tenant units include 1876th Communications Sqdn.; Frank J. Seiler Research Lab (AFSC); DoD Medical Exam Review Board; Det. 470 of AF Audit Agency; 557th Flying Training Sqdn. (ATC). Area 18,000 acres. Altitude 7,280 ft. Military 2,758; civilians 1,795. Payroll $134 million. Housing: 622 officer; 621 enlisted; 18 transient. 70-bed hospital.

Vance AFB, Okla. 73702; 3 mi. SSW of Enid. Phone (405) 237-2121; AUTOVON 962-7110. ATC base. 71st Flying Training Wing, undergraduate pilot training. Base activated Nov. 1941; named for Lt. Col. Leon R. Vance, Jr., native of Enid, 1939 West Point graduate, Medal of Honor recipient, killed July 26, 1944, when the air-evac plane returning him to the U.S. went down in the Atlantic near Iceland. Area 1,811 acres. Altitude 1,307 ft. Military 1,400; civilians 1,300. Payroll $47 million. Housing: 119 officer; 111 NCO; 1 transient. Clinic.

Vandenberg AFB, Calif. 93437; 8 mi. NNW of Lompoc. Phone (805) 866-1611; AUTOVON 276-1110. SAC base. Site of 1st Strategic Aerospace Div. (SAC); Space and Missile Test Organization (AFSC); Western Space and Missile Test Center (AFSC); Shuttle Activation Task Force (AFSC). Host command conducts missile crew training and provides facilities and support for operational ICBM tests. Vandenberg is the only base that launches operational ballistic missiles in the SAC deterrent force. WSMC is responsible for conducting R&D testing of USAF space and ballistic missile programs, and unmanned polar-orbiting space operations of DoD, USAF, NASA, contractors, and others. This includes development, testing, and evaluation of the MX and the Space Transportation System. MX testing is scheduled to begin in early 1983. Site Alteration Task Force (SATAF) is responsible for facility construction, equipment installation, and validation for future Vandenberg Space Shuttle launches beginning in late 1985. Approximately 1,493 launches have taken place from Vandenberg since Dec. 1958. Originally Army's Camp Cooke. Activated Oct. 1941. Base taken over by USAF June 7, 1957; renamed for Gen. Hoyt S. Vandenberg, USAF's second chief of staff, died April 2, 1954. Area 98,400 acres. Altitude 400 ft. Military 6,415; civilians 7,905. Payroll $207 million. Housing: 393 officer; 1,575 enlisted; 172 mobile trailer spaces; 20 transient. 45-bed hospital.

Warren AFB *(see Francis E. Warren AFB).*

Wheeler AFB, Hawaii 96854; near center of the island of Oahu, adjacent to Army's Schofield Barracks. Phone (808) 655-1414; AUTOVON 430-0111. PACAF base. 15th Air Base Sqdn., base host unit; 326th Air Div. (Air Defense Control Center), 22d TASS, a subordinate unit, and the 169th ACWS (Hawaii Air National Guard—Air Defense Direction Center); U.S. Army flying activities from Schofield Barracks; and several other tenant units. Base activated Feb. 1922; named for Maj. Sheldon H. Wheeler, who became CO of Luke Field, Hawaii, in 1919 and was killed there July 13, 1921, when his biplane crashed during aerial exhibition. Area 1,369 acres. Altitude 845 ft. Military 720; civilians 112. Payroll included in entry for Hickam AFB. Housing: 102 officer; 390 NCO. Dispensary.

Whiteman AFB, Mo. 65305; 1.5 mi. S of Knob Noster. Phone (816) 687-1110;

AUTOVON 975-1110. SAC base. 351st Strategic Missile Wing. Base activated in 1942; named for 2d Lt. George A. Whiteman, shot down while taking off in a fighter from Wheeler Field, Hawaii, on Dec. 7, 1941, the first Army Air Forces airman to be shot down in WW II. Area 3,384 acres, plus missile complex of about 10,000 sq. mi. Altitude 869 ft. Military 3,178; civilians 373. Payroll $54.7 million. Housing: 209 officer; 783 NCO; 57 transient (incl. 19 VOQ, 4 guest houses, and 34 VAQ). 10-bed hospital.

Williams AFB, Ariz. 85224; 16 mi. SE of Mesa. Phone (602) 988-2611; AUTOVON 474-1001. ATC base. 82d Flying Training Wing, largest undergraduate pilot training base; also provides F-5 combat crew training for foreign students. Home of AFSC Human Resources Lab/Flying Training Div., doing extensive research on flight simulators. Base activated July 1941; named for 1st Lt. Charles D. Williams, killed in crash of a bomber near Fort De Russy, Hawaii, July 6,1927. Area 4,762 acres. Altitude 1,385 ft. Military 3,300; civilians 1,070. Payroll $72 million. Housing: 310 officer; 496 NCO; 40 transient. 25-bed hospital.

Wright-Patterson AFB, Ohio 45433; 10 mi. ENE of Dayton. Phone (513) 257-1110; AUTOVON 787-1110. AFLC base. Hq. Air Force Logistics Command; Hq. Aeronautical Systems Div. (AFSC); 4950th Test Wing (AFSC); Foreign Technology Div. (AFSC); AF Institute of Technology; USAF Medical Center, Wright-Patterson; U.S. Air Force Museum; AF Acquisition Logistics Div.; AFLC International Logistics Center; 2750th Air Base Wing (AFLC); 906th Tactical Fighter Gp. (AFRES); plus more than 75 other DoD activities and government agencies. Originally separate, Wright Field and Patterson Field were merged and redesignated Wright-Patterson AFB on Jan. 13, 1948; named for aviation pioneers Orville and Wilbur Wright and for 1st Lt. Frank S. Patterson, killed June 19, 1918, in the crash of a DH-4. The Wright brothers did much of their early flying on Huffman Prairie, now in Area C of present base. Area 8,174 acres. Altitude 824 ft. Military 7,900; civilians 16,000; contracted service and contractor employees 8,000. Payroll $695 million. Housing: 1,090 officer; 1,245 NCO; 40 transient. 285-bed hospital.

Wurtsmith AFB, Mich. 48753; 3 mi. NW of Oscoda. Phone (517) 739-2011; AUTOVON 623-1110. SAC base. 40th Air Div.; 379th Bomb Wing. Base activated 1924 as Camp Skeel, gunnery camp for Selfridge Field; became Oscoda Army Air Field during WW II; renamed in 1953 for Maj. Gen. Paul B. Wurtsmith, killed Sept. 13, 1946, in a B-25 crash near Asheville, N.C.; assigned to SAC April 1, 1960. Area 5,200 acres. Altitude 634 ft. Military 3,152; civilians 399. Payroll $43 million. Housing: 290 officer; 1,065 NCO; 59 transient. 20-bed hospital.

Appendix B

Air Force Oversea Installations

We are a strong nation. But we cannot live to ourselves and remain strong.
—George C. Marshall

The information in this appendix is taken from Air Force Fact Sheet 74-1, and sets forth all major oversea installations of the U.S. Air Force outside the continental United States. The information provided includes related facts about each air base such as parent command, major operations, and government housing available. On-base dependent schools overseas are listed where available.

ACTIVE MAJOR INSTALLATIONS OVERSEAS
(By Host Country)

Azores (Portugal)

LAJES Field on northeast tip of Terceira Island, one of the nine islands in Azores Archipelago, lies 2,300 miles east of New York City and 1,000 miles west of Lisbon. MAC base. Major operations: military airlift. Housing: 40 appropriated fund and 108 surplus commodity. Medical facilities: USAF hospital. Schools: Kindergarten through 12th grade.

Panama

ALBROOK AS, located in Canal Zone of Panama on the Pacific side of the Isthmus. USAFSO base. Major operations: U.S. Air Force Southern Division, TAC; Inter-American Air Force Academy; Tropical Survival School. Housing: 470 appropriated fund. Named for 1st Lt. Frank P. Albrook, WW I pilot who in 1922 commanded first military unit at Balboa Field, later redesignated in his honor.

HOWARD AB, 10 miles across the Panama Canal from Albrook AFB, 6 miles southwest of Balboa. USAFSO base. Major operations: support base for Southern Command. Housing: 712 appropriated fund. Medical facilities: USAF clinic. Named for Maj. Charles H. Howard, WW I pilot whose postwar career included two tours of Canal Zone duty.

Germany

BITBURG AB, 3 miles southeast of Bitburg, 135 miles west of Wiesbaden. USAFE base. Major operations: tactical fighter. Medical facilities: USAF hospital. Schools: Kindergarten through 12th grade; special classes available.

ERDING AS, near Landshuterstrasse. USAFE base. Major operations: fighter-interceptor. Housing: 24 foreign source.

HAHN AB, 60 miles west of Wiesbaden. USAFE base. Major operations: tactical fighter. Housing: 672 foreign source. Medical facilities: USAF hospital. Schools: Kindergarten through 10th grade.

LINDSEY AS, in Wiesbaden. USAFE base. Major operations: support base for USAFE. Schools: 1st through 6th grade; special classes available. Named for Capt. Darrell Robins Lindsey, WW II B-26 pilot, posthumously awarded the Medal of Honor.

RAMSTEIN AB, 2 miles southeast of Landstuhl, 7 miles west of Kaiserslautern. USAFE base. Major operations: Headquarters U.S. Air Forces in Europe; tactical reconnaissance; support base for HQ USAFE. Housing: 2,412 foreign source. Medical facilities: USAF clinic. Schools: Kindergarten through 9th grade; special classes available.

RHEIN MAIN AB, 5 miles south of Frankfurt. USAFE base. Major operations tactical airlift base; MAC operations. Housing: 6 leased and 852 foreign source. Medical facilities: USAF clinic. Schools: Kindergarten through 9th grade; special classes available.

SEMBACH AB, 1 mile south southeast of Sembach, 10 miles from Kaiserslautern. USAFE base. Major operations: Headquarters 17th Air Force; support base for USAFE. Housing: 506 foreign source. Medical facilities: USAF clinic. Schools: Kindergarten through 8th grade; special classes.

SPANGDAHLEM AB, 1 mile northwest of Spangdahlem. USAFE base. Major operations: tactical fighter. Housing: 1,294 foreign source. Medical facilities: USAF clinic. Schools: Kindergarten through 8th grade.

WIESBADEN AB, 5 miles east of Wiesbaden. USAFE base. Major operations: Headquarters 601st Tactical Control Wing; MAC weather operations. Housing: 2,541 foreign source. Medical facilities: USAF hospital. Schools: Kindergarten through 8th grade, 9th through 12th grade with 5-day dormitory and special classes.

ZWEIBRUCKEN AB, 30 miles south of Ramstein. USAFE base. Major operations: tactical reconnaissance. Medical facilities: USAF clinic.

Greenland (Denmark)

SONDRESTROM AB, 70 miles west of Holstensbourg, at the head of Sondrestrom Fjord. ADC base. Major operations: support base for ADC. Medical facilities: USAF clinic.

THULE AB, on northwest coast of Greenland, 2,180 miles from Seattle and 2,185 miles from New York City, 931 miles below the North Pole. Major operations: ADC units. Medical facilities: USAF hospital.

Guam (U.S.)

ANDERSEN AFB, on north end of island, 11 miles from Agana. SAC base. Major operations: Headquarters 8th Air Force; heavy bomber and tanker; weather reconnaissance. Housing: 403 appropriated fund and 1,050 Capehart. Medical facilities: USAF clinic and USN hospital. Named for Brig. Gen. James R. Andersen, lost on flight from Kwajalein to Hawaii in 1945.

Iceland

KEFLAVIK Airport, 22 miles west of Reykjavik. ADC is tenant on base. Major operations: fighter-interceptor. Schools: Kindergarten through 12th grade.

Italy

AVIANO AB, adjacent to town of Aviano, at foot of the Italian Alps, 50 miles north of Venice. USAFE base. Major operations: tactical group. Housing: 4 leased. Medical facilities: USAF hospital. Schools: Kindergarten through 10th grade.

Japan

YOKOTA AB, 25 miles northwest of Tokyo. PACAF base. Major operations: PACAF support base. Housing: 11 appropriated fund, 4,616 foreign source, 144 surplus commodity, and 88 other. Schools: Kindergarten through 8th grade.

Korea

KUNSAN AB, 8 miles southwest of Kunsan. PACAF base. Major operations: tactical fighter base. Medical facilities: USAF hosp1tal.

KWANGJU AB, 5 miles east of Song Jong-Ni. PACAF base. Major operations combat support base. Medical facilities: USAF clinic.

OSAN AB, 5 miles south southeast of Osan, 45 miles south of Seoul. PACAF base. Major operations: 314th Air Division. Medical facilities: USAF hospital.

TAEGU AB, within city limits of Taegu. PACAF base. Major operations: combat support base. Medical facilities: USAF clinic. Schools: Kindergarten through 9th grade.

Netherlands

CAMP NEW AMSTERDAM AB, part of Soesterberg AB, RNAF, 9 miles east northeast of Utrecht. USAFE base. Major operations: fighter-interceptor base. Schools: Kindergarten through 9th grade.

Okinawa (Japan)

KADENA AB, 2 miles northwest of Misato, 12 miles north of Naha. PACAF base. Major operations: F-4 Tactical Fighter Wing; PACAF air division; SAC strategic operations. Housing: 1,934 appropriated fund and 209 other. Medical facilities: USAF clinic and USA hospital. Schools: Kindergarten through 8th grade.

Philippines

CLARK AB, 3 miles northwest of Angeles, 50 miles northwest of Manila on the island of Luzon. PACAF base. Major operations: Headquarters 13th Air Force; tactical fighter wing; tactical operations; Pacific Jungle Survival School. Housing: 1,270 appropriated fund and 225 other. Medical facilities: USAF hospital. Schools: Kindergarten through 12th grade. Named for Maj. Howard Clark, first airman to fly in Hawaiian islands; died in crash in Panama in 1919.

Spain

MORON AB, 30 miles southeast of Seville. USAFE base. Major operations: support base for USAFE. Housing: 36 surplus commodity.

TORREJON AB, 2 miles northeast of Torrejon de Ardoz, 15 miles northwest of Madrid. USAFE base. Major operations: Headquarters 16th Air Force; tactical fighter operations. Housing: 862 leased and 67 surplus commodity. Medical facilities: USAF hospital. Schools: Kindergarten through 12th grade; special classes available.

ZARAGOZA AB, 8 miles west of Zaragoza. USAFE base. Major operations: tactical fighter training. Housing: 156 surplus commodity. Medical facilities: USAF hospital. Schools: Kindergarten through 12th grade with 7-day dormitory available.

Turkey

INCIRLIK AB, 10 miles from Adana. USAFE base. Major operations: tactical fighter base. Housing: 150 surplus commodity, 375 trailers, and 200 units approved for construction in 1973. Medical facilities: USAF hospital in Ankara. Schools: Kindergarten through 10th grade; special classes available.

United Kingdom

RAF ALCONBURY, 3 miles northwest of Huntingdon, 60 miles north of London. USAFE base. Major operations: tactical reconnaissance wing. Medical facilities: USAF clinic. Schools: Kindergarten through 6th grade.

RAF BENTWATERS, 87 miles northeast of London. USAFE base. Major operations: tactical fighter wing. Medical facilities: USAF clinic. Schools: Kindergarten through 6th grade; special classes available.

RAF LAKENHEATH, 70 miles northeast of London. USAFE base. Major operations: tactical fighter operations. Medical facilities: USAF hospital. Schools: Kindergarten through 12th grade; special classes and 5-day dormitory available.

RAF MILDENHALL, 25 miles northeast of Cambridge. USAFE base. Major operations: tactical airlift; Headquarters 3d Air Force.

RAF UPPER HEYFORD, 15 miles north of Oxford. USAFE base. Major operations: tactical fighter operations. Medical facilities: USAF hospital. Schools: Kindergarten through 12th grade; special classes available.

RAF WEST RUISLIP, USAFE base. Major operations: USAFE support base. Schools: Kindergarten through 6th grade; special classes available.

RAF WOODBRIDGE, 4 miles west of Woodbridge, Suffolk County, USAFE base. Major operations: tactical fighter. Schools: Kindergarten through 10th grade; special classes available.

INDEX

O

P

U

V

W